The Molecular Biology and Biochemistry of Fruit Ripening

The Molecular Biology and Biochemistry of Fruit Ripening

Edited by
GRAHAM B. SEYMOUR
MERVIN POOLE
JAMES J. GIOVANNONI
GREGORY A. TUCKER

WILEY-BLACKWELL
A John Wiley & Sons, Inc., Publication

Wiley-Blackwell is an imprint of John Wiley & Sons, formed by the merger of Wiley's global Scientific, Technical and Medical business with Blackwell Publishing.

Editorial offices: 2121 State Avenue, Ames, Iowa 50014-8300, USA
The Atrium, Southern Gate, Chichester, West Sussex, PO19 8SQ, UK
9600 Garsington Road, Oxford, OX4 2DQ, UK

For details of our global editorial offices, for customer services and for information about how to apply for permission to reuse the copyright material in this book please see our website at www.wiley.com/wiley-blackwell.

Library of Congress Cataloging-in-Publication Data is available upon request

A catalogue record for this book is available from the British Library.

Wiley also publishes its books in a variety of electronic formats. Some content that appears in print may not be available in electronic books.

Cover design by Nicole Teut

Set in 11/13 pt Times by Aptara® Inc., New Delhi, India
Printed and bound in Malaysia by Vivar Printing Sdn Bhd

1 2013

Contents

List of Contributors ix
Preface xi

Chapter 1 Biochemistry of Fruit Ripening 1
Sonia Osorio and Alisdair R. Fernie

Introduction 1
Central Carbon Metabolism 4
Ethylene in Ripening 7
Polyamines 9
Volatiles 10
Cell Wall Metabolism 11
Concluding Remarks 13
References 13

Chapter 2 Fruit—An Angiosperm Innovation 21
Sandra Knapp and Amy Litt

Introduction 21
Fruit in the Fossil Record 30
Fruit Variation and Angiosperm Phylogeny 32
Fruit Development 33
Fruit as a Driver of Angiosperm Diversity 36
Acknowledgments 38
References 38

Chapter 3 Ethylene and the Control of Fruit Ripening 43
Don Grierson

Introduction 43
Ethylene and Climacteric and Nonclimacteric Fruits 46
A Molecular Explanation for System-1 and System-2 Ethylene 48
Ethylene and Ripening Gene Networks in Flower and Fruit Development 53
Ethylene Perception and Signaling 54
Ethylene Response Factors 60

Ethylene and Ripening Gene Expression 60
Conclusions 67
Acknowledgments 68
References 68

Chapter 4 Carotenoid Biosynthesis and Chlorophyll Degradation 75
Peter M. Bramley

Introduction 75
Distribution of Carotenoids and Chlorophylls in Fruit 75
Chlorophyll Degradation and Recycling 78
Carotenoids and Carotenoid Metabolites 82
Future Perspectives 100
Acknowledgments 102
Bibliography 102

Chapter 5 Phenylpropanoid Metabolism and Biosynthesis of Anthocyanins 117
Laura Jaakola

Introduction 117
Cinnamic Acids 119
Monolignols, Lignans, and Lignin 120
Coumarins 120
Stilbenoids 122
Flavonoids 122
Engineering Elevated Levels of Flavonoids and Other Phenylpropanoids 128
Conclusion 129
References 129

Chapter 6 Biosynthesis of Volatile Compounds 135
Antonio Granell and José Luis Rambla

Introduction 135
Metabolic Pathways 136
Identification of Quantitative Trait Loci for Volatiles 152
Metabolic Engineering of the Fruit Volatile Pathways 153
Future Perspectives 154
References 155

Chapter 7 Cell Wall Architecture and Metabolism in Ripening Fruit and the Complex Relationship with Softening 163
Eliel Ruiz-May and Jocelyn K.C. Rose

Introduction 163
Building Blocks of Fruit Cell Walls 164

The Architecture of Fruit Cell Walls 168
Cell Wall Dynamics in Ripening Fruit 171
The Cuticular Cell Wall and Fruit Softening 177
Summary 179
Acknowledgments 180
References 180

Chapter 8 Regulatory Networks Controlling Ripening 189
Betsy Ampopho, Natalie Chapman, Graham B. Seymour, and James J. Giovannoni

Hormonal Control 189
Genetic Networks 191
Epigenetic Regulation 200
References 201

Index 207

List of Contributors

Betsy Ampopho
Boyce Thompson Institute for Plant Science Research
Cornell University
Ithaca, New York, NY, USA.

Peter M Bramley
School of Biological Sciences
Royal Holloway
University of London
Egham, Surrey, United Kingdom.

Natalie Chapman
Plant and Crop Science Division
University of Nottingham
Sutton Bonington, Loughborough, Leics, United Kingdom.

Alisdair R. Fernie
Department of Molecular Physiology
Max-Planck-Institute for Molecular Plant Physiology
Potsdam-Golm, Germany.

James J Giovannoni
Department of Agriculture–Agricultural Research Service
Boyce Thompson Institute for Plant Science Research
Cornell University
Ithaca, New York, NY, USA.

Antonio Granell
Instituto de Biología Molecular y Celular de Plantas
Consejo Superior de Investigaciones Científicas
Universidad Politécnica de Valencia
Valencia, Spain.

Don Grierson
Laboratory of Molecular Physiology and Biotechnology
Zhejiang University
Zhejiang, China.

Division of Plant and Crop Sciences
School of Biosciences
University of Nottingham
Sutton Bonington Campus
Loughborough, Leicestershire, United Kingdom.

Laura Jaakola Department of Biology
University of Oulu
Oulu, Finland.

Sandra Knapp Department of Botany
The Natural History Museum
London, United Kingdom.

Amy Litt The New York Botanical Garden
Bronx, New York, NY, USA.

Sonia Osorio Max-Planck-Institute for Molecular Plant Physiology
Potsdam-Golm, Germany.

Mervin Poole Plant Science Division
University of Nottingham
Sutton Bonington Campus
Loughborough, Leics, United Kingdom.

Jose Luis Rambla Instituto de Biología Molecular y Celular de Plantas
Consejo Superior de Investigaciones Científicas
Universidad Politécnica de Valencia
Valencia, Spain.

Jocelyn K.C. Rose Department of Plant Biology
Cornell University
Ithaca, New York, NY, USA.

Eliel Ruiz-May Department of Plant Biology
Cornell University
Ithaca, New York, NY, USA.

Graham B. Seymour Plant and Crop Science Division
University of Nottingham
Sutton Bonington
Loughborough, Leics, United Kingdom.

Gregory A. Tucker School of Biosciences
University of Nottingham
Sutton Bonington Campus
Loughborough, Leics, United Kingdom.

Preface

Evolution has fashioned multiple means of protecting seed and dispersing them upon maturation. None is as fascinating nor as consequential to humankind as the ripe and delectable fleshy fruit. Ripe fruits comprise a significant and expanding proportion of human and animal diets, which the medical community contends should only be increased. In addition to being visual delights with seductive tastes and aromas, ripe fruits deliver a diverse array of antioxidants and nutrients to those who consume them, in addition to healthy doses of carbohydrates and fiber. The chemistry of fruits comprises attributes that producers, processors, and distributors alike seek to understand, optimize, and deliver to increasingly health-conscious consumers expecting high quality and diversity of choices. Plant scientists have endeavored to unravel the mysteries of fleshy fruit biology and the underlying molecular and biochemical processes that contribute to fruit ripening and the resulting desirable attributes of fruits and fruit products.

This book offers a useful overview of fruit ontology and evolution emphasizing the exponential growth in advances and discoveries in ripening-related chemistry and associated regulatory processes accumulated in the last decade. The reader will appreciate the broad and deep impact of comprehensive genomics and metabolomics in addition to the computational tools necessary to decipher the resulting data on the progress of the field. As a consequence of these all-encompassing approaches, fruit biology has advanced from the investigation of single genes and enzymatic reactions to the development of nuanced molecular regulatory models overseeing complex biochemical pathways leading to numerous metabolic outputs. Looking at the physiological and molecular symphony of events impacting textural changes of the ripening fruit, the array of novel phenolic metabolites, or the network of genes and signaling processes regulating ethylene hormone response, it becomes strikingly clear that recent technical advances have moved ripening biology forward at an astounding rate. This book captures the advances of the field and couches them in an evolutionary context and a fundamental knowledge of fruit biology, making it an excellent primer for those interested in the field and a comprehensive reference for those familiar with it. *The Molecular Biology and Biochemistry of Fruit Ripening* is essential reading for any student of plant science and those especially interested in fruit biology and its relationship to human diet and nutrition.

1 Biochemistry of Fruit Ripening

Sonia Osorio and Alisdair R. Fernie

Introduction

This chapter is intended to provide an overview of the key metabolic and regulatory pathways involved in fruit ripening, and the reader is referred to more detailed discussions of specific topics in subsequent chapters.

The quality of fruit is determined by a wide range of desirable characteristics such as nutritional value, flavor, processing qualities, and shelf life. Fruit is an important source of supplementary diet, providing minerals, vitamins, fibers, and antioxidants. In particular, they are generally rich sources of potassium, folate, vitamins C, E, and K as well as other phytonutrients such as carotenoids (beta-carotene being a provitamin A) and polyphenols such as flavonols (Saltmarsh et al., 2003). A similar, but perhaps more disparate, group of nutrients is associated with vegetables. Thus nutritionists tend to include fruits and vegetables together as a single "food group," and it is in this manner that their potential nutritional benefits are normally investigated and reported. Over the past few decades, the increased consumption of fruits and vegetables has been linked to a reduction in a range of chronic diseases (Buttriss, 2012). This has led the WHO to issue a recommendation for the consumption of at least 400 g of fruits and vegetables per day. This in turn has prompted many countries to issue their own recommendations regarding the consumption of fruits and vegetables. In Britain this has given rise to the five-a-day recommendation. A portion in the United Kingdom is deemed to be around 80 g; so five-a-day corresponds to about 400 g per day. Other countries have opted for different recommendations (Buttriss, 2012), but all recognize the need for increased consumption.

The rationale for the five-a-day and other recommendations to increase fruit and vegetable consumption comes from the potential link between high intake of fruits and vegetables and low incidence of a range of diseases. There have been many studies carried out over the last few decades. The early studies tended to have a predominance of case-control approaches while recently more cohort studies, which are considered to be more robust, have been carried out. This has given rise to many critical and systematic reviews, examining this cumulative

The Molecular Biology and Biochemistry of Fruit Ripening, First Edition.
Edited by Graham B. Seymour, Mervin Poole, James J. Giovannoni and Gregory A. Tucker.

evidence base, over the years which have sometimes drawn disparate conclusions regarding the strength of the links between consumption and disease prevention (Buttriss, 2012). One of the most recent (Boeing et al., 2012) has concluded that there is convincing evidence for a link with hypertension, chronic heart disease, and stroke and probable evidence for a link with cancer in general. However, there might also be probable evidence for an association between specific metabolites and certain cancer states such as between carotenoids and cancers of the mouth and pharynx and beta-carotene and esophageal cancer and lycopene and prostate cancer (WRCF and American Institute for Cancer Research, 2007). There is also a possible link that increased fruit and vegetable consumption may prevent body weight gain. This reduces the propensity to obesity and as such could act as an indirect reduction in type 2 diabetes, although there is no direct link (Boeing et al., 2012). Boeing et al. (2012) also concluded there is possible evidence that increased consumption of fruits and vegetables may be linked to a reduced risk of eye disease, dementia, and osteoporosis. In almost all of these studies, fruits and vegetables are classed together as a single "nutrient group." It is thus not possible in most cases to assign relative importance to either fruits or vegetables. Similarly, there is very little differentiation between the very wide range of botanical species included under the banner of fruits and vegetables and it is entirely possible that beneficial effects, as related to individual disease states, may derive from metabolites found specifically in individual species.

Several studies have sought to attribute the potential beneficial effects of fruits and vegetables to specific metabolites or groups of metabolites. One such which has received a significant amount of interest is the antioxidants. Fruit is particularly rich in ascorbate or vitamin C which represents one of the major water-soluble antioxidants in our diet and also in carotenoids such as beta-carotene (provitamin A) and lycopene which are fat-soluble antioxidants (Chapter 4). However, intervention studies using vitamin C or indeed any of the other major antioxidants, such as beta-carotene, often fail to elicit similar protective effects, especially in respect of cancer (Stanner et al., 2004). Polyphenols are another group of potential antioxidants that have attracted much attention in the past. The stilbene—resveratrol—which is found in grapes, for example, has been associated with potential beneficial effects in a number of diseases (Baur and Sinclair, 2006). Similarly, the anthocyanins (Chapter 5), which are common pigments in many fruits, have again been implicated with therapeutic properties (Zafra-Stone et al., 2007). It is possible that these individual molecules may be having quite specific nutrient–gene expression effects. It is difficult to study these effects *in vivo*, as bioavailability and metabolism both in the gut and postabsorption can be confounding factors.

Although there are recommendations across many countries regarding the consumption of fruits and vegetables, in general, the actual intake falls below these recommendations (Buttriss, 2012). However, trends in consumption are on the increase driven potentially by increasing nutritional awareness on the part of the consumer and an increasing diversity of available produce. Fruit is available either fresh or processed in a number of ways the most obvious being in the form of juices or more recently smoothies. The list of fruits and vegetables traded throughout the world is both long and diverse. The FAO lists over 100 "lines" of which 60 are individual fruits or vegetables or related groups of these commodities. The remaining "lines" are juices and processed or prepared material. However, the top five traded products are all fruits and these are banana, tomato, apple, grape, and orange. In 1982–1984 these five between

Table 1.1 Global production, consumption, and net export of the five major (million tons) fruit commodities in 2002–2004. Data from European Commission Directorate-General for Agriculture and Rural Development (2007).

Commodity	Production	Consumption	Net Export
Banana	71	58	12.9
Tomato	119	103	2.1
Apple	59	56	3
Grape	64	59	1.7
Orange	63	53	2.5

them accounted for around half of global trade in fruits and vegetables; by 2002–2004, this had fallen to around 40% (European Commission Directorate-General for Agriculture and Rural Development, 2007). This probably reflects a growing trend toward diversification in the fruit market, especially in respect of tropical fruit. These figures represent traded commodities and in no way reflect global production of these commodities. In fact only about 5–10% of global production is actually traded. The EU commissioned a report in 2007 to examine trends in global production, consumption, and export of fruits and vegetables between 1980–1982 and 2002–2004. This demonstrated that fruits and vegetables represented one of the fastest growing areas of growth within the agricultural markets with total global production increasing by around 94% during this period. Global fruit production in 2004 was estimated at 0.5 billion tonnes. The growth in fruit production, at 2.2% per annum, was about half that for vegetables during this period. The report breaks these figures down into data for the most commonly traded commodities and the results for production, consumption, and net export in 2002–2004 are summarized in Table 1.1. Not all of the five major fruit commodities increased equally during this period. Banana and tomato production both doubled; apple and orange production both went up by about 50% while grape stagnated or even declined slightly during this period.

Global consumption of fruits and vegetables rose by 52% between 1992–2004 and 2002–2004 (European Commission Directorate-General for Agriculture and Rural Development, 2007). This means that global fruit and vegetable consumption rose by around 4.5% per annum during this period. This exceeded the population growth during the same period and as such suggested an increased consumption per capita of the population. Again the results for the consumption amongst the five major traded crops were variable with increases of banana, tomato being higher at 3.9% per annum and 4.5% per annum, respectively, while grapes (1.6% per annum) and oranges (1.9% per annum) were lower.

The net export figures reported above do not include trade between individual EU countries; however, even taking this into account, it is clear that only a small proportion of fruit production enters international trade. A major problem with trade in fresh fruit is the perishable nature of most of the commodities. This requires rapid transport or sophisticated means of reducing or modifying the fruits' metabolism. This can be readily achieved for some fruits, such as apple, by refrigeration; however, several fruits, such as mango, are subject to chilling injury that limits this approach. Other methods that are employed are the application of controlled or modified atmospheres (Jayas and Jeyamkondan, 2002). Generally an increase in carbon dioxide accompanied by a reduction in oxygen, will serve to reduce ethylene synthesis and respiration rate. The application of chemicals such as 1-MCP, an ethylene analog, can also significantly reduce ripening rates (Blankenship and Dole, 2003). Genetically modifying the fruit, for instance to reduce ethylene production, can also lead to an increase in shelf life (Picton et al., 1993).

Fruit ripening is highly coordinated, genetically programmed, and an irreversible developmental process involving specific biochemical and physiological attributes that lead to the development of a soft and edible fruit with desirable quality attributes (Giovannoni, 2001). The main changes associated with ripening include color (loss of green color and increase in nonphotosynthetic pigments that vary depending on species and cultivar), firmness (softening by cell-wall-degrading activities), taste (increase in sugar and decline in organic acids), and odor (production of volatile compounds providing the characteristic aroma). While the majority of this chapter will concentrate on central carbon metabolism, it is also intended to document progress in the understanding of metabolic regulation of the secondary metabolites of importance to fruit quality. These include vitamins, volatiles, flavonoids, pigments, and the major hormones. The interrelationship of these compound types is presented in Figure 1.1. Understanding the mechanistic basis of the events that underlie the ripening process will be critical for developing more effective methods for its control.

Central Carbon Metabolism

Sucrose, glucose, and fructose are the most abundant carbohydrates and are widely distributed food components derived from plants. The sweetness of fruits is the central characteristic determining fruit quality and it is determined by the total sugar content and by their ratios among those sugars. Accumulation of sucrose, glucose, and fructose in fruits such as melons, watermelons (Brown and Summers, 1985), strawberries (Fait et al., 2008) and peach (Lo Bianco and Rieger, 2002) is evident during ripening; however, in domesticated tomato (*Solanum lycopersicum*) only a high accumulation of the two hexoses is observed, whereas some wild tomato species (i.e., *Solanum chmielewskii*) accumulate mostly sucrose (Yelle et al., 1991). The variance in relative levels of sucrose and hexoses is most likely due to the relative activities of the enzymes responsible for the degradation of sucrose, invertase, and sucrose synthase.

The importance of the supply to, and the subsequent mobilization of sucrose in, plant heterotrophic organs has been the subject of intensive research effort over many years (Miller and Chourey, 1992; Zrenner et al., 1996; Wobus and Weber, 1999; Heyer et al., 2004; Roitsch and Gonzalez, 2004; Biemelt and Sonnewald, 2006; Sergeeva et al., 2006; Lytovchenko et al., 2007). While the mechanisms of sucrose loading into the phloem have been intensively studied over a similar time period (Riesmeier et al., 1993; Burkle et al., 1998; Meyer et al., 2004; Sauer et al., 2004), those by which it is unloaded into the sink organ (the developing organs attract nutrients) have only been clarified relatively recently and only for a subset of plants studied (Bret-Harte and Silk, 1994; Viola et al., 2001; Kuhn et al., 2003; Carpaneto et al., 2005). Recently, in the tomato fruit, the path of sucrose unloading in early developmental stages has been characterized as apoplastic. The study used tomato introgression lines containing

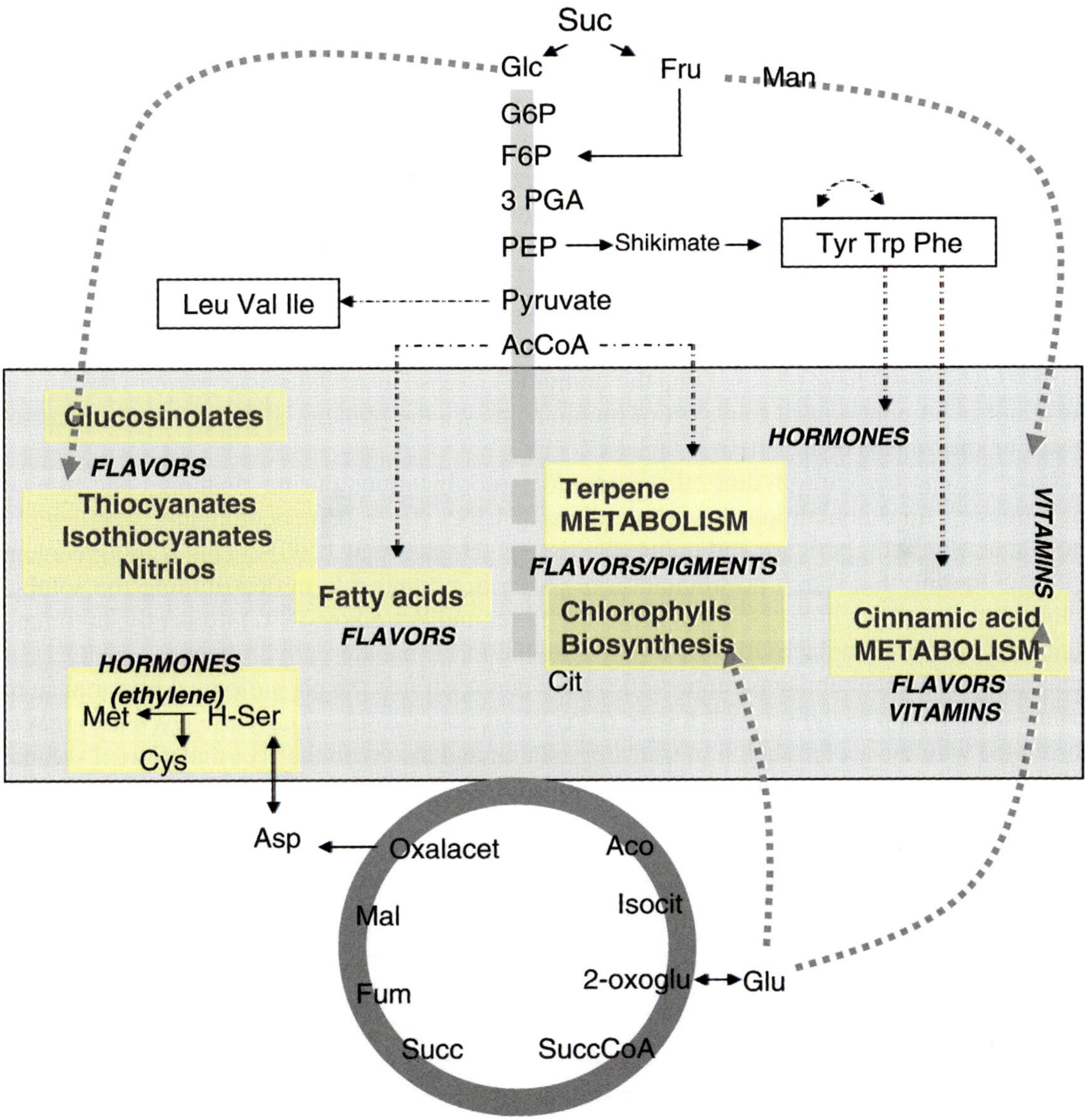

Figure 1.1 Interrelationships of primary and secondary metabolism pathways leading to the biosynthesis of aroma volatiles, hormones, pigments and vitamins (adapted from Carrari and Fernie (2006)).

an exotic allele of LIN5, a cell wall invertase that is exclusively expressed in flower (mainly ovary but also petal and stamen) and in young fruit (Godt and Roitsch, 1997; Fridman and Zamir, 2003), and it has been demonstrated that alterations in the efficiency of this enzyme result in significantly increased partitioning of photosynthate to the fruit and hence an enhanced agronomic yield (Fridman et al., 2004; Baxter et al., 2005; Schauer et al., 2006). Utilizing the reverse genetic approach, Zanor et al. (2009a) reported that *LIN5* antisense plants had decreased glucose and fructose in the fruit proving *in planta* the importance of LIN5 in the control of the total soluble solids content. The transformants were characterized by an altered flower and fruit morphology, displaying increased numbers of petals and sepals per flower, an increased rate of fruit abortion, and a reduction in fruit size. Evaluation of the mature fruit revealed that the transformants had a reduction of seed number per plant as well as altered levels of phytohormones. Interestingly, a role for apoplastic invertase in the control of sink size has been postulated previously in other species; the apoplastic invertase-deficient *miniature1* mutant of maize exhibits a dramatically decreased seed size as well as altered levels of phytohormones (Miller and Chourey, 1992; Sonnewald et al., 1997; LeClere et al., 2008). This raises interesting questions regarding the regulation of carbon partitioning in fruits. Recently, a metabolic and transcriptional study using introgression lines resulting from a cross between *S. lycopersicum* and *S. chmielewskii* have revealed that the dramatic increase in amino acid content in the fruit is the result of an upregulated transport of amino acids via the phloem, although the mechanism is still unknown (Do et al., 2010).

Starch is another carbohydrate that undergoes modifications during ripening. The tomato introgression lines containing the exotic allele of LIN5 (IL 9-2-5) accumulated significantly more starch in both, pericarp and columella tissues (Baxter et al., 2005). This is in agreement with the finding that starch accumulation plays an important role in determining the soluble solids content or Brix index of mature fruit (Schaffer and Petreikov, 1997). Recently, in tomato fruits, reduction of the activities of either mitochondrial malate dehydrogenase (mMDH) or fumarase via targeted antisense approaches have demonstrated the physiological importance of malate metabolism in the activation state of ADP-glucose pyrophosphorylase (AGPase) that is correlated with the accumulation of transitory starch and also with the accumulation of soluble solids at harvest (Centeno et al., 2011).

Organic acid manipulation is highly valuable from a metabolic engineering perspective because the organic acid to sugar ratio defines quality parameters at harvest time in fruits. However, their study has received much less attention than that of the sugars to date. Malate is the predominant acid in many fruits, both climacteric, including tomato (Kortstee et al., 2007), apple (Beruter, 2004), and nonclimacteric, including pineapple (Saradhuldhat and Paull, 2007), cherry (Usenik et al., 2008), strawberry (Moing et al., 2001), and grape (Kliewer et al., 1967). Interestingly, levels of both citrate and malate were also highly correlated to many important regulators of ripening in an independent study that was focused on early fruit development (Mounet et al., 2009). Patterns of malate accumulation differ between plant species and even cultivars (Kliewer et al., 1967). In fruits, patterns of malate accumulation and degradation cannot be explained by the classification of species as climacteric or nonclimacteric, nor can they be attributed to changes in overall respiration rates. Some climacteric fruits such as plum and tomato appear to utilize malate during the respiratory burst (Goodenough et al., 1985; Kortstee et al., 2007), while others such as banana and mango continue to accumulate malate throughout ripening, even at the climacteric stage (Selvaraj and Kumar, 1989a; Agravante et al., 1991). Nonclimacteric fruits also display widely varying malate accumulation and degradation events (Moing et al., 2001; Saradhuldhat and Paull, 2007); some fruits, including mango, kiwifruit, and strawberry display no net loss of malate throughout ripening (Selvaraj and Kumar, 1989a; Walton and De Jong, 1990; Moing et al., 2001). For this reason, the metabolism of malate has been a strong focus of research on grapes and tomato fruits, in which the acid plays a more metabolically active role (Goodenough et al., 1985). In grapefruit, malate is increasing in earlier stages and then is decreasing during ripening (Ruffner and Hawker, 1977). In earlier stages, malate is accumulated mostly through the metabolism of sugars (Hale, 1962) and during ripening, malate is a vital source of carbon for different pathways: TCA cycle and respiration, gluconeogenesis, amino acid interconversion, ethanol fermentation, and production of secondary compounds such as anthocyanins and flavonols (Ruffner, 1982; Famiani et al., 2000). Work with tomato fruit suggests that in early development, the majority of malate oxidation occurs through the TCA cycle.

The structure of the TCA cycle is well known in plants; however, until recently its regulation was poorly characterized. In our laboratory, several studies have been pursued to determine the role of mitochondrial TCA cycle in plants. Biochemical analysis of the *Aco1* mutant revealed that it exhibited a decreased flux through the TCA cycle, decreased levels of TCA cycle intermediates, enhanced carbon assimilation, and dramatically increased fruit weight (Carrari et al., 2003). Nunes-Nesi et al. (2005) produced tomato plants with reduced mMDH activity. Plants showed an increment in fruit dry weight likely due to the enhanced photosynthetic activity and carbon assimilation in the leaves, which also led to increased accumulation of starch and sugars, as well as some organic acids (succinate, ascorbate, and dehydroascorbate). Reduction of fumarase activity has been investigated in tomato plants (Nunes-Nesi et al., 2007), which led to lower fruit yield and total dry weight. Those plants showed opposite characteristics to plants that were impaired for mMDH activity. Additionally, biochemical analyses of antisense tomato mitochondrial NAD-dependent isocitrate dehydrogenase plants revealed clear reduction in flux through the TCA cycle, decreased levels of TCA cycle intermediates, and relatively few changes in photosynthetic parameters; however, fruit size and yield were reduced (Sienkiewicz-Porzucek et al., 2010). All those studies have been performed on the illuminated leaf; recently, it has characterized tomato plants independently exhibiting a fruit-specific decreased expression of genes encoding consecutive enzymes of the TCA cycle, fumarase, and mMDH (Centeno et al., 2011). Detailed biochemical characterization revealed that the changes in starch concentration, and consequently soluble solids content, were likely due to a redox regulation of AGPase. Those plants showed also a little effect on the total fruit yield as well as unanticipated changes in postharvest shelf life and susceptibility to bacterial infection. Despite the fact that much research work is needed to understand the exact mechanism for the increment in the fruit dry matter, manipulation of central organic acids is clearly a promising approach to enhance fruit yield (Nunes-Nesi et al., 2011).

Ethylene in Ripening

Based on the respiratory pattern and ethylene biosynthesis during ripening, fruits have been classified either as "climacteric" or "nonclimacteric." Climacteric fruits such as tomato show an increase in respiration rate and ethylene formation. Nonclimacteric fruits do not increase respiration, although they produce a little ethylene during ripening and do not respond to external ethylene treatment (Giovannoni, 2001). This difference is one of the main reasons that the majority of biochemical research has concentrated on this hormone. The role of ethylene in ripening of climacteric fruits has been known for more than 50 years (see Chapter 3). Since then, considerable effort has been focused on the studies of ethylene biosynthesis (S-adenosylmethionine, SAM; SAM synthetase; 1-aminocyclopropane carboxylic acid; ACC synthase; and ACC oxidase), ethylene perception (ethylene receptors, ETRs); signal transduction (ethylene response factor, ERFs); and ethylene-regulated genes such as cell-wall-disassembling genes (endopolygalacturonase; pectin methyl esterase, PME; and pectate lyase).

The *Arabidopsis* model system has served as starting point in the knowledge of the steps involved in ethylene perception and signal transduction; however, more efforts in understanding the ethylene response during fruit ripening have focused on the characterization of tomato homologs (Giovannoni, 2007). In this vein, six ethylene receptors have been isolated in tomato (*ETHYLENE RECEPTOR1, LeETR1; ETHYLENE RECEPTOR2, LeETR2; ETHYLENE RECEPTOR5, LeETR5; NEVER-RIPE, NR; ETHYLENE RECEPTOR4, LeETR4;* and *ETHYLENE RECEPTOR6, LeETR6*) compared to five members in *Arabidopsis* (*ETHYLENE RECEPTOR1, ETR1; ETHYLENE RECEPTOR2, ETR2; ETHYLENE RESPONSE SENSOR1, ERS1; ETHYLENE RESPONSE SENSOR2, ERS2; and ETHYLENE INSENSITIVE4, EIN4*) (Bleecker, 1999; Chang and Stadler, 2001). Five of the six tomato receptors have shown to bind ethylene (Klee and Tieman, 2002; Klee, 2002) but expression studies have been shown different profiles. Transcript levels of *LeETR1*, *LeETR2*, and *LeETR5* change little upon treatment of ethylene in fruit, where *NR*, *LeETR4*, and *LeETR6* are strongly induced during ripening (Kevany et al., 2007). Interestingly, analysis of transgenic plants with reduced *LeETR4* and *LeETR6*, caused an early ripening phenotype (Kevany et al., 2007; Kevany et al., 2008). On the other hand, *NR* mutation resulted in not fully ripened fruit (Wilkinson et al., 1995; Yen et al., 1995). Nevertheless, analysis of transgenic plants with reduction in *NR* levels suggested that this gene was not necessary for ripening to proceed (Hackett et al., 2000), suggesting that the other fruit-specific member of the receptor family has compensatory upregulation (Tieman et al., 2000). Overexpression of the *NR* receptor in tomato resulted in reduced sensitivity in seedlings and mature plants (Ciardi et al., 2000). This is in agreement with models where ethylene receptors act as negative regulators of ethylene signaling (Klee and Tieman, 2002; Klee, 2002). Consistent with this model, an exposure of immature fruits to ethylene caused a reduction in the amount of ethylene receptor protein and earlier ripening (Kevany et al., 2007). Recently, further ethylene-inducible (*CONSTITUTIVE TRIPLE RESPONSES MAP kinase kinase, CTR*) family of four genes have been identified in tomato (*LeCTR1, LeCTR2, LeCTR3, and LeCTR4*). Like *NR, LeETR4,* and *LeETR6, LeCRT1* is also upregulated during ripening (Adams-Phillips et al., 2004). Recently, studies of two-hybrid yeast interaction assay of tomato ethylene receptor and LeCTR proteins have demonstrated that those proteins are capable of interacting with NR (Zhong et al., 2008), reinforcing the idea that ethylene receptors transmit the signal to the downstream CTRs.

Recently, genomics approaches have provided insight into primary ripening control upstream of ethylene (Chapter 8). Tomato pleiotropic ripening mutations, *ripening inhibitor* (*rin*), *non-ripening* (*nor*), and *Colorless nonripening* (C*nr)* have added great insights in this regard. The *rin, nor,* and *Cnr* mutations are affected in all aspects of the tomato fruit ripening process that are unable to respond to ripening-associated ethylene genes (Vrebalov et al., 2002; Manning et al., 2006). Furthermore, in fruits from those mutants, the ripening-associated ethylene genes are induced by exogenous ethylene indicating that all three genes operate upstream of ethylene biosynthesis and are involved in process controlled exclusively by ethylene. The three mutant loci encode putative transcription factors. The *rin* encoded a partially deleted MADS-box protein of the SEPALLATA clade (Hileman et al., 2006), where *Cnr* is an epigenetic change that alters the promoter methylation of SQUAMOSA promoter binding (SPB) proteins. Manning et al. (2006) and J. Vrebalov and J. Giovannoni (unpublished results) suggest that the *nor* loci encodes a transcription factor, although not a member of MADS-box family. The observed ethylene-independent aspect of ripening suggests that RIN, NOR, and CNR proteins are candidates for conserved molecular mechanisms of fruit in both the climacteric and nonclimacteric categories.

Biochemical evidence suggests that ethylene production may be influenced or regulated by interactions between its biosynthesis and other metabolic pathways. One such example is provided by the fact that SAM is the substrate for both the polyamine pathway and the nucleic acid methylation; the competition for substrate was demonstrated by the finding that the overexpression of a SAM hydrolase has been associated with inhibited ethylene production during ripening (Good et al., 1994). On the other hand, the methionine cycle directly links ethylene biosynthesis to the central pathways of primary metabolism.

Polyamines

The most common plant polyamines are the diamine putrescine and the higher polyamines spermidine and spermine and it is known to be implicated in different biological processes, including cell division, cell elongation, embryogenesis, root formation, floral development, fruit development and ripening, pollen tube growth and senescence, and in response to biotic and abiotic stress (Kaur-Sawhney et al., 2003). In plants, putrescine is synthesized from arginine, a reaction catalyzed by arginine decarboxylase, or from ornithine by ornithine decarboxylase. Spermidine is synthesized from putrescine and SAM. SAM as a key intermediate for ethylene (Good et al., 1994; Fluhr and Mattoo, 1996; Giovannoni, 2004) has the potential to commit the flux of SAM either into polyamine biosynthesis, ethylene biosynthesis, or both. The overexpression of a SAM hydrolase has been associated with inhibited ethylene production during ripening (Good et al., 1994) which led to suggestions that changes in the levels of polyamines and ethylene may influence specific physiological processes in the plant (Kaur-Sawhney et al., 2003).

Mattoo et al. (2007) produced tomato fruits with increased SAM decarboxylase, in an attempt to over-accumulate spermidine and spermine whose levels decline during normal ripening process in tomato (Mehta et al., 2002). In the metabolite levels, those fruits showed prominent changes which influence multiple cellular pathways in diverse subcellular compartments such as mitochondria, cytoplasm, chloroplasts, and chromoplasts during fruit ripening. Red fruits showed upregulation of phosphoenolpyruvate carboxylase (PEPC) and cytosolic isocitrate dehydrogenase (ICDHc) as well as increase in the levels of glutamate, glutamine, asparagine, and organic acids; those of aspartate, valine, glucose, and sucrose showed a decrease compared to the wild type. The authors suggested that spermidine and spermine are perceived as signals of carbon metabolism in order to optimize C and N budgets following similar N regulatory aspects as in roots or leaves (Corruzi and Zhou, 2001; Foyer and Noctor, 2002). Also these data revealed a role of polyamines in mitochondrial metabolic regulation suggested by upregulation of the mitochondrial cytochrome oxidase transcripts, higher respiratory activity as well as higher content of citrate, malate, and fumarate in the ripe transgenic fruits (Mattoo et al., 2006). Polyamines are also postulated to regulate stress responses as is shown in transgenic rice plants overexpressing arginine decarboxylase (Capell et al., 2004). Those plants resulted in activation of SAM decarboxylase and higher levels of spermidine and spermine which triggered drought tolerance. Further support for this role has been provided by a spermine mutant of Arabidopsis that displayed salt sensitivity (Yamaguchi et al., 2006). Various mechanisms have been invoked to explain the effects of polyamines; however, much research work is needed to understand how the plant cells sense threshold levels of polyamines, and what downstream signaling pathways are involved.

Volatiles

Metabolism in the fruit involves the conversion of high-molecular-weight precursors to smaller compounds that help to obtain viable seeds and to attract seed-dispersing species (Chapter 6). The flavor of fruit is generally determined from tens and hundreds of constituents, most of them generated during the ripening phase of the fruit growth and development process. The content of sugars and organic acids and the ratios between them play a significant role in the overall flavor of fruit. Indeed, sugar content has previously been regarded as the major quantitative factor determining this parameter (Park et al., 2006). Amino acids are other soluble components that contribute significantly to fruit flavor. In the case of tomato fruit, flavor—a valuable trait—is the sum of the interaction between sugars (principally glucose and fructose), acids (citric, malic, and ascorbic), and glutamate and approximately 400 volatile compounds (Petro-Turza, 1987; Buttery, 1993; Buttery and Ling, 1993; Fulton et al., 2002), although a smaller set of only 15–20 are made in sufficient quantities to have an impact on human perception (Baldwin et al., 2000). Any study on the metabolic pathways leading to their synthesis must be considered in the context of this developmental process. Thus, it is known that the rapid growth phase of the fruits act as strong sinks that import massive amounts of photoassimilates from photosynthesizing organs. The translocation of metabolites occurs in the phloem. Sucrose is the metabolite mostly translocated, although in some species other compounds are predominant as polyalcohols like mannitol or sorbitol, and even oligosaccharides. These translocated compounds, which are the result of the primary metabolism, are the precursors of most of the metabolites that account for the fruit flavor, generally classified as secondary metabolites. Thus, the synthesis of these compounds is necessarily supported by the supply of the primary photoassimilates.

Flavor perception is often described as a combination of taste and smell. Some of these primary metabolites can be essential components of taste since they might be, depending on the species, main components of the harvested fruits, being recognized by sweet taste receptors. Recently, a metabolomic approach was used to describe the phenotypic variation of a broad range of primary and volatile metabolites, across a series of tomato lines, resulting from crosses between a cherry tomato and three independent large fruit cultivar (Levovil, VilB, and VilD)

(Zanor et al., 2009b). The results of the most highly abundant primary metabolite analysis of cherry and large-fruited tomato lines were largely in accordance with those obtained from previous studies (Causse et al., 2002). The low sugar and high malate content of the Levovil parental and the corresponding very low sugar/acid ratio could explain the lower acceptance of the fruit by the food panel tasters, especially given that malate is perceived as sourer tasting than citrate (Marsh et al., 2003). In addition to the changes observed in sugars and acids in cherry tomatoes, the glutamate level, known to be sensed as the fifth basic taste (umami) which evokes a savory feeling, was found to be considerably higher in the cherry variety than in the large-fruited varieties. This finding is, additionally, in accordance with the fact that cherry tomatoes were found to be tastier than the other parental lines used in this study. Additionally, in this study considerable correlation within the levels of primary metabolites and volatile compounds, respectively, were also observed. However, there was relatively little association between the levels of primary metabolites and volatile compounds, implying that they are not tightly linked to one another with the exception of sucrose which showed a strong association with a number of volatile compounds (Zanor et al., 2009b).

A broad profiling of tomato volatiles on a tomato introgression line population harboring introgressions of the wild species *Solanum pennellii* yielded over 100 QTL that are reproducibly altered in one or more volatiles contributing to flavor (Tieman et al., 2006b). These QTL have been used as tools to identify the genes responsible for controlling the synthesis of many volatile compounds. Very few genes involved in the biosynthetic pathways of tomato flavor volatiles have been identified, although the detection of *malodorous*, a wild species allele that affects tomato aroma, allowed the identification of a QTL that is linked with a markedly undesirable flavor within the *S. pennellii* IL8-2 (Tadmor et al., 2002). A complementary approach, utilizing broad genetic crosses, has been used to identify QTL for organoleptic properties of tomatoes (Causse et al., 2002). The lines identified as preferable by consumer could now be comprehensively characterized with respect to volatile and nonvolatile compounds alike. By using a combination of metabolic and flux profiling alongside reverse genetic studies on IL8-2, it was possible to confirm the biological pathway of a set of phenylalanine-derived volatiles, 2-phenylacetaldehyde and 2-phenylethanol, important aromatic compounds in tomato (Tieman et al., 2006a). A combined metabolic, genomic, and biochemical analysis of glandular trichomes from the wild tomato species *Solanum Habrochaites* identified a key enzyme in the biosynthesis of methyl ketones that serve this purpose (Fridman et al., 2005). In recent years, there have been dramatic improvements in the knowledge of volatiles; however, there is still work to be done before it can be claimed that the understanding of their biosynthesis is comprehensive.

Cell Wall Metabolism

Fruit growth and ripening are complex developmental processes that involve many events contributing to the textural and constitutional changes in the fruits and determining their final composition. The metabolic changes during ripening include alteration of cell structure (Chapter 7), involves changes in cell wall thickness, permeability of plasma membrane, hydration of cell wall, decrease in the structural integrity, and increase in intracellular spaces (Redgwell et al., 1997). Cell wall disassembly rate and extent are crucial for the maintenance of fruit quality and integrity (Matas et al., 2009). For this reason, maintenance of firmness has long been the target for breeders in many crops to minimize postharvest decay.

The major textural changes resulting in the softening of fruit are due to enzyme-mediated alterations in the structure and composition of cell wall, partial or complete solubilization of the major classes of cell wall polysaccharides such as pectins and cellulose (Seymour et al., 1987; Tucker and Grierson, 1987; Redgwell et al., 1992), and hydrolysis of starch and other storage polysaccharides (Fuchs et al., 1980; Selvaraj and Kumar, 1989b). The activity of these enzymes is directly linked to the shelf life of the fruits and it is why those genes have been frequent targets for genetic engineering (Goulao and Oliveira, 2007; Vicente et al., 2007). Among cell wall hydrolases, pectin-degrading enzymes are mostly implicated in fruit softening. Increased solubilization of the pectin substances, progressive loss of tissue firmness, and a rapid rise in the polygalacturonase (PG) activity accompany normal ripening in many fruits (Brady, 1987; Fisher and Bennett, 1991). A positive correlation between PG activity and initiation of softening is known in a number of fruits like guava (El-Zoghbi, 1994), papaya (Paull and Chen, 1983), mango (Roe and Bruemmer, 1981), strawberry (Garcia-Gago et al., 2009; Quesada et al., 2009). However, experiments with transgenic tomatoes have shown that even though PG is important for the degradation of pectins, it is not sole determinant of tissue softening during ripening (Gray et al., 1992). PME catalyzes the de-esterification of pectin, and its activity together with PG increase remarkably during ripening in peach, tomato, pear, and strawberry (Tucker and Grierson, 1987; Osorio et al., 2010). The loss of neutral sugar side chains from pectin is one of the most important features occurring during ripening. Substantial variation in the cell wall composition among fruits exists and their metabolism in relation to softening also varies (Gross and Sams, 1984), that is, neutral sugar side chains are not lost in ripening plum and cucumber fruits (Gross and Sams, 1984). The mutant *rin* containing little or no PG activity showed a substantial loss of galactose from cell wall, suggesting that this loss is not due to the action of PG. This evidence suggests that other cell wall hydrolases play an important role in the texture softening during ripening (Gray et al., 1992). In parallel to these changes in the cell wall, in many fruits a dramatic increase in susceptibility to necrotrophic pathogens has been reported (Prusky, 1996). It is now accepted that cell wall disassembly can be a key component of this susceptibility (Flors et al., 2007; Cantu et al., 2008).

The regulation of texture and shelf life is clearly far more complex than was previously envisaged (see Chapter 7), and so new approaches are needed; a better understanding of the relationship between changes in the texture properties of specific fruit tissues as well as intact fruit "firmness" and shelf life. Polysaccharide degradation is not the sole determinant of fruit softening and other ripening-related physiological processes also play critical roles. The cuticle has a number of biological functions that could have an important impact on fruit quality and shelf life that include the ability to maintain fruit skin integrity (Hovav et al., 2007), restrict cuticular transpiration (Leide et al., 2007), and limit microbial infection. Other reports also highlight other processes that contribute to fruit softening such as turgor pressure (Saladie et al., 2007; Thomas et al., 2008; Wada et al., 2008) and the possible associated developmental changes in apoplastic solute accumulation (Wada et al., 2008).

Concluding Remarks

Metabolomics allows the identification of changes in chemical composition with agronomic value. The shift from single metabolite measurements to platforms that can provide information on hundreds of metabolites has led to the development of better models to describe the links both between metabolites and between metaboloms. It is likely that the combination of molecular marker sequence analysis, PCR amplification and sequencing, analysis of allelic variation, and evaluation of co-responses between gene expression and metabolite composition traits will allow the detection of both expression QTL (wherein the mechanism underlying the metabolic change is an alteration in transcript and by implication, in protein amount), as well as change in function in which the level of expression is unaltered. It is hoped that in the future, this approach will allow a comprehensive understanding of genetic and metabolic networks that govern fruit metabolism and its effect on compositional quality.

References

Adams-Phillips, L., Barry, C., Kannan, P., Leclercq, J., Bouzayen, M., and Giovannoni, J. (2004) Evidence that CTR1-mediated ethylene signal transduction in tomato is encoded by a multigene family whose members display distinct regulatory features. *Plant Molecular Biology*, 54, 387–404.

Agravante, J.U., Matsui, T., and Kitagawa, H. (1991) Sugars and organic acids in ethanol-treated and ethylene-treated banana fruits. *Journal of the Japanese Society for Food Science and Technology*, 38, 441–444.

Baldwin, E.A., Scott, J.W., Shewmaker, C.K., and Schuch, W. (2000) Flavor trivia and tomato aroma: biochemistry and possible mechanisms for control of important aroma components. *HortScience*, 35, 1013–1022.

Baur, J.A., and Sinclair, D.A. (2006) Therapeutic potential of resveratrol: the in vivo evidence. *Nature Reviews. Drug Discovery*, 5, 493–506.

Baxter, C.J., Carrari, F., Bauke, A., Overy, S., Hill, S.A., Quick, P.W., Fernie, A.R., and Sweetlove, L.J. (2005) Fruit carbohydrate metabolism in an introgression line of tomato with increased fruit soluble solids. *Plant and Cell Physiology*, 46, 425–437.

Beruter, J. (2004) Carbohydrate metabolism in two apple genotypes that differ in malate accumulation. *Journal of Plant Physiology*, 161, 1011–1029.

Biemelt, S., and Sonnewald, U. (2006) Plant-microbe interactions to probe regulation of plant carbon metabolism. *Journal of Plant Physiology*, 163, 307–318.

Blankenship, S.M., and Dole, J.M. (2003) 1-Methylcyclopropene: a review. *Postharvest Biology and Technology*, 28, 1–25.

Bleecker, A.B. (1999) Ethylene perception and signaling: an evolutionary perspective. *Trends in Plant Science*, 4, 269–274.

Boeing, H., Bechthold, A., Bub, A., Ellinger, S., Haller, D., Kroke, A., Leschik-Bonnet, E., Muller, M.J., Oberritter, H., Schulze, M., Stehle, P, and Watzl, B. (2012) Critical Review: vegetables and fruit in the prevention of chronic diseases. *European Journal of Nutrition*, 51, 637–663.

Brady, C.J. (1987) Fruit ripening. *Annual Review of Plant Physiology*, 38, 155–178.

Bret-Harte, M.S., and Silk, W.K. (1994) Nonvascular, symplasmic diffusion of sucrose cannot satisfy the carbon demands of growth in the primary root tip of Zea mays L. *Plant Physiology*, 105, 19–33.

Brown, A.C., and Summers, W.L. (1985) Carbohydrate accumulation and color development in watermelon. *Journal of the American Society for Horticultural Science*, 110, 683–687.

Burkle, L., Hibberd, J.M., Quick, W.P., Kuhn, C., Hirner, B., and Frommer, W.B. (1998) The H + -sucrose cotransporter NtSUT1 is essential for sugar export from tobacco leaves. *Plant Physiology*, 118, 59–68.

Buttery, R.G. (1993) Quantitative and sensory aspects of flavour of tomato and other vegetables and fruit. In: *Flavor Science: Sensible Principles and Techniques* (eds. T.E. Acree, and R. Teranishi), pp. 259–286. American Chemical Society, Washington, DC.

Buttery, R.G., and Ling, L.C. (1993) Volatiles of tomato fruit and plant parts: relationship and biogenesis. In: *Bioactive Volatile Compounds from Plants* (eds. R. Teranishsi, R.G. Buttery, H. Sugisawa), pp. 23–34. ACS Books, Washington, D.C.

Buttriss, J. (2012) *Plant Foods and Health in Phytonutrients,* (eds. A. Salter, H. Wiseman, and G. Tucker), pp. 1–51. Wiley-Blackwell, Oxford.

Cantu, D., Vicente, A.R., Greve, L.C., Dewey, F.M., Bennett, A.B., Labavitch, J.M., and Powell, A.L. (2008) The intersection between cell wall disassembly, ripening, and fruit susceptibility to Botrytis cinerea. *Proceedings of the National Academy of Sciences of the United States of America*, 105, 859–864.

Capell, T., Bassie, L., and Christou, P. (2004) Modulation of the polyamine biosynthetic pathway in transgenic rice confers tolerance to drought stress. *Proceedings of the National Academy of Sciences of the United States of America*, 101, 9909–9914.

Carpaneto, A., Geiger, D., Bamberg, E., Sauer, N., Fromm, J., and Hedrich, R. (2005) Phloem-localized, proton-coupled sucrose carrier ZmSUT1 mediates sucrose efflux under the control of the sucrose gradient and the proton motive force. *The Journal of Biological Chemistry*, 280, 21437–21443.

Carrari, F., Fernie, A.R. (2006) Metabolic regulation underlying tomato fruit development. *Journal of Experimental Botany*, 57(9):1883–1897.

Carrari, F., Nunes-Nesi, A., Gibon, Y., Lytovchenko, A., Loureiro, M.E., and Fernie, A.R. (2003) Reduced expression of aconitase results in an enhanced rate of photosynthesis and marked shifts in carbon partitioning in illuminated leaves of wild species tomato. *Plant Physiology*, 133, 1322–1335.

Causse, M., Saliba-Colombani, V., Lecomte, L., Duffe, P., Rousselle, P., and Buret, M. (2002) QTL analysis of fruit quality in fresh market tomato: a few chromosome regions control the variation of sensory and instrumental traits. *Journal of Experimental Botany*, 53, 2089–2098.

Centeno, D.C., Osorio, S., Nunes-Nesi, A., Bertolo, A.L.F., Carneiro, R.T., Araújo, W.L., Steinhauser, M.C., Michalska, J., Rohrmann, J., Geigenberger, P., Oliver, S.N., Stitt, M., Carrari, F., Rose, J.K, and Fernie, A.R. (2011) On the crucial role of malate in transitory starch metabolism, ripening, total soluble solid content at harvest and post-harvest softening in tomato fruit. *The Plant Cell*, 23, 162–184.

Chang, C., and Stadler, R. (2001) Ethylene hormone receptor action in *Arabidopsis. BioEssays*, 23, 619–627.

Ciardi, J.A., Tieman, D.M., Lund, S.T., Jones, J.B., Stall, R.E., and Klee, H.J. (2000) Response to Xanthomonas campestris pv. vesicatoria in tomato involves regulation of ethylene receptor gene expression. *Plant Physiology*, 123, 81–92.

Corruzi, G.M., and Zhou, L. (2001) Carbon and nitrogen sensing and signaling in plants: emerging 'matrix effects'. *Current Opinion in Plant Biology*, 4, 247–253.

Do, P.T., Prudent, M., Sulpice, R., Causse, M., and Fernie, A.R. (2010) The influence of fruit load on the tomato pericarp metabolome in a Solanum chmielewskii introgression line population. *Plant Physiology*, 154, 1128–1142.

El-Zoghbi, M. (1994) Biochemical changes in some tropical fruits during ripening. *Food Chemistry*, 49, 33–37.

European Commission Directorate-General for Agriculture and Rural Development (2007) http://ec.europa.eu/agriculture/analysis/tradepol/worldmarkets/fruitveg/072007_en.pdf.

Fait, A., Hanhineva, K., Beleggia, R., Dai, N., Rogachev, I., Nikiforova, V.J., Fernie, A.R., and Aharoni, A. (2008) Reconfiguration of the achene and receptacle metabolic networks during strawberry fruit development. *Plant Physiology*, 148, 730–750.

Famiani, F., Walker, R.P., Tecsi, L., Chen, Z.H., Proietti, P., and Leegood, R.C. (2000) An immunohistochemical study of the compartmentation of metabolism during the development of grape (Vitis vinifera L.) berries. *Journal of Experimental Botany*, 51, 675–683.

Fisher, R.L., and Bennett, A.B. (1991) Role of cell wall hydrolases in fruit ripening. *Annual Review of Plant Physiology and Plant Molecular Biology*, 42, 675–703.

Flors, V., Leyva Mde, L., Vicedo, B., Finiti, I., Real, M.D., Garcia-Agustin, P., Bennett, A.B., and Gonzalez-Bosch, C. (2007) Absence of the endo-beta-1,4-glucanases Cel1 and Cel2 reduces susceptibility to Botrytis cinerea in tomato. *The Plant Journal: for cell and molecular biology*, 52, 1027–1040.

Fluhr, R., and Mattoo, A.K. (1996) Ethylene-biosynthesis and perception. *Critical Reviews in Plant Sciences*, 15, 479–523.

Foyer, C.H., and Noctor, G. (2002) *Photosynthetic Nitrogen Assimilation and Associated Carbon and Respiratory Metabolism*, Kluwer Academic Publishers, Boston, MA.

Fridman, E., Carrari, F., Liu, Y.S., Fernie, A.R., and Zamir, D. (2004) Zooming in on a quantitative trait for tomato yield using interspecific introgressions. *Science*, 305, 1786–1789.

Fridman, E., Wang, J., Iijima, Y., Froehlich, J.E., Gang, D.R., Ohlrogge, J., and Pichersky, E. (2005) Metabolic, genomic, and biochemical analyses of glandular trichomes from the wild tomato species Lycopersicon hirsutum identify a key enzyme in the biosynthesis of methylketones. *Plant Cell*, 17, 1252–1267.

Fridman, E., and Zamir, D. (2003) Functional divergence of a syntenic invertase gene family in tomato, potato, and Arabidopsis. *Plant Physiology*, 131, 603–609.

Fuchs, Y., Pesis, E., and Zauberman, G. (1980) Changes in amylase activity, starch and sugar contents in mango fruit pulp. *Horticulture Science*, 13, 155–160.

Fulton, T.M., Buchell, P., Voirol, E., Lopez, J., Petiard, V., and Tanksley, S.D. (2002) Quantitative trait loci (QTL) affecting sugars, organic acids and other biochemical properties possible contributing to flavour, identified in four advanced backcross populations of tomato. *Euphytica*, 127, 163–177.

Garcia-Gago, J.A., Pose, S., Munoz-Blanco, J., Quesada, M.A., and Mercado, J.A. (2009) The polygalacturonase FaPG1 gene plays a key role in strawberry fruit softening. *Plant Signaling and Behavior*, 4, 766–768.

Giovannoni, J. (2001) Molecular Biology of Fruit Maturation and Ripening. *Annual Review of Plant Physiology and Plant Molecular Biology*, 52, 725–749.

Giovannoni, J. (2004) Genetic regulation of fruit development and ripening. *The Plant Cell*, 16, 170–180.

Giovannoni, J.J. (2007) Fruit ripening mutants yield insights into ripening control. *Current Opinion in Plant Biology*, 10, 283–289.

Godt, D.E., and Roitsch, T. (1997) Regulation and tissue-specific distribution of mRNAs for three extracellular invertase isoenzymes of tomato suggests an important function in establishing and maintaining sink metabolism. *Plant Physiology*, 115, 273–282.

Good, X., Kellogg, J.A., Wagoner, W., Langhoff, D., Matsumura, W., and Bestwick, R.K. (1994) Reduced ethylene synthesis by transgenic tomatoes expressing S-adenosylmethionine hydrolase. *Plant Molecular Biology*, 26, 781–790.

Goodenough, P.W., Prosser, I.M., and Young, K. (1985) NADP-linked malic enzyme and malate metabolism in ageing tomato fruit. *Phytochemistry*, 24, 1157–1162.

Goulao, L.F., and Oliveira, C.M. (2007) Cell wall modifications during fruit ripening: when a fruit is not a fruit. *Trends in Food Science & Technology*, 19, 4–25.

Gray, J., Picton, S., Shabbeer, J., Schuch, W., and Grierson, D. (1992) Molecular biology of fruit ripening and its manipulation with antisense genes. *Plant Molecular Biology*, 19, 69–87.

Gross, K.C., and Sams, C.E. (1984) Changes in cell wall neutral sugar composition during fruit ripening: a species survey. *Phytochemistry*, 23, 2457–2461.

Hackett, R.M., Ho, C.W., Lin, Z., Foote, H.C., Fray, R.G., and Grierson, D. (2000) Antisense inhibition of the Nr gene restores normal ripening to the tomato Never-ripe mutant, consistent with the ethylene receptor-inhibition model. *Plant Physiology*, 124, 1079–1086.

Hale, C.R. (1962) Synthesis of organic acids in fruit of grape. *Nature*, 195, 917–918.

Heyer, A.G., Raap, M., Schroeer, B., Marty, B., and Willmitzer, L. (2004) Cell wall invertase expression at the apical meristem alters floral, architectural, and reproductive traits in Arabidopsis thaliana. *The Plant Journal: for cell and molecular biology*, 39, 161–169.

Hileman, L.C., Sundstrom, J.F., Litt, A., Chen, M., Shumba, T., and Irish, V.F. (2006) Molecular and phylogenetic analyses of the MADS-box gene family in tomato. *Molecular Biology and Evolution*, 23, 2245–2258.

Hovav, R., Chehanovsky, N., Moy, M., Jetter, R., and Schaffer, A.A. (2007) The identification of a gene (Cwp1), silenced during Solanum evolution, which causes cuticle microfissuring and dehydration when expressed in tomato fruit. *The Plant Journal: for cell and molecular biology*, 52, 627–639.

Jayas, D.S., and Jeyamkondan, S. (2002) Modified Atmosphere Storage of Grains, Meats, Fruits and Vegetables. *Biosystems Engineering*, 82, 235–251.

Kaur-Sawhney, R., Tiburcio, A.F., Altabella, T., and Galston, A.W. (2003) Polyamines in plants: an overview. *Journal of Cell and Molecular Biology*, 2, 1–12.

Kevany, B.M., Taylor, M.G., and Klee, H.J. (2008) Fruit-specific suppression of the ethylene receptor LeETR4 results in early-ripening tomato fruit. *Plant Biotechnology Journal*, 6, 295–300.

Kevany, B.M., Tieman, D.M., Taylor, M.G., Cin, V.D., and Klee, H.J. (2007) Ethylene receptor degradation controls the timing of ripening in tomato fruit. *The Plant Journal: for cell and molecular biology*, 51, 458–467.

Klee, H.J. (2002) Control of ethylene-mediated processes in tomato at the level of receptors. *Journal of Experimental Botany*, 53, 2057–2063.

Klee, H., and Tieman, D. (2002) The tomato ethylene receptor gene family: form and function. *Physiologia Plantarum*, 115, 336–341.

Kliewer, W.M., Howarth, L., and Omori, M. (1967) Concentrations of tartaric acid and malic acid and their salts in vitis vinifera grapes. *American Journal of Enology and Viticulture*, 18, 42–54.

Kortstee, A.J., Appeldoorn, N.J., Oortwijn, M.E., and Visser, R.G. (2007) Differences in regulation of carbohydrate metabolism during early fruit development between domesticated tomato and two wild relatives. *Planta*, 226, 929–939.

Kuhn, C., Hajirezaei, M.R., Fernie, A.R., Roessner-Tunali, U., Czechowski, T., Hirner, B., and Frommer, W.B. (2003) The sucrose transporter StSUT1 localizes to sieve elements in potato tuber phloem and influences tuber physiology and development. *Plant Physiology*, 131, 102–113.

LeClere, S., Schmelz, E.A., and Chourey, P.S. (2008) Cell wall invertase-deficient miniature1 kernels have altered phytohormone levels. *Phytochemistry*, 69, 692–699.

Leide, J., Hildebrandt, U., Reussing, K., Riederer, M., and Vogg, G. (2007) The developmental pattern of tomato fruit wax accumulation and its impact on cuticular transpiration barrier properties: effects of a deficiency in a beta-ketoacyl-coenzyme A synthase (LeCER6). *Plant Physiology*, 144, 1667–1679.

Lo Bianco, R., and Rieger, M. (2002) Partitioning of sorbitol and sucrose catabolism within peach fruit. *Journal of the American Society for Horticultural Science*, 127, 115–121.

Lytovchenko, A., Sonnewald, U., and Fernie, A.R. (2007) The complex network of non-cellulosic carbohydrate metabolism. *Current Opinion in Plant Biology*, 10, 227–235.

Manning, K., Tor, M., Poole, M., Hong, Y., Thompson, A.J., King, G.J., Giovannoni, J.J., and Seymour, G.B. (2006) A naturally occurring epigenetic mutation in a gene encoding an SBP-box transcription factor inhibits tomato fruit ripening. *Nature Genetics*, 38, 948–952.

Marsh, K.B., Rossiter, K., Lau, K., Walker, S., Gunson, A., and MacRae, E.A. (2003) Using fruit pulps to explore flavour in kiwifruit. *Acta Horticulturae*, 610, 229–238.

Matas, A.J., Gapper, N.E., Chung, M.Y., Giovannoni, J.J., and Rose, J.K. (2009) Biology and genetic engineering of fruit maturation for enhanced quality and shelf-life. *Current Opinion in Biotechnology*, 20, 197–203.

Mattoo, A.K., Sobolev, A.P., Neelam, A., Goyal, R.K., Handa, A.K., and Segre, A.L. (2006) Nuclear magnetic resonance spectroscopy-based metabolite profiling of transgenic tomato fruit engineered to accumulate spermidine and spermine reveals enhanced anabolic and nitrogen-carbon interactions. *Plant Physiology*, 142, 1759–1770.

Mattoo, A.K., Chung, S.H., Goyal, R.K., Fatima, T., Solomos, T., Srivastava, A., Handa, A.K. (2007) Overaccumulation of higher polyamines in ripening transgenic tomato fruit revives metabolic memory, upregulates anabolism-related genes, and positively impacts nutritional quality. *Journal of AOAC INTERNATIONAL*, 90(5): 1456–1464.

Mehta, R.A., Cassol, T., Li, N., Ali, N., Handa, A.K., and Mattoo, A.K. (2002) Engineered polyamine accumulation in tomato enhances phytonutrient content, juice quality, and vine life. *Nature Biotechnology*, 20, 613–618.

Meyer, S., Lauterbach, C., Niedermeier, M., Barth, I., Sjolund, R.D., and Sauer, N. (2004) Wounding enhances expression of AtSUC3, a sucrose transporter from Arabidopsis sieve elements and sink tissues. *Plant Physiology*, 134, 684–693.

Miller, M.E., and Chourey, P.S. (1992) The Maize Invertase-Deficient miniature-1 Seed Mutation Is Associated with Aberrant Pedicel and Endosperm Development. *The Plant Cell*, 4, 297–305.

Moing, A., Renaud, C., Gaudillere, M., Raymond, P., Roudeillac, P., and Denoyes-Rothan, B. (2001) Biochemical changes during fruit development of four strawberry cultivars. *Journal of the American Society for Horticultural Science*, 126, 394–403.

Mounet, F., Moing, A., Garcia, V., Petit, J., Maucourt, M., Deborde, C., Bernillon, S., Le Gall, G., Colquhoun, I., Defernez, M., Giraudel, J.L., Rolin, D., Rothan, C., and Lemaire-Chamley, M. (2009) Gene and metabolite regulatory network analysis of early developing fruit tissues highlights new candidate genes for the control of tomato fruit composition and development. *Plant Physiology*, 149, 1505–1528.

Nunes-Nesi, A., Araujo, W.L., and Fernie, A. (2011) Targeting mitochondrial metabolism and machinery as a means to enhance photosynthesis. *Plant Physiology*, 155, 101–107.

Nunes-Nesi, A., Carrari, F., Gibon, Y., Sulpice, R., Lytovchenko, A., Fisahn, J., Graham, J., Ratcliffe, R.G., Sweetlove, L.J., and Fernie, A.R. (2007) Deficiency of mitochondrial fumarase activity in tomato plants impairs photosynthesis via an effect on stomatal function. *Plant Journal*, 50, 1093–106.

Nunes-Nesi, A., Carrari, F., Lytovchenko, A., Smith, A.M., Loureiro, M.E., Ratcliffe, R.G., Sweetlove, L.J., and Fernie, A.R. (2005) Enhanced photosynthetic performance and growth as a consequence of decreasing mitochondrial malate dehydrogenase activity in transgenic tomato plants. *Plant Physiology*, 137, 611–622.

Osorio, S., Bombarely, A., Giavalisco, P., Usadel, B., Stephens, C., Aragüez, I., Medina-Escobar, N., Botella, M.A., Fernie, A.R., and Valpuesta, V. (2010) Demethylation of oligogalacturonides by FaPE1 in the fruits of the wild strawberry Fragaria vesca triggers metabolic and transcriptional changes associated to defence and development of the fruit. *Journal of Experimental Botany*, 62, 2855–2873.

Park, Y.S., Jung, S.T., Kang, S.G., Drzewiecki, J., Namiesnik, J., Haruenkit, R., Barasch, D., Trakhtenberg, S., and Gorinstein, S. (2006) In vitro studies of polyphenols, antioxidants and other dietary indices in kiwifruit (Actinidia deliciosa). *International Journal of Food Sciences and Nutrition*, 57, 107–122.

Paull, R.E., and Chen, N.J. (1983) Postharvest Variation in Cell Wall-Degrading Enzymes of Papaya (Carica papaya L.) during Fruit Ripening. *Plant Physiology*, 72, 382–385.

Petro-Turza, M. (1987) Flavor of tomato and tomato products. *Food Review International*, 2, 309–351.

Picton, S., Barton, SL., Bouzeyen, M., Hamilton, AJ, and Grierson, D. (1993) Altered fruit ripening and leaf senescence in tomatoes expressing an antisense ethylene-forming enzyme transgene. *The Plant Journal: for cell and molecular biology*, 3, 469–481.

Prusky, D. (1996) Pathogen quiescence in postharvest diseases. *Annual Review of Phytopathology*, 34, 413–434.

Quesada, M.A., Blanco-Portales, R., Pose, S., Garcia-Gago, J.A., Jimenez-Bermudez, S., Munoz-Serrano, A., Caballero, J.L., Pliego-Alfaro, F., Mercado, J.A., and Munoz-Blanco, J. (2009) Antisense down-regulation of the FaPG1 gene reveals an unexpected central role for polygalacturonase in strawberry fruit softening. *Plant Physiology*, 150, 1022–1032.

Redgwell, R.J., MacRae, E.A., Hallet, I., Fischer, M., Perry, J., and Harker, R. (1997) *In vivo* and *in vitro* swelling of cell walls during fruit ripening. *Planta*, 203, 162–173.

Redgwell, R.J., Melton, L.D., and Brasch, D.J. (1992) Cell wall dissolution in ripening kiwifruit (Actinidia deliciosa): solubilization of the pectic polymers. *Plant Physiology*, 98, 71–81.

Riesmeier, J.W., Hirner, B., and Frommer, W.B. (1993) Potato sucrose transporter expression in minor veins indicates a role in phloem loading. *The Plant Cell*, 5, 1591–1598.

Roe, B., and Bruemmer, J.H. (1981) Changes in pectin substances and enzymes during ripening and storage of "keitt" mangos. *Journal of Food Science*, 46, 186–189.

Roitsch, T., and Gonzalez, M.C. (2004) Function and regulation of plant invertases: sweet sensations. *Trends in Plant Science*, 9, 606–613.

Ruffner, H.P. (1982) Metabolism of tartaric and malic acids in Vitis: a review – Part A. *Vitis,* 21, 247–259 and 346–358.

Ruffner, H.P., and Hawker, J.S. (1977) Control of glycolysis in ripening berries of Vitis vinifera. *Phytochemistry*, 16, 1171–1175.

Saladie, M., Matas, A.J., Isaacson, T., Jenks, M.A., Goodwin, S.M., Niklas, K.J., Xiaolin, R., Labavitch, J.M., Shackel, K.A., Fernie, A.R., Lytovchenko, A., O'Neill, M.A., Watkins, C.B., and Rose, J.K. (2007) A reevaluation of the key factors that influence tomato fruit softening and integrity. *Plant Physiology*, 144, 1012–1028.

Saltmarsh, M., Crozier, AC, and Ratcliffe, B. (2003) *Fruit and Vegetables in Plants: Diet and Health,* (ed. G. Goldberg), pp. 107–133. Blackwood Publishing.

Saradhuldhat, P., and Paull, R.E. (2007) Pineapple organic acid metabolism and accumulation during fruit development. *Scientia Horticulturae*, 112, 297–303.

Sauer, N., Ludwig, A., Knoblauch, A., Rothe, P., Gahrtz, M., and Klebl, F. (2004) AtSUC8 and AtSUC9 encode functional sucrose transporters, but the closely related AtSUC6 and AtSUC7 genes encode aberrant proteins in different Arabidopsis ecotypes. *The Plant Journal: for cell and molecular biology*, 40, 120–130.

Schaffer, A.A., and Petreikov, M. (1997) Sucrose-to-Starch Metabolism in Tomato Fruit Undergoing Transient Starch Accumulation. *Plant Physiology*, 113, 739–746.

Schauer, N., Semel, Y., Roessner, U., Gur, A., Balbo, I., Carrari, F., Pleban, T., Perez-Melis, A., Bruedigam, C., Kopka, J., Willmitzer, L., Zamir, D., and Fernie, A.R. (2006) Comprehensive metabolic profiling and phenotyping of interspecific introgression lines for tomato improvement. *Nature Biotechnology*, 24, 447–454.

Selvaraj, Y., and Kumar, R. (1989a) Pyruvate kinase activity in ripening mando (Mangifera indica L.) fruit. *Journal Of Food Science and Technology Mysore*, 26, 228–229.

Selvaraj, Y., and Kumar, R. (1989b) Studies on fruit softening enzymes and polyphenol oxidase activity in ripening mando (*Mangifera indica* L.) fruit. *Journal of Food Science and Technology*, 26, 218–222.

Sergeeva, L.I., Keurentjes, J.J., Bentsink, L., Vonk, J., van der Plas, L.H., Koornneef, M., and Vreugdenhil, D. (2006) Vacuolar invertase regulates elongation of Arabidopsis thaliana roots as revealed by QTL and mutant analysis. *Proceedings of the National Academy of Sciences of the United States of America*, 103, 2994–2999.

Seymour, G.B., Harding, S.E., Taylor, M.G., Hobson, G.E., and Tucker, G.A. (1987) Polyuronide solubilization during ripening of normal and mutant tomato fruit. *Phytochemistry*, 26, 1871–1875.

Sienkiewicz-Porzucek, A., Sulpice, R., Osorio, S., Krahnert, I., Leisse, A., Urbanczyk-Wochniak, E., Hodges, M., Fernie, A.R., and Nunes-Nesi, A. (2010) Mild Reductions in Mitochondrial NAD-Dependent Isocitrate Dehydrogenase Activity Result in Altered Nitrate Assimilation and Pigmentation But Do Not Impact Growth. *Molecular Plant*, 3, 156–173.

Sonnewald, U., Hajirezaei, M.R., Kossmann, J., Heyer, A., Trethewey, R.N., and Willmitzer, L. (1997) Increased potato tuber size resulting from apoplastic expression of a yeast invertase. *Nature Biotechnology*, 15, 794–797.

Stanner, S.A., Hughes, J., Kelly, C.N.M., and Buttriss, J. (2004) A review of the epidemiological evidence for the "antioxidant hypothesis". *Public Health Nutrition*, 7, 407–422.

Tadmor, Y., Fridman, E., Gur, A., Larkov, O., Lastochkin, E., Ravid, U., Zamir, D., and Lewinsohn, E. (2002) Identification of malodorous, a wild species allele affecting tomato aroma that was selected against during domestication. *Journal of Agricultural and Food Chemistry*, 50, 2005–2009.

Thomas, T.R., Shackel, K.A., and Matthews, M.A. (2008) Mesocarp cell turgor in Vitis vinifera L. berries throughout development and its relation to firmness, growth, and the onset of ripening. *Planta*, 228, 1067–1076.

Tieman, D., Taylor, M., Schauer, N., Fernie, A.R., Hanson, A.D., and Klee, H.J. (2006a) Tomato aromatic amino acid decarboxylases participate in synthesis of the flavor volatiles 2-phenylethanol and 2-phenylacetaldehyde. *Proceedings of the National Academy of Sciences of the United States of America*, 103, 8287–8292.

Tieman, D.M., Taylor, M.G., Ciardi, J.A., and Klee, H.J. (2000) The tomato ethylene receptors NR and LeETR4 are negative regulators of ethylene response and exhibit functional compensation within a multigene family. *Proceedings of the National Academy of Sciences of the United States of America*, 97, 5663–5668.

Tieman, D.M., Zeigler, M., Schmelz, E.A., Taylor, M.G., Bliss, P., Kirst, M., and Klee, H.J. (2006b) Identification of loci affecting flavour volatile emissions in tomato fruits. *Journal of Experimental Botany*, 57, 887–896.

Tucker, G.A., and Grierson, D. (1987) Fruit ripening. In: *The Biochemistry of Plants* (ed. D. Davies), pp. 265–319. Academic Press Inc., New York, NY.

Usenik, V., Fabcic, J., and Stampar, F. (2008) Sugars, organic acid. Phenolic composition and antioxidant activity of sweet cherry (Prunus avium L.). *Food Chemistry*, 107, 185–192.

Vicente, A.R., Saladié, M., Rose, J.K.C., and Labavitch, J.M. (2007) The linkage between cell wall metabolism and fruit softening: looking to the future. *Journal of the Science of Food and Agriculture*, 87, 1435–1448.

Viola, R., Roberts, A.G., Haupt, S., Gazzani, S., Hancock, R.D., Marmiroli, N., Machray, G.C., and Oparka, K.J. (2001) Tuberization in potato involves a switch from apoplastic to symplastic phloem unloading. *The Plant Cell*, 13, 385–398.

Vrebalov, J., Ruezinsky, D., Padmanabhan, V., White, R., Medrano, D., Drake, R., Schuch, W., and Giovannoni, J. (2002) A MADS-box gene necessary for fruit ripening at the tomato ripening-inhibitor (rin) locus. *Science*, 296, 343–346.

Wada, H., Shackel, K.A., and Matthews, M.A. (2008) Fruit ripening in Vitis vinifera: apoplastic solute accumulation accounts for pre-veraison turgor loss in berries. *Planta*, 227, 1351–1361.

Walton, E.F., and De Jong, T.M. (1990) Growth and compositional changes in kiwifruit berries from three Californian locations. *Annals of Botany*, 66, 285–298.

Wilkinson, J.Q., Lanahan, M.B., Yen, H.C., Giovannoni, J.J., and Klee, H.J. (1995) An ethylene-inducible component of signal transduction encoded by never-ripe. *Science*, 270, 1807–1809.

Wobus, U., and Weber, H. (1999) Sugars as signal molecules in plant seed development. *Biological Chemistry*, 380, 937–944.

WRCF and American Institute for Cancer Research (2007) *Food, Nutrition, Physical Activity and the Prevention of Cancer: a Global Perspective*, American Institute for Cancer Research, Washington, DC.

Yamaguchi, K., Takahashi, Y., Berberich, T., Imai, A., Miyazaki, A., Takahashi, T., Michael, A., and Kusano, T. (2006) The polyamine spermine protects against high salt stress in Arabidopsis thaliana. *FEBS Letters*, 580, 6783–6788.

Yelle, S., Chetelat, R.T., Dorais, M., Deverna, J.W., and Bennett, A.B. (1991) Sink metabolism in tomato fruit : IV. Genetic and biochemical analysis of sucrose accumulation. *Plant Physiology*, 95, 1026–1035.

Yen, H.C., Lee, S., Tanksley, S.D., Lanahan, M.B., Klee, H.J., and Giovannoni, J.J. (1995) The tomato Never-ripe locus regulates ethylene-inducible gene expression and is linked to a homolog of the Arabidopsis ETR1 gene. *Plant Physiology*, 107, 1343–1353.

Zafra-Stone, S., Yasmin, T., Bagchi, M., Chatterjee, A., Vinson, J A., and Bagchi, D. (2007) Berry anthocyanins as novel antioxidants in human health and disease prevention. *Molecular Nutrition and Food Research*, 51, 675–683.

Zanor, M.I., Osorio, S., Nunes-Nesi, A., Carrari, F., Lohse, M., Usadel, B., Kuhn, C., Bleiss, W., Giavalisco, P., Willmitzer, L., Sulpice, R., Zhou, Y.H., and Fernie, A.R. (2009a) RNA interference of LIN5 in tomato confirms its role in controlling Brix content, uncovers the influence of sugars on the levels of fruit hormones, and demonstrates the importance of sucrose cleavage for normal fruit development and fertility. *Plant Physiology*, 150, 1204–1218.

Zanor, M.I., Rambla, J.L., Chaib, J., Steppa, A., Medina, A., Granell, A., Fernie, A.R., and Causse, M. (2009b) Metabolic characterization of loci affecting sensory attributes in tomato allows an assessment of the influence of the levels of primary metabolites and volatile organic contents. *Journal of Experimental Botany*, 60, 2139–2154.

Zhong, S., Lin, Z., and Grierson, D. (2008) Tomato ethylene receptor-CTR interactions: visualization of NEVER-RIPE interactions with multiple CTRs at the endoplasmic reticulum. *Journal of Experimental Botany*, 59, 965–972.

Zrenner, R., Schuler, K., and Sonnewald, U. (1996) Soluble acid invertase determines the hexose-to-sucrose ratio in cold-stored potato tubers. *Planta*, 198, 246–252.

2 Fruit—An Angiosperm Innovation

Sandra Knapp and Amy Litt

Introduction

The possession of fruits is part of the definition of the angiosperms; although commonly referred to as the flowering plants, their very Latin name, angio—hidden, sperm—seed, refers to structures surrounding and protecting the next generation—the seeds. The angiosperms, or flowering plants (subclass Magnoliidae, sensu Chase and Reveal, 2009) comprise some 250–400,000 species in four major clades (APG, 2009) that are the dominant components of vegetation in most tropical, subtropical, and temperate zones worldwide. Angiosperms are tiny herbs to large canopy trees, occupy terrestrial and aquatic (including marine) habitats, and are the source of all major crop plants. As such, human beings have a long and intimate association with angiosperms, and most particularly with angiosperm fruits and seeds. The majority of staple crop plants (such as rice, maize, wheat, peas, and beans) are derived from seeds, and thus from angiosperm fruit.

A huge variety of fruit types is found in flowering plants (Spujt, 1994); as many as 150 different fruit types have been described in the literature—Corner (1964) has characterized fruits as "bewildering in their diversity." The eighteenth-century German botanist Joseph Gaertner (1788) defined a fruit as a structure derived from a mature ovary containing seeds, but many structures we might define as fruit are in fact composed of not only these tissue types. Various floral and vegetative parts can also be parts of a fruit—for example, fleshy parts of strawberries or figs are formed from the receptacle (base) of the flower, the pineapple "fruit" is composed of the coalesced parts of many flowers, and a rose hip is an expanded receptacle filled with achenes. Van der Pijl (1982) used the concept of the dispersal unit as his definition of a fruit; this links the fruit to their role in the ecology of the plant life cycle as the mechanism by which seeds are dispersed. Dispersal units occur in land plants other than angiosperms as well; some Carboniferous seed ferns had fleshy structures that are thought to have been an attractant for reptiles (Tiffney, 1986, see below) and the seed of the "living fossil" *Gingko biloba* is enveloped in a soft fleshy covering called the sarcotesta. These propagules are not considered true fruits, despite functional similarities. True fruits are limited to the angiosperms and are derived from the

---

The Molecular Biology and Biochemistry of Fruit Ripening, First Edition.
Edited by Graham B. Seymour, Mervin Poole, James J. Giovannoni and Gregory A. Tucker.

gynoecium: the collection of carpels in a flower, the carpel being that structure that encloses the seed (see below for developmental definitions). The origin of the carpel itself is not at all clear, but several new ideas rooted in phylogenetics and in developmental anatomy are currently being tested (see Frohlich and Chase, 2007 for a review). Among these are the "gametoheterotopy" theory of Meyen (1988) in which the carpel originates as a homeotic transformation in pollen-bearing structures that resulted in ovules on a flat surface that then folded, and the "mostly male" theory (Frohlich and Parker, 2000) in which female structures were expressed ectopically on male parts, creating the carpel precursor. Fruits are interpreted as those structures that have allowed for or that triggered angiosperm diversification and dominance through enabling the efficient dispersal of seeds (Lorts et al., 2008).

In discussing the evolution of fruit types, authors have usually either used many terms (Spujt, 1994) or have tried to simplify fruit types in the group in question (see Knapp, 2002; Lorts et al., 2008). The myriad fruit types recognized in angiosperms (see Figs. 2.1, 2.2, 2.3, 2.4, 2.5) can be simplified using combinations of the following characteristics: dehiscence or nondehiscence, dry or fleshy exterior, and free (apocarpous) or fused (syncarpous) carpels (see Table 2.1). A capsule, such as a milkweed or poppy fruit, is a syncarpous dehiscent dry fruit, a nut, like hazelnut or acorn, is unicarpellate, dry, and indehiscent, while a berry, like a tomato or grape, is syncarpous, indehiscent, and fleshy. Fruits of plants like columbines or delphiniums are apocarpous (with free carpels), dry, and dehiscent; these are termed follicles (see also examples in Table 2.1 and Fig. 2.2). Syncarpy occurs in 83% of extant angiosperm species, and has been interpreted as facilitating pollen tube competition and thus fitness (Endress, 1982); syncarpy has arisen in both the monocots and the eudicots independently (Endress, 2011). Endress (1982) has suggested that syncarpy provides for more possibilities for adaptive radiation in fruit type than does apocarpy through greater possibility for development of fleshiness and/or dehiscence mechanisms.

Our understanding of angiosperm relationships has improved greatly due to the advent of DNA sequencing and its use in phylogeny reconstruction. Previous ideas about the polarity of morphological characteristics were based on the assumption that particular types of floral or fruit morphology were "primitive" and thus were the ancestral states (Corner, 1964; Cronquist, 1981). Thus flowers/fruits with free parts (many free petals, many free carpels, like *Magnolia*) were thought to be ancestral, and those with fused parts (tubular flowers, fused carpels, like morning glories, *Ipomoea*) were advanced. The traditional belief has been that reproductive features are best suited for assessing relationships (Baker, 1959). They have been viewed as "phylogenetically conservative" and thus given great weight in taxonomy (Spujt, 1994), and have been considered the best indicators of evolutionary trends (Corner, 1949). Overreliance on the apparent stability and evolutionary importance of reproductive characters in the past has confounded rigorous study of patterns of fruit evolution. New views on the relationships of the orders and families of the flowering plants based on DNA sequence analysis (APG, 1998, 2003, 2009) have allowed us to better examine the evolution of fruit types in a phylogenetic context and to separate ideas of evolution from functional states.

The current view of angiosperm relationships recognizes three large groupings that are referred to as the ANA (ANITA) grade (*Amborella*, the sister group to all other angiosperms, waterlilies, and Austrobaileyaceae), the magnoliids (magnolias and their relatives including

Table 2.1 Some commonly encountered fruit type names with some commonly encountered examples (see also Figures 2.1, 2.2, 2.3, 2.4, and 2.5). A great many more complex fruit types occur, such as schizocarps (also called nutlets or mericarps), in which a syncarpous ovary develops into a fruit that splits at maturity simulating a fruit derived from an apocarpous gynoecium; aggregate fruit types are also common. For a comprehensive dictionary of fruit types and a more complete treatment of the many synonyms for these simplistic renditions, see Spujt, 1994.

	Definition	Examples	Notes
Dry indehiscent fruits			
Achene	One-seeded, seed attached to fruit wall at one point only	Many members of the Asteraceae; sunflower, lettuce; strawberries (true fruit is the pip on the outside)	
Grain (caryopsis)	One-seeded, seed coat fused to seed	Wheat, rice, most grasses	
Nut	One-seeded, pericarp hard	Hazelnuts, acorns	
Samara	One-seeded, winged	Maples, dipterocarps	
Dry dehiscent fruits			
Capsule	Syncarpous gynoecium of two or more carpels, opening in a variety of ways	Tobacco, poppies, violet	The types of capsules are defined on how and where they open
Follicle	Apocarpous gynoecium of one carpel, splits along one suture	Columbine, magnolia	
Legume	Apocarpous gynoecium of one carpel, splits along two sutures	Fabaceae such as peas and beans	All fruits of the pea family are generally referred to as legumes, but can be variously modified
Silique or silicle	Syncarpous gynoecium of two or more carpels, splits along two sutures, has a persistent partition	Brassicaceae such as *Arabidopsis* or canola	Silique was traditionally defined as a silicle wider than long, but now used more generally
Fleshy indehiscent fruits			
Berry	Syncarpous gynoecium of one-several fused carpels, one-many seeds imbedded in fleshy mass	Grape, tomato, pepper, orange (hesperidium)	Also called a bacca (Latin)
Drupe/pyrene	Seed(s) enclosed in a stony endocarp (see Bobrov et al., 2005 for *Amborella*)	Peach, almond, holly	Drupelets are aggregate fruits composed of many tiny drupes (e.g., raspberries); pyrenes are those fruits with two or more seeds enclosed in a hard endocarp

Piperaceae), and the mesangiosperms (monocots + eudicots) (APG, 2009; Chase and Reveal, 2009; see Fig. 2.6). The traditionally recognized dicotyledons (comprising the ANA grade, magnoliids, and eudicots of the APG III system) are not monophyletic, and previous scenarios for fruit evolution based on the divergence of monocotyledons and dicotyledons are no longer relevant. The core eudicots (a monophyletic clade within the eudicots, see Fig. 2.6) comprise two major lineages, the rosids (fabids + malvids) and the asterids (lamiids + campanulids), which between them account for the bulk of angiosperm species diversity (APG, 2009). For orders

Figure 2.1 (a) *Amborella trichopoda* (Amborellaceae), aggregation of fruitlets (drupes) from free carpels, photo M. Christenhusz, (b) *Gyrocarpus americanus* (Hernandiaceae; Laurales), nuts with elongate apical wings, photo M. Peña-Chocarro, (c) *Magnolia virginiana* (Magnoliaceaee; Magnoliales), a group of single-seeded follicles from an apocarpous gynoecium, photo R. Moran, (d) *Annona squamosa* (Annonaceae; Magnoliales), an aggregate fruit of fused separate carpels, photo M. Peña-Chocarro, (e) *Drimys granadensis* (Winteraceae; Canellales), fleshy berries from separate carpels, photo F. Michelangeli, (f) *Zea mays* (Poaceae; Poales), each kernel of maize is a caryopsis, derived from a single flower, photo A. Litt, (g) *Cocos nucifera* (Arecaceae; Arecales), large nuts with liquid endosperm, photo M. Peña-Chocarro.

Figure 2.2 (a) *Paris mairei* (Liliaceae; Liliales), capsule with brightly colored seeds, photo S. Knapp, (b) *Musa velutina* (Musaceae; Zingiberales), fleshy berry, photo S. Knapp, (c) *Aquilegia coerulea* (Ranunculaceae; Ranunculales), a group of many-seeded follicles from an apocarpous gynoecium, photo E. Kramer, (d) *Nelumbo nucifera* (Nelumbonaceae; Proteales), nuts held in an expanded hypanthium, photo S. Knapp, (e) *Protea sp.* (Proteaceae; Proteales), a group of capsules with plumed seeds, photo S. Knapp, (f) *Buxus latistyla* (Buxaceae; Buxales), woody capsule, photo D.W. Stephenson, (g) *Ecballium elaterium* (Cucurbitaceae; Cucurbitales), a berry (pepo), with unusual pressure-related dehiscence, photo D.W. Stephenson, (h) *Rosa omeiensis* (Rosaceae; Rosales), a rose hip is an expanded hypanthium containing many achenes, photo S. Knapp.

Figure 2.3 (a) *Pterogyne nitens* (Fabaceae; Fabales), a legume/samara, photo M. Peña-Chocarro, (b) *Euonymus macropterus* (Celastraceae; Celastrales), a capsule with seeds enclosed in a fleshy orange aril (outgrowth of the seed), photo M. Christenhusz, (c) *Brugueira gymnorhiza* (Rhizophoraceae; Malpighiales), a viviparous "fruit" (hypocotyl emerging directly from calyx), photo S. Knapp, (d) *Bulnesia sarmientoi* (Zygophyllaceae; Zygophyllales), winged capsules, photo M. Peña-Chocarro, (e) *Helicteres isora* (Malvaceae; Malvales), capsules with twisted carpels, photo R. Prakash, (f) *Cochlospermum vitifolium* (Bixaceae; Malvales), capsule with plumed seeds, photo M. Peña-Chocarro, (g) *Iberis umbellata* (Brassicaceae; Brassicales), an obtriangular silicle, photo S. Knapp, (h) *Cedrela mexicana* (Meliaceae; Sapindales), woody capsules with winged seeds, photo M. Peña-Chocarro.

Figure 2.4 (a) *Syzigium jambos* (Myrtaceae; Myrtales), a fleshy berry, photo S. Knapp, (b) *Eucalyptus* sp. (Myrtaceae; Myrtales), aggregation of woody capsules, photo S. Knapp, (c) *Geranium maculatum* (Geraniaceae; Geraniales), a capsule with elongate column and single-seeded carpels, photo S. Mori, (d) *Ampelopsis brevipedunculata* (Vitaceae; Vitales), a fleshy berry, photo M. Christenhusz, (e) *Heisteria povedae* (Olacaceae; Santalales), a drupe with brightly colored attractive calyx, photo S. Knapp, (f) *Cornus capitata* (Cornaceae; Cornales), a fused aggregate fleshy fruit, photo S. Knapp, (g) *Lecythis pisonis* (Lecythidaceae; Myrtales), a massive woody capsule opening via a "lid" (this type of capsule is called a pyxidium), photo C. Gracie, (h) *Diospyros digyna* (Ebenaceae; Ericales), fleshy multi-seeded berries, photo M. Peña-Chocarro.

Figure 2.5 (a) *Genipa americana* (Rubiaceae; Gentianales), a woody berry, photo M. Peña-Chocarro, (b) *Galposis speciosa* (Lamiaceae; Lamiales), a group of nutlets (one-seeded carpels that split apart at maturity), photo M. Christenhusz, (c) *Mandragora officinalis* (Solanaceae; Solanales), a fleshy berry, photo S. Knapp, (d) *Lithospermum arvense* (Boraginaceae; unplaced as to order), dry nutlets (one-seeded carpels that split apart at maturity), photo M. Christenhusz, (e) *Ilex* (Aquifoliaceae; Aquifoliales), a pyrene with two hard seeds, photo R. Moran, (f) *Trapopogon porrifolius* (Asteraceae; Asterales), achenes with plumed apical structure, photo A. Matthews, (g) *Calendula arvensis* (Asteraceae; Asterales), achenes of several shapes in a single head, photo M. Christenhusz, (h) *Myrrhis odorata* (Apiaceae; Apiales), mericarps (one-seeded carpels that split apart at maturity), photo M. Christenhusz.

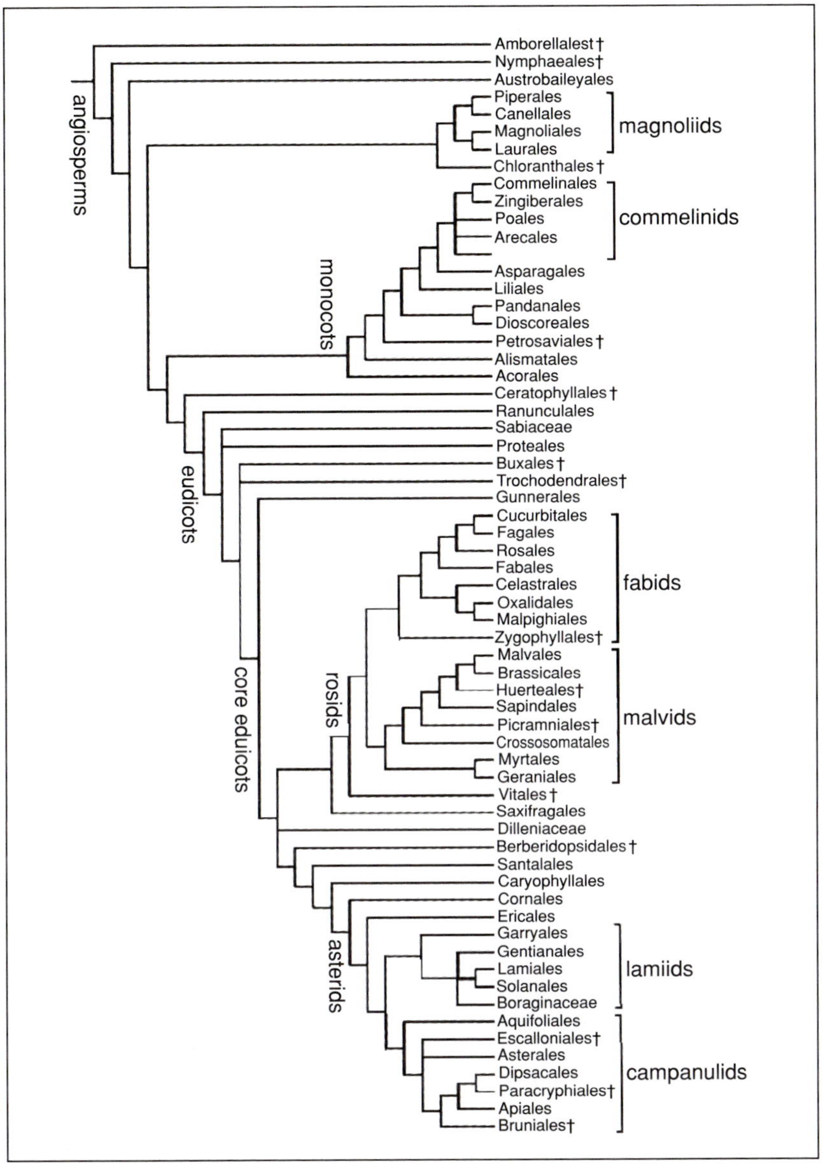

Figure 2.6 The APG III angiosperm phylogeny; for explanation of methods and literature supporting these relationships, see APG (2009). (Published with permission from APG 2009).

and families included in each of these clades, the reader is referred to Figure 1 of APG III and other web resources concerned with angiosperm phylogeny (see APG, 2009; Angiosperm Phylogeny Website, http://www.mobot.org/mobot/research/apweb/; http://www2.biologie.fu-berlin.de/sysbot/poster/poster1.pdf).

It is our intention in this chapter to provide a broad-brush review of a few of what we consider important elements in thinking about fruit evolution: (1) fruit variation and evolution as seen in the fossil record (i.e., through time), (2) fruit variation as it is related to current ideas of angiosperm phylogeny, (3) the basic molecular and structural mechanisms underlying fruit development, and (4) the possible role of fruits in angiosperm diversification. This is not a comprehensive review of the literature, although we have attempted to include a broad range of references; we hope this will stimulate thinking about fruit development and ripening that is more broadly based on angiosperm phylogeny and on the role of fruits in natural ecosystems.

Fruit in the Fossil Record

The apparent recent origin and rapid diversification of the angiosperms were for Charles Darwin "an abominable mystery" (in a letter to J.D. Hooker dated July 22, 1879, see Friedman, 2009). The angiosperm fossil record extends back to the lower Cretaceous Period, with the earliest unambiguous fossils from pollen floras in southern England (see Hughes, 1994); these are dated conservatively around 132 million years ago (mya). By the middle part of the Cretaceous, angiosperms were common and diverse, indicating that the first major diversification and ecological radiation occurred over a relatively short time frame (Friis et al., 2010). Relatively derived lineages are represented in these early fossils, thus suggesting that the rapid radiation was underestimated or that angiosperms originated and diversified earlier than was previously thought. Methods for dating phylogenetic trees (see references in Wikström et al., 2001; Smith et al., 2010) derived from molecular sequence data have provided new views on the question of angiosperm age. Evidence from these dated molecular phylogenies has pushed back estimates of the origin of the crown group of angiosperms to the early to middle Jurassic (179–158 mya, Wikström et al., 2001) or even into the Triassic (182–257 mya, Smith et al., 2010); both these estimates predate the earliest angiosperm fossils. Origin of the derived eudicot clade which today accounts for most of angiosperm diversity is also estimated to predate the earliest eudicot fossils (see Smith et al., 2010). The combination of dated molecular phylogenies and a huge increase in the fossil evidence for early angiosperms (see Friis et al., 2010) has resulted in a clearer picture of early angiosperm evolution, both in terms of age and character diversity.

Using mostly fruit types as evidence, Corner (1949) proposed his "Durian theory," in which he suggested the earliest flowering plants were large tropical forest trees with dehiscent red capsules and large black arillate seeds. It is clear, however, from a wealth of angiosperm mesofossils (flowers, fruits, and leaves, see Friis et al., 2006 for a review) from the early Cretaceous, that early angiosperms were mostly small and herbaceous, rather than trees, and they were weeds of disturbed habitats (Wing and Boucher, 1998). The early Cretaceous fossil *Leefructus* (Sun et al., 2011) is one such small herbaceous plant and is considered to have affinities with Ranunculales (the leaves and fruit are visually similar to *Delphinium* or *Aquilegia*, see Figs. 3 and 4 of Sun et al., 2011), a member of the eudicot clade (APG, 2009). The compression fossil from China described as *Archaefructus* was initially proposed as the sister group to extant angiosperms (Sun et al., 2002), but further examination has suggested it represents a specialized crown-group member with an inflorescence of small flowers rather than a single large flower on an extended axis (Friis et al., 2003; Endress and Doyle, 2009); this interpretation is supported by the morphology of a second species of *Archaefructus* (Ji et al., 2004).

Flowers of these early Cretaceous angiosperms are mostly quite small, as are fruits (Friis et al., 2010); fruit volume varies between 0.1 and 8.3 mm^3 (the size of a small cherry) and seeds from 0.02 to 6.9 mm^3 (Eriksson et al., 2000a). Diverse fruit types are represented in early Cretaceous fossils; the vast majority are indehiscent one-seeded nuts or drupes with one or few seeds, but some apocarpous dehiscent fruits are found, as are many fruits with fleshy outer layers (Friis et al., 2010). One quarter of all fruits found in one site were fleshy drupes or berries (Eriksson et al., 2000b). Fruit size increased in the Tertiary concomitantly with seed size (Eriksson et al., 2000a), with a drastic change at the Cretaceous–Tertiary boundary (Tiffney, 1984; Wing and Boucher, 1998). Fleshiness evolved before large seed size, and it has been suggested that the origin of fleshiness was related to defense against pathogens rather than to dispersal, being later co-opted for biotic dispersal (Mack, 2000; Tiffney, 2004).

Size change in both fruit and seeds through the fossil record has been attributed to co-evolution with seed-dispersing animals (Tiffney, 1984). This mutualistic relationship between fruits and animals dispersing large seeds is thought to have been a major factor in the development of closed canopy, forested habitats (large seeds are considered characteristic of closed canopy plant communities). An alternative is that changes in habitat through climate change, coupled perhaps with the demise of large herbivores (e.g., dinosaurs), drove the evolution of larger fruits and seeds, and that the availability of seed dispersers reinforced this trend (Eriksson et al., 2000a); thus dispersers tracked, not led, the changes in size. Distinguishing between these two hypotheses is difficult due to bias in the fossil record (Eriksson et al., 2000a) and uncertainty over ecological dynamics in the past (Tiffney, 2004), but a series of studies using both extant and fossil fruits has strongly supported the latter hypothesis.

Bolmgren and Eriksson (2005) tested 50 phylogenetically independent lineages in the European flora and examined shifts in size and fleshiness using phylogenetically independent contrasts in order to test the role of niche shifts in the evolution of fleshy fruits. They found significant correlations between fruit type and habitat conditions, suggesting that phylogenetically independent origins of fleshiness were related to changing vegetation, and that shifts to fleshiness have been occurring continuously over the last 70 million years. Their results also suggested that frugivore-mediated selection on fruit characteristics was enhanced when plant population sizes were small and in forest understory (low light) conditions (Bolmgren and Erikson, 2005). Lorts et al. (2008) compared Australian tropical forests and Indian deserts and found greater numbers of fleshy fruits in the tropical forests; this suggested to them that fruit evolution was driven by dispersers, but their data were not tested statistically and so are of limited utility. The lack of strong correlation between fleshy fruits and selection pressure by frugivores has been interpreted to indicate that the fruit/frugivore interaction is diffuse and probably phylogenetically constrained (Herrera, 1982), but it is clear that the situation is not so simple. A more plausible suggestion is that fleshy fruit evolution has been an important and continually recurring theme throughout flowering plant evolution (Bolmgren and Eriksson, 2005), and that many factors contribute to the evolution of fruit types, not dispersers alone.

Fruit Variation and Angiosperm Phylogeny

Bolmgren and Eriksson (2005) suggest that fleshy fruit evolution is a recurring theme in angiosperm history and thus should not be phylogenetically constrained. Lorts et al. (2008) mapped crude fruit type categories (dry dehiscent, dry indehiscent, and fleshy) onto the angiosperm phylogeny at the ordinal level (using APG II, APG, 2003) and found no association (assessed visually) between fruit type and major clade. They conclude that there is no phylogenetic constraint on fruit type evolution across the major lineages (orders) of angiosperms, with most orders having all three of their broadly defined fruit types (see Fig. 1 of Lorts et al., 2008). On an angiosperm-wide level, it is clear that there is massive variation in fruit morphology and the "primitive" fruit type is not known. Basal lineages do not necessarily have what have long been considered "primitive" fruit types; *Amborella trichopoda*, for example, the sister to all other angiosperms (APG, 2009) has a group of fleshy fruitlets, each derived from a separate carpel, that have been interpreted as drupes (see Bobrov et al., 2005).

More detailed analyses done with smaller clades (e.g., at the family or subfamilial level) have demonstrated that fruit types are highly homoplasious (i.e., the product of convergent or parallel evolution) and have evolved multiple times in many lineages (see references in Knapp, 2002; most phylogenetic analyses of relationships at all taxonomic levels have some discussion of fruit evolution, only a handful are discussed here). In the family Bignoniaceae, for example, indehiscent fruit with large seeds have convergently evolved in three lineages, and each has a sister group relationship with genera with the more typical bignoniaceous dehiscent fruit with winged seeds (Zjhra et al., 2004). The independent evolution of indehiscence is supported by structural differences in these fruits; other studies at lower taxonomic levels have shown that within the Neotropical group the taxa with indehiscent fruit form monophyletic groups (Grose and Olmstead, 2007). In the Rubiaceae, fleshy fruits have arisen 12 times independently (Bremer and Eriksson, 1992; Bremer et al., 1995), and in some parts of the family fruit types are highly variable and can exhibit almost any combination of characters of dehiscence, shape, and seed morphology (Motley et al., 2005). Motley et al. (2005) showed that in the Catesbeae-Chiococceae complex, indehiscent fleshy fruits (drupes and berries) have evolved multiple times, and that the fruit type in the group was not correlated with distribution, indicating that fleshiness did not confer any advantage in dispersibility.

Traditional classifications in many families of flowering plants have depended heavily on fruit characters (see above). In Melastomataceae berries have evolved from capsules at least four times (Clausing and Renner, 2001) and fleshiness has been derived from many different parts of the developing ovary (Clausing et al., 2000), highlighting the necessity of determining the homology of characters before making hypotheses concerning fruit type evolution. In Solanaceae, fleshy fruits have multiple origins (Knapp, 2002), but the possession of berries appears to define a large, monophyletic clade, thus indicating some phylogenetic component to fruit evolution in the family. Within this clade, however, multiple reversions to dry dehiscent fruits have occurred, primarily associated with dry habitats (see Knapp, 2002).

It is clear from an examination of the fruit in relation to phylogeny at a variety of levels that fruit evolution is highly labile and that fleshiness has been an important and continually recurring theme throughout flowering plant evolution (Bolmgren and Eriksson, 2005). The use of fruit type as a character is perhaps analogous to the use of pollen types (Punt, 1971) or pollination syndromes (Faegri and van der Pijl, 1979). Both of these have been shown to be useful for a variety of discussions, but to be inappropriate for the analysis of evolution of these organs in flowering plants (Blackmore et al., 1991; Ollerton, 1996; Knapp, 2010). In looking at fruit evolution it might be more useful to use fruit types as descriptive terms, but to examine the organ/tissue changes separately in order to look at evolutionary patterns as has been done extensively in the pollen literature (Persson et al., 1994). Fruits, although derived from a single whorl in the angiosperm flower (see below), are complex structures and very slightly different developmental trajectories can lead to quite different outcomes (Evans and Dickinson, 1999).

Fruit Development

Different types of fruits are characterized by different tissue types and proportions, for instance, berries often remain primarily undifferentiated parenchyma whereas capsules become lignified in defined regions. These distinguishing features become manifest for the most part after fertilization, but the molecular processes that underlie the different developmental trajectories are not well understood. Fruits develop from the carpels, the innermost organs of the flower, and in fact fruit development may be seen as a continuation of a developmental sequence that begins with carpel initiation in the developing floral bud and ends with fruit maturation and seed dispersal. Carpel identity is specified by genes that provide the C-function of the ABC model (Coen and Meyerowitz, 1991); the activity of C-function genes is required for the formation of both of the reproductive whorls of the flower, the stamens and carpels. The best characterized of these stamen- and carpel-identity genes belong to the *AGAMOUS* clade of MADS-box transcription factors, which includes the *Arabidopsis* gene *AGAMOUS* (*AG*) and the tomato gene *TAG1*. Loss of function of *AG*-lineage genes may result in complete loss of carpel identity (Bowman et al., 1989; Yanofsky et al., 1990; Pnueli et al., 1994), although other genes are also able to confer carpel identity such as *SHATTERPROOF1 (SHP1)* and *SHATTERPROOF2 (SHP2)* (Pinyopich et al., 2003), redundant paralogs that are also members of the *AG* gene clade, the YABBY transcription factor *CRABS CLAW*, and the bHLH transcription factor *SPATULA* in *Arabidopsis* (Alvarez and Smyth, 1999). The carpel wall, which will form the fleshy wall of a berry or the dry woody wall of a capsule, is usually relatively undifferentiated during flower development, consisting of an inner and outer epidermis and interior vascularized parenchymatous tissue, although some distinctive features that are characteristic of the fruit may be already manifest in the carpels, such as the externally visible junction of the replum and valves in the *Arabidopsis* silique or the spiny projections of the *Datura* fruit. In addition, some aspects of fruit morphology, such as fruit shape and size, are at least partly specified during carpel development (e.g., Ku et al., 2000).

The ovary and ovules have usually reached maturity when the flower opens, although the style often continues to elongate until it reaches full length and becomes receptive. It may be several days after anthesis before fruit development formally commences, upon fertilization of the ovules in the carpels. Fertilization triggers major changes in the wall of the fruit (pericarp), as well as in the seed, that vary depending on fruit type. The best-studied examples of fruit development are the silique of *Arabidopsis*, an example of a dry dehiscent fruit (see Table 2.1), and the berry of tomato, an example of a fleshy indehiscent fruit. In both of these cases, and in fact in most species that have been studied, fertilization initiates a period characterized by cell division (Gillaspy et al., 1993; Vivian-Smith and Koltunow, 1999; Tanksley, 2004; Pabón-Mora and Litt, 2011). These divisions can be both periclinal and anticlinal, in which case the fruit will grow in overall size and the pericarp will increase in thickness; in general, this pattern is characteristic of fleshy fruits. Alternatively, the divisions can be largely anticlinal, in which case there is overall growth in fruit size but no increase in pericarp thickness. This latter pattern is seen more often in dry dehiscent fruits, in which the pericarp remains thin with a limited number of cell layers. In the *Arabidopsis* silique (stages 14–16 as defined by Ferrandiz et al., 1999), division is largely anticlinal and perpendicular to the long axis of the silique. The result is elongation of the developing fruit; however, the six to seven pericarp cell layers are maintained. At this stage in *Arabidopsis*, cell expansion occurs as well, contributing to an increase in silique diameter as well as length (Vivian-Smith and Koltunow, 1999; Vivian-Smith et al., 2001). In contrast, in the capsules of Solanaceae (such as *Nicotiana*, see Knapp, 2002) there is little or no cell expansion so that most increase in fruit length and diameter is due to increased cell numbers; cell division is exclusively anticlinal thus there is no increase in the number of pericarp cell layers (Pabón-Mora and Litt, 2011). In tomato, grape, and other fleshy fruits, extensive cell division in all planes occurs during this stage leading to increase in fruit length and width, and pericarp thickness (Gillaspy et al., 1993; Hardie et al., 1996; Tanksley, 2004).

Cell division tapers off after a few days or up to a few weeks depending on species, and is followed by various processes depending on fruit type. In the silique of *Arabidopsis*, differentiation of the dehiscence zone, where the valves will detach from the replum, begins at this stage, including lignification of valve margin layers and formation of the separation layer (stage 17a and b of Ferrandiz et al., 1999). Likewise in Solanaceae capsules, lignification of the endocarp begins in approximately the same time frame (Pabón-Mora and Litt, 2011). In contrast, in fleshy berries, an often prolonged phase of cell expansion occurs after cell division, accompanied by endoreduplication (Gillaspy et al., 1993; Tanksley, 2004). Cell expansion is largely responsible for the dramatic increase in size seen in fleshy fruits such as tomato. The final stage of fruit development occurs after cell expansion or lignification and differentiation are complete. Drying and dehiscence occur in the silique and the Solanaceae capsule. At this stage cell death occurs, including the cells of the mesocarp of the Solanaceae capsules and the separation layer of the silique of *Arabidopsis* (Smyth et al., 1990; Ferrandiz et al., 1999; Pabón-Mora and Litt, 2011); in the latter case this causes the valves to detach from the replum, exposing the seeds. In capsules and perhaps in siliques as well, drying of the pericarp causes shrinkage of nonlignified layers; juxtaposed against lignified layers which are unable to shrink, this causes tension that results in dehiscence. In fleshy fruits, cell expansion ceases and ripening begins (Gillaspy et al., 1993; Tanksley, 2004). This process involves dramatic changes in the pericarp including accumulation of sugars and other nutrients, color change, softening of cell walls, and other processes.

In both *Arabidopsis* and tomato, the molecular processes that regulate early stages of post-fertilization fruit development are not as well characterized as those that direct the later stages.

In the pericarp of the *Arabidopsis* silique, cell division and expansion appear to be controlled by a variety of genes that are up- or downregulated at fertilization. For instance, *FRUIT WITHOUT FERTILIZATION* (*FWF*) and *ABERRANT TESTA SHAPE* (*ATS*) appear to negatively regulate cell division, whereas genes involved in GA biosynthesis and signalling largely promote cell expansion and division (Vivian-Smith and Koltunow, 1999; Vivian-Smith et al., 2001). In tomato, several loci have been identified that affect cell division and expansion, for instance mutations in fw2.2, which represses cell division during the phase that follows fertilization, appear to have played an important role in the increase in size of domesticated tomatoes (Frary et al., 2000; Cong et al., 2002) and downregulation of the MADS-box gene *TAGL1*, the tomato ortholog of *SHP1/SHP2*, produces a fruit with a thinner pericarp composed of fewer cell layers than wild type (Vrebalov et al., 2009). Cell expansion, which is correlated in tomato with dramatic levels of endoreduplication (Cheniclet et al., 2005), appears to be controlled by such genes as *SlFSM1*, a MYB-like gene that negatively regulates cell expansion in the inner mesocarp (Machemer et al., 2011), and the kinase *WEE1* (Gonzalez et al., 2007), which plays a role in triggering endoreduplication.

In the *Arabidopsis* silique, differentiation of the dehiscence zone is largely regulated by the interaction among *FRUITFULL* (*FUL*), a MADS-box transcription factor, *SHP1/SHP2*, and two bHLH transcription factors, *INDEHISCENT* (*IND*) and *ALCATRAZ* (*ALC*). *IND*, along with *ALC* and the *SHP* genes, promotes lignification of the valve margins and differentiation of the separation layer; their expression is repressed by *FUL* activity in the valves and *REPLUMLESS* (*RPL*), a homeodomain transcription factor, in the replum, which thereby keeps those structures unlignified (Ferrandiz et al., 2000; Liljegren et al., 2000a, 2000b; Rajani and Sundaresan, 2001; Roeder et al., 2003; Liljegren et al., 2004). The siliques of *alc*, *ind*, and *shp1 shp2* double mutants fail to dehisce due to lack of differentiation of the tissues involved in dehiscence; the *ful* mutant is defective in valve elongation, due to ectopic lignification, and bursts open prematurely due to crowding of the seeds (Gu et al., 1998).

The pericarp of the tomato berry remains relatively undifferentiated compared to that of *Arabidopsis* silique; aside from the inner and outer epidermis and vascular tissue, the main distinction among cells is differential expansion. Thus there are no complex interactions setting up boundaries and dehiscence zones as in the silique. However, once the process of cell expansion has ceased, ripening is initiated, and with it a complex series of cellular and metabolic changes that result in alterations in color, texture, metabolite content, and other features.

Although the ripening process has been extensively studied in tomato as well as other fleshy fruits such as melon and strawberry, we know far less about the molecular processes that govern early development, less still about the parallel processes in other fruit types such as capsules or samaras, and virtually nothing about the way these processes are altered during the evolution of new fruit morphologies. Although there have been numerous shifts between dry and fleshy fruits during angiosperm evolution, we know little about the molecular changes that have accompanied these changes. In Solanaceae, for instance, basal lineages predominately produce capsules, but the derived Solanoideae clade is characterized mainly by berries. Studies indicate that species in this family share similar gene complements and even sequences, suggesting that changes in regulation may underlie different morphologies, and this is likely to be a common theme in fruit evolution. In Solanaceae, for instance, one of the characteristics that distinguishes berries from capsules is the increase in the number of cell layers in the pericarp of the former. Two transcription factors, *TAGL1* (Vrebalov et al., 2009) and *SlFUL2* (Meyer, Pabón-Mora, and Litt, unpublished data), have recently been shown to play a role in this process in tomato; when they are downregulated, the number of cell layers is dramatically reduced. Thus, changes in the regulation of these genes resulting in extensive cell proliferation in the pericarp may have occurred in correlation with the origin of the Solanoideae, and may have been a key factor in the origin of the thick-walled berry that characterizes this clade. It is likely that similar shifts in gene regulation and function, particularly transcription factors and components of signal transduction pathways, underlie the lability of fruit morphology seen throughout angiosperm evolution.

Fruit as a Driver of Angiosperm Diversity

It is tempting to assume that fruit diversity is the result of adaptation to frugivores; adaptationist scenarios about tight obligate relationships between fruits and their dispersers abound in the literature, but they have largely been shown to be less compelling than originally thought. The slow extinction of the tambalacoque (*Sideroxylon grandiflorum*) of Mauritius due to the extirpation of its disperser the dodo (*Raphus cucullatus*) (Temple, 1977), the recalcitrant germination of Galapagos tomatoes overcome by passage through the gut of the giant Galapagos tortoise (Rick and Bowman, 1961), and the apparent anachronism of large fruits with a "megafaunal dispersal syndrome" (Janzen and Martin, 1981) are all "just-so-stories" that were for a long-time staples of introductory classes. Closer examination of all three of these compelling scenarios has shown that tambalacoque seeds do not need scarification by dodos to germinate and in fact these trees are found in mixed age stands in Mauritius (Witmer and Cheke, 1991), Galapagos tortoises do not normally eat tomatoes (Darwin et al., 2003), and the "fruits the gomphotheres ate" in fact have a large number of extant seed dispersers (Howe, 1985). These three case studies, so long used as examples of close adaptation between fruits and frugivores, clearly illustrate the need to see beyond a set of adaptationist scenarios and examine the details, often complex and difficult to disentangle, that contribute to the evolution of fruit morphology.

The strong role of habitat and climate in the origin and evolution of fruit characters such as fleshiness (Eriksson et al., 2000a) has been shown clearly in both the fossil record and through correlation studies with extant angiosperms. Syncarpy may represent a key innovation for angiosperm fruits in that it facilitated the development of pollen presentation mechanisms that allow pollen competition and pollen tube protection (Endress, 2011). Fruit color is another characteristic that on the face of it seems to be adaptative for frugivore attraction, but studies examining color as a signal have demonstrated that, like other fruit characters, evolution of fruit color is mediated through diffuse interactions between the plants, antagonists, and mutualists (see Schaefer and Schaefer, 2007). Carotenoids, for example, are not perceived by birds (Schaefer et al., 2008); and variability in insect predation, not dispersal agents, accounts for orange/yellow/red color differences in the arillate seeds of *Acacia* in Australia (Whitney and Stanton, 2004). Antagonists have also played a clear role in the evolution of fruit chemistry

(Levey et al., 2007); special (secondary) metabolites can have a deterrent effect on microbes, whose effects on fruits limit seed dispersal and thus are detrimental to plant fitness. An apparent paradox is that most wild fleshy fruits look delicious but taste terrible; astringency (bitterness) is the rule, not the exception for most species of fleshy-fruited angiosperms (Herrera, 1982). Wild peppers with a polymorphism in capsaicin content (responsible for pungency) differed in their microbial defenses (higher capsaicin content provided better protection), had no difference in attractiveness to birds who are the primary consumers and dispersers of wild chillis, and caused differences in gut retention time of seeds (Tewksbury et al., 2006). Fruit chemistry thus, like fleshiness and color, is more complex than it appears at first glance and relationships between many players drive evolution of this fruit characteristic.

Valido et al. (2011) examined phenotypic integration in 111 fleshy fruited species in the Mediterranean region. They analysed the magnitude and co-variation of morphological, nutritional, and visual traits across a community phylogeny and found that fleshy fruits consumed by birds were more integrated, that is, the traits examined co-varied more than those consumed by mammals. Visual and nutritional traits had positive phylogenetic co-variance, suggesting that fruit color signals will be an interesting area in which to explore the intersection between gene expression and natural selection on a local level.

Plant–insect interactions have played a crucial role in the diversification and ecological dominance of angiosperms, despite some of these interactions being quite diffuse. Insect pollination occurs in 78% of angiosperms, suggesting this is a key driver of flower and thus angiosperm evolution (Ollerton et al., 2011). Both flowers and fruits have been implicated in the rapid radiation of angiosperms, but the relative importance of each is difficult to assess from the fossil record. In an analysis of three floras, Whitney (2009) examined the morphological space occupied by species in terms of flower and fruit diversity. If flowers and fruits have been exposed to different selection pressures via their animal interactions (Schaefer et al., 2004), then one would expect a difference in character amplitude (degree of morphological difference) between flowers and fruits. In the datasets tested, flowers of animal-pollinated, animal-dispersed species diverge more than do fruits of the same taxa in morphological space, consistent with the hypothesis that diversifying selection is stronger for flower than fruit morphology (Whitney, 2009). This then suggests that interactions of flowers, rather than fruits, are the major driver of angiosperm diversification. This study highlights the importance of assessing traits as characters, rather than as syndromes, in order to dissect the possible evolutionary correlations evident in the highly variable morphology we see in today's angiosperm reproductive features.

Data from the fossil record and from studies of extant angiosperms indicate that we need to stop automatically assuming that fleshy fruits have evolved to attract dispersers, but instead that fleshy fruit evolution may be driven by other ecological factors. The scale at which relationships are analysed also needs to be carefully taken into account. The diffuse, rather than tightly mutualistic, nature of the relationship between plants and animals mediated by fruits means that factors contributing to the origin and diversification are likely to be diverse and not strictly adaptative. Using the comparative framework provided by phylogeny and the detailed dissection of development provided by increased understanding of molecular regulatory mechanisms, insights gained from the study of fruit traits in single species whose importance is agronomic or genetic will have wider evolutionary relevance. A new view of fruit not as types or dispersal syndromes will allow us to compartmentalize fruit characters into their component parts in order to study their control, development, and ecology more closely and clearly. The sequencing of whole genomes of several angiosperms with very different types of fruits will allow us to combine the power of phylogenetic, ecological, and molecular methods in understanding the fruit—the characteristic that makes flowering plants angiosperms.

Acknowledgments

We thank Graham Seymour and Jim Giovannoni for inviting us to contribute this chapter; SK was supported by the National Science Foundation's Planetary Biodiversity Inventory programme (DEB-0316614); we thank Maarten Christenhusz, Scott Mori, Fabian Michelangeli, Mamen Peña-Chocarro, Carol Gracie, Dennis W. Stevenson, Robbin Moran, Ranee Prakash, and http://www.plantsystematics.org for permission to use their photographs of fruits.

References

Alvarez, J., and Smyth, D.R. (1999) CRABS CLAW and SPATULA, two *Arabidopsis* genes that control carpel development in parallel with AGAMOUS. *Development*, 126(11), 2377–2386.

Angiosperm Phylogeny Group. (1998) An ordinal classification for the families of flowering plants. *Annals of the Missouri Botanical Garden*, 85, 531–553.

Angiosperm Phylogeny Group. (2003) An update of the Angiosperm Phylogeny Group classification for the orders and families of flowering plants: APG II. *Botanical Journal of the Linnean Society*, 141, 399–436.

Angiosperm Phylogeny Group. (2009) An update of the Angiosperm Phylogeny Group classification for the orders and families of flowering plants: APG III. *Botanical Journal of the Linnean Society*, 161, 105–121.

Baker, H.G. (1959) Reproductive methods as factors in speciation in flowering plants. *Cold Spring Harbor Symposia on Quantitative Biology*, 24, 177–190.

Blackmore, S., Scotland, R., and Stafford, P.J. (1991) The comparative method in palynology. *Palaeoecology of Africa and the Surrounding Islands*, 22, 3–10.

Bobrov, A.V.F.C., Endress, P.K., Melikov, A.P., Romanov, M.S., Sorokin, A.N., and Palmarola-Bejerano, A. (2005) Fruit structure of *Amborella trichopoda* (Amborellaceae). *Botanical Journal of the Linnean Society*, 148, 265–274.

Bolmgren, K., and Eriksson, O. (2005) Fleshy fruits – origins, niche shifts, and diversification. *Oikos*, 109, 255–272.

Bowman, J.L., Smyth, D.R., and Meyerowitz, E.M. (1989) Genes directing flower development in *Arabidopsis*. *Plant Cell*, 1, 37–52.

Bremer, B., and Eriksson, O. (1992) Evolution of fruit characters and dispersal modes in the tropical family Rubiaceae. *Biological Journal of the Linnean Society*, 47, 79–95.

Chase, M.W., and Reveal, J. (2009) A phylogenetic classification of the land plants to accompany APG III. *Botanical Journal of the Linnean Society*, 161, 122–127.

Cheniclet, C., Rong, W.Y., Causse, M., Frangne, N., Bolling, L., Carde, J.P., and Renaudin, J.P. (2005) Cell expansion and endoreduplication show a large genetic variability in pericarp and contribute strongly to tomato fruit growth. *Plant Physiology*, 139(4), 1984–1994.

Clausing, G., Meyer, K., and Renner, S.S. (2000) Correlations among fruit traits and evolution of different fruits within Melastomataceae. *Botanical Journal of the Linnean Society*, 133, 303–326.

Clausing, G., and Renner, S.S. (2001) Molecular phylogenetics of Melastomataceae and Memcylacaeae: implications for character evolution. *American Journal of Botany*, 88, 486–498.

Coen, E.S., and Meyerowitz, E.M. (1991) The war of the whorls: genetic interactions controlling flower development. *Nature*, 353, 31–37.

Cong, B., Liu, J., and Tanksley, S.D. (2002) Natural alleles at a tomato fruit size quantitative trait locus differ by heterochronic regulatory mutations. *Proceedings of the National Academy of Sciences of the United States of America*, 99(21), 13606–13611.

Corner, E.J.H. (1949) The Durian theory or the origin of the modern tree. *Annals of Botany*, 13, 367–414.

Corner, E.J.H. (1964) *The Life of Plants*. Weidenfeld & Nicholson, London.

Cronquist, A. (1981) *An Integrated System of Classification of Flowering Plants*. Columbia University Press, New York.

Darwin, S.C., Knapp, S., and Peralta, I.E. (2003). Tomatoes in the Galápagos Islands: morphology of native and introduced species of *Solanum* section *Lycopersicon* (Solanaceae). *Systematics and Biodiversity*, 1, 29–54.

Endress, P.K. (2011) Evolutionary diversification of the flowers in angiosperms. *American Journal of Botany*, 98, 370–396.

Endress, P.K. (1982) Syncarpy and alternative modes of escaping disadvantages of apocarpy in primitive angiosperms. *Taxon*, 31, 48–52.

Endress, P.K., and Doyle, J.A. (2009) Reconstructing the ancestral flower and its initial specializations. *American Journal of Botany*, 96, 22–66.

Eriksson, O., Friis, E.M., and Löfgren, P. (2000a) Seed size, fruit size, and dispersal systems in angiosperms from the early Cretaceous to the late Tertiary. *American Naturalist*, 156, 47–58.

Eriksson, O., Friis, E.M., Pederson, K.R., and Crane, P.R. (2000b) Seed size and dispersal mechanisms of early Cretaceous angiosperms for Famalicão, Portugal. *International Journal of Plant Sciences*, 161, 319–329.

Evans, R.C., and Dickinson, T.A. (1999) Floral ontogeny and morphology in subfamily Amygdaloideae T. and G. (Rosaceae). *International Journal of Plant Sciences*, 160, 955–979.

Faegri, K., and van der Pijl, L. (1979) *The Principles of Pollination Ecology,* 3rd revised ed. Pergamon Press, Oxford.

Ferrandiz, C., Liljegren, S.J., and Yanofsky, M.F. (2000) Negative regulation of the SHATTERPROOF genes by FRUITFULL during Arabidopsis fruit development. *Science*, 289, 436–438.

Ferrandiz, C., Pelaz, S., and Yanofsky, M.F. (1999) Control of carpel and fruit development in Arabidopsis. *Annual Review of Biochemistry*, 68, 321–354.

Frary, A., Nesbitt, T.C., Grandillo, S., van der Knaap, E., Cong, B., Liu, J., Meller, J., Elber, R., Alpert, K.B., and Tanksley, S.D. (2000) fw2.2: a quantitative trait locus key to the evolution of tomato fruit size. *Science*, 289(5476), 85–88.

Friedman, W.E. (2009) The meaning of Darwin's "abominable mystery". *American Journal of Botany*, 96, 5–21.

Friis, E.M., Doyle, J.A., Endress, P.K., and Leng, Q. (2003) *Archaefructus* – angiosperm precursor or specialized early angiosperm? *Trends in Plant Science*, 8, 369–373.

Friis, E.M., Pedersen, K.R., and Crane, P.R. (2006) Cretaceous angiosperm flowers: innovation and evolution in plant reproduction. *Palaeogeography, Palaeoclimatology, Palaeoecology*, 232, 251–293.

Friis, E.M., Pedersen, K.R., and Crane, P.R. (2010) Diversity in obscurity: fossil flowers and the early history of angiosperms. *Philosophical Transactions of the Royal Society B: Biological Sciences*, 365, 369–382.

Frohlich, M.W., and Chase, M.W. (2007) After a dozen years of progress the origin of angiosperms is still a great mystery. *Nature*, 450, 1184–1189.

Frohlich, M.W., and Parker, D.S. (2000) The Mostly Male theory of flower evolutionary origins: from genes to fossils. *Systematic Botany*, 25, 155–170.

Gaertner, J. (1788) (1788–1792) *De Fructibus et Seminibus Plantarum*. Vol. 4. Academiae Carolinae, Stuttgart.

Gillaspy, G., Ben-David, H., and Gruissem, W. (1993) Fruits: a developmental perspective. *Plant Cell*, 5(10), 1439–1451.

Gonzalez, N., Gevaudant, F., Hernould, M., Chevalier, C., and Mouras, A. (2007) The cell cycle-associated protein kinase WEE1 regulates cell size in relation to endoreduplication in developing tomato fruit. *Plant Journal*, 51(4), 642–655.

Grose, S.O., and Olmstead, R.G. (2007) Evolution of a charismatic Neotropical clade: molecular phylogeny of *Tabebuia* s. l., Crescentieae, and allied genera (Bignoniaceae). *Systematic Botany*, 32, 650–659.

Gu, Q., Ferrandiz, C., Yanofsky, M.F., and Martienssen, R. (1998) The *FRUITFULL MADS*-box gene mediates cell differentiation during *Arabidopsis* fruit development. *Development*, 125, 1509–1517.

Hardie, W.J., O'Brien, T.P., and Jaudzems, V.G. (1996) Morphology, anatomy and development of the pericarp after anthesis in grape, *Vitis vinifera* L. *Australian Journal of Grape and Wine Research*, 2(2), 97–141.

Herrera, C.M. (1982) Defense of ripe fruits from pests: its significance in relation to plant-disperser interactions. *American Naturalist*, 120, 218–241.

Howe, H.F. (1985) Gomphothere fruits: a critique. *American Naturalist*, 125, 853–865.

Hughes, N.F. (1994) *The Enigma of Angiosperm Origins*. Cambridge University Press, Cambridge.

Janzen, D.H., and Martin, P.S. (1981) Neotropical anachronisms: the fruits the gomphotheres ate. *Science*, 215, 19–27.

Ji, Q., Li, H., Bowe, L.M., Liu, Y., and Taylor, D.W. (2004) Early Cretaceous *Archaefructus eoflora* sp. nov. with bisexual flowers from Beiopao, Liaoning, western China. *Acta Geologica Sinica*, 78, 883–896.

Knapp, S. (2010) On "various contrivances": pollination, phylogeny and flower form in the Solanaceae. *Philosophical Transactions of the Royal Society B: Biological Sciences*, 365, 449–460.

Knapp, S. (2002) Tobacco to tomatoes: a phylogenetic perspective on fruit diversity in the Solanaceae. *Journal of Experimental Botany*, 53, 2001–2022.

Ku, H.M., Grandillo, S., and Tanksley, S. (2000) fs8.1, a major QTL, sets the pattern of tomato carpel shape well before anthesis. *Theoretial and Applied Genetics*, 101, 873–878.

Levey, D.J., Tewksbury, J.J., Izhaki, I., Tsahar, E., and Haak, D.C. (2007) Evolutionary ecology of secondary compounds in ripe fruit: case studies with capsaicin and emodin. In: *Seed Dispersal: Theory and Application in a Changing World* (eds. A.J. Dennis, E.W. Schupp, R.J. Green, and D.A. Westcott), pp. 59–77. CAB International, Wallingford, UK.

Liljegren, S.J., Ditta, G.S., Eshed, Y., Savidge, B., Bowman, J.L., and Yanofsky, M.F. (2000a) SHATTERPROOF MADS-box genes control seed dispersal in Arabidopsis. *Nature*, 404, 766–770.

Liljegren, S.J., Kempin, S., Chen, A., Roeder, A., Guimil, S., Khammungkhune, T., and Yanofsky, M.F. (2000b) A bHLH gene, INDEHISCENT1, is required for fruit dehiscence and mediates the fruitfull phenotype. In: *11th International Conference on Arabidopsis Research*. University of Wisconsin, Madison Press, Madison, WI.

Liljegren, S.J., Roeder, A.H., Kempin, S.A., Gremski, K., Ostergaard, L., Guimil, S., Reyes, D.K., and Yanofsky, M.F. (2004) Control of fruit patterning in Arabidopsis by INDEHISCENT. *Cell*, 116(6), 843–853.

Lorts, C.M., Briggeman, T., and Sang, T. (2008) Evolution of fruit types and seed dispersal: a phylogenetic and ecological snapshot. *Journal of Systematics and Evolution*, 46, 396–404.

Machemer, K., Shaiman, O., Salts, Y., Shabtai, S., Sobolev, I., Belausov, E., Grotewold, E., and Barg, R. (2011) Interplay of MYB factors in differential cell expansion and consequences for tomato fruit development. *Plant Journal*, 68, 337–350. doi: 10.1111/j.1365-313X.2011.04690.x.

Mack, A.L. (2000) Did fleshy fruit pulp evolve as a defense against seed loss rather than as a dispersal mechanism? *Journal of Bioscience*, 25, 93–97.

Meyen, S.V. (1988) Origin of the angiosperm gynoecium by gametoheterotopy. *Botanical Journal of the Linnean Society*, 97, 171–178.

Motley, T.J., Wurdack, K.J., and Delprete, P.G. (2005) Molecular systematics of the Catesbeae-Chicocceae complex (Rubiaceae): flower and fruit evolution and biogeographic implications. *American Journal of Botany*, 92, 316–329.

Ollerton, J. (1996) Reconciling ecological processes with phylogenetic patterns, the apparent paradox of plant-pollinator systems. *Journal of Ecology*, 84, 767–769.

Ollerton, J., Winfree, R., and Tarrant, S. (2011) How many flowering plants are pollinated by animals? *Oikos*, 120, 321–326.

Pabón-Mora, N., and Litt, A. (2011) Comparative anatomical and developmental analysis of dry and fleshy fruits in Solanaceae. *American Journal of Botany*, 98, 1415–1436.

Persson, V., Knapp, S., and Blackmore, S. (1994) Pollen morphology and systematics of the tribe Juanulloeae AT Hunziker (Solanaceae). *Review of Palaeobotany and Palynology*, 83, 1–30.

Pinyopich, A., Ditta, G.S., Savidge, B., Liljegren, S.J., Baumann, E., Wisman, E., and Yanosky, M.F. (2003) Assessing the redundancy of MADS-box genes during carpel and ovule development. *Nature*, 424, 85–88.

Pnueli, L., Hareven, D., Rounsley, S.D., Yanofsky, M.F., and Lifschitz, E. (1994) Isolation of the tomato AGAMOUS gene TAG1 and analysis of its homeotic role in transgenic plants. *Plant Cell*, 6, 163–173.

Punt, W. (1971) Pollen morphology in *Norantea, Souroubea* and *Ruyschia (Marcgraviacea). Pollen et Spores*, 13: 199–232.

Rajani, S., and Sundaresan, V. (2001) The Arabidopsis myc/bHLH gene ALCATRAZ enables cell separation in fruit dehiscence. *Current Biology*, 11(24), 1914–1922.

Rick, C.M., and Bowman, R.I. (1961) Galápagos tomatoes and tortoises. *Evolution*, 15, 407–417.

Roeder, A.H., Ferrandiz, C., and Yanofsky, M.F. (2003) The role of the REPLUMLESS homeodomain protein in patterning the *Arabidopsis* fruit. *Current Biology*, 13(18), 1630–1635.

Schaefer, H.M., Schaefer, V., and Levey, D.J. (2004) How plant-animal interactions signal new insights in communication. *Trends in Ecology and Evolution*, 19, 577–584.

Schaefer, H.M., McGraw, K., and Catoni, C. (2008) Birds use fruit colour as honest signal of dietary antioxidant rewards. *Functional Ecology*, 22, 303–310.

Schaefer, H.M., and Schaefer, V. (2007) The evolution of visual fruit signals: concepts and constraints. In: *Seed Dispersal: Theory and Application in a Changing World* (eds. A.J. Dennis, E.W. Schupp, R.J. Green, and D.A. Westcott), pp. 37–58. CAB International, Wallingford, UK.

Smith, S.A., Beaulieu, J.M., and Donoghue, M.J. (2010) An uncorrelated relaxed-clock analysis suggests an earlier origin for flowering plants. *Proceedings of the National Academy of Sciences of the United States of America*, 107, 5897–5902.

Smyth, D.R., Bowman, J.L., and Meyerowitz, E.M. (1990) Early flower development in *Arabidopsis. Plant Cell*, 2, 755–767.

Spujt, R.W. (1994) A systematic treatment of fruit types. *Memoirs of the New York Botanical Garden*, 70, 1–181.

Sun, G., Dilcher, D.L., Wang, H., and Chen, Z. (2011) A eudicot from the Early Cretaceous of China. *Nature*, 471, 625–628.

Sun, G., Ji, Q., Dilcher, D.L., Zheng, S., Nixon, K.C., and Wang, X. (2002) Archaeofructaceae – a new basal angiosperm family. *Science*, 296, 899–904.

Tanksley, S.D. (2004) The genetic, developmental, and molecular bases of fruit size and shape variation in tomato. *Plant Cell*, 16 (Suppl.), S181–189.

Temple, S.A. (1977) Plant-animal mutualism: co-evolution with dodo leads to near extinction of plant. *Science*, 197, 885–886.

Tewksbury, J.J., Manchego, C., Haak, D., and Levey, D.J. (2006) Where did the chili get its spice? Biogeography of capsaicinoid production in cancestral wild chili species. *Journal of Chemical Ecology*, 32, 547–564.

Tiffney, B.H. (1984) Seed size, dispersal syndromes, and the rise of the angiosperms: evidence and hypothesis. *Annals of the Missouri Botanical Garden*, 71, 551–576.

Tiffney, B.H. (1986) Evolution of seed dispersal syndromes according to the fossil record. In: *Seed Dispersal* (ed. D.R. Murray). Academic Press, Sydney.

Tiffney, B.H. (2004) Vertebrate dispersal of seed plants through time. *Annual Review of Ecology, Evolution, and Systematics*, 35, 1–19.

Valido, A., Schafer, H.M., and Jordano, P. (2011) Colour, design and reward: phenotypic integration of fleshy fruit displays. *Journal of Evolutionary Biology*, 24, 751–780,

Van der Pijl, L. (1982) *Principles of Dispersal in Higher Plants*. 3rd ed. Springer-Verlag, New York.

Vivian-Smith, A., and Koltunow, A.M. (1999) Genetic analysis of growth-regulator-induced parthenocarpy in Arabidopsis. *Plant Physiology*, 121(2), 437–451.

Vivian-Smith, A., Luo, M., Chaudhury, A., and Koltunow, A. (2001) Fruit development is actively restricted in the absence of fertilization in *Arabidopsis*. *Development*, 128(12), 2321–2331.

Vrebalov, J., Pan, I.L., Arroyo, A.J.M., McQuinn, R., Chung, M., Poole, M., Rose, J.K.C., Seymour, G., Grandillo, S., Giovannoni, J., and Irish, V.F. (2009) Fleshy fruit expansion and ripening are regulated by the tomato SHATTERPROOF gene TAGL1. *Plant Cell*, 21(10), 3041–3062.

Whitney, K.D. (2009) Comparative evolution of flower and fruit morphology. *Proceedings of the Royal Society B: Biological Sciences*, 276, 2941–2947.

Whitney, K.D., and Stanton, M.L. (2004) Insect seed predators as novel agents of selection on fruit color. *Ecology*, 85, 2153–2160.

Wikström, N., Savolainen, V., and Chase, M.W. (2001) Evolution of the angiosperms: calibrating the family tree. *Proceedings of the Royal Society B: Biological Sciences*, 268, 2211–2220.

Wing, S.L., and Boucher, L.D. (1998) Ecological aspects of the Cretaceous flowering plant radiation. *Annual Review of Earth and Planetary Science*, 26, 379–421.

Witmer, M.C., and Cheke, A.S. (1991) The dodo and the tambalaque tree: an obligate mutualism reconsidered. *Oikos*, 61, 133–137.

Yanofsky, M.F., Ma, H., Bowman, J.L., Drews, G.N., Feldmann, K.A., and Meyerowitz, E.M. (1990) The protein encoded by the *Arabidopsis* homeotic gene *agamous* resembles transcription factors. *Nature*, 346, 35–39.

Zjhra, M.L., Sytsma, K.J., and Olmstead, R.G. (2004) Delimitation of the Malagasy tribe Coleeae and implications for fruit evolution in Bignoniaceae inferred from a chloroplast DNA phylogeny. *Plant Systematics and Evolution*, 245, 55–67.

3 Ethylene and the Control of Fruit Ripening

Don Grierson

Introduction

Ripening of fleshy fruits has evolved as a mechanism to aid seed dispersal. This adaptation is not confined to one particular group of plants but occurs in anatomically and taxonomically different types of fruits (See Chapter 2), suggesting it may have arisen independently several times during evolution. The unripe fruit initially provides protection for the developing seeds, and is generally hard and unattractive to predators. During the growth phase, the fruit accumulates water, nutrients, and reserves from the parent plant, and synthesizes cellular components. Most growing fruits are green and the cells in the outer layers carry out photosynthesis, which contributes to dry matter deposition. After seed development and fruit growth have been completed, the properties of the fruit cells change so that they become more attractive for potential consumers such as animals and humans. The value to the plant of investing resources in the ripe fruit is that after the fruit is eaten the seeds may be discarded some distance away or, alternatively, if eaten they survive the digestion process and are deposited elsewhere.

It may seem surprising, but 50 years ago ripening and senescence were thought of as degradative processes that culminate in general disruption of cellular organization. There certainly is some cellular disruption in over-ripe fruit, but prior to that, ripening is driven by an active and regulated developmental program. A crucial finding was that leaf senescence could be inhibited by the protein synthesis inhibitor cycloheximide. Further investigation showed that there was an increase in incorporation of radioactive precursors into fruit RNA and proteins associated with the climacteric rise during fruit ripening, and using *in vitro* techniques it was shown that this is associated with the production of new mRNAs.

During the last 25 years the twin approaches of ripening gene identification and expression studies, particularly in tomato, and intensive study of mutants of the ethylene response pathway in Arabidopsis, have made a major contribution to our understanding of the molecular basis of fruit quality and ripening control. Ripening requires complex biochemical changes, involving switching off the fruit developmental program and the initiation of new gene expression programs that cause ripening changes (see Giovannoni, 2004, 2007). These changes often include

The Molecular Biology and Biochemistry of Fruit Ripening, First Edition.
Edited by Graham B. Seymour, Mervin Poole, James J. Giovannoni and Gregory A. Tucker.

alterations in fruit texture, accumulation of sugars, and the destruction of toxic compounds such as alkaloids, thus making the fruit more palatable. Frequently, ripe fruit also becomes more attractive to predators by producing aroma volatiles, and change color by accumulating red, blue, or purple anthocyanins in the cell vacuoles or by converting their chloroplasts to chromoplasts, which accumulate yellow and red carotenoids. The variation in attributes of different types of fruits can be largely explained if it is assumed that different biochemical processes are involved in accumulation of reserves in different species, and a range of different genes have been recruited to the ripening program during evolution. For example, specific mechanisms of softening are related to differences in the battery of enzymes involved in cell wall synthesis and cell wall disruption in different fruits. Nutritional value varies with the types of reserve compounds (fats, carbohydrates, proteins), vitamins, and other micronutrients stored by the fruit, and flavor and aroma are determined by the types of volatile secondary products generated during ripening, or during mastication.

The existence of tomato-ripening mutants confirmed the genetic basis of ripening. The *Neverripe (Nr)* and *ripening inhibitor (rin)* fruit shown in Figure 3.1 were photographed 6 months after control fruit ripened and remained in good condition for a year or more, provided they were prevented from becoming dehydrated. After 1 year the *rin* seeds began to germinate but the flesh had still not changed their color, although there had been some softening. This indicates that ripening and the eventual deterioration and rotting of normal fruit is a regulated process; it is not inevitable, but is genetically determined. As we shall see, ethylene plays an important role in this process. A range of fruit color mutants, in which no other aspect of ripening appeared to be affected, indicated that the ripening pathway had several independent branches (Fig. 3.2), and this was confirmed later when it was shown, for example, that the production of individual enzymes involved in cell wall metabolism (e.g., polygalacturonase (PG), pectin esterase (PE), pigment synthesis (phytoene synthase (PSY)), and

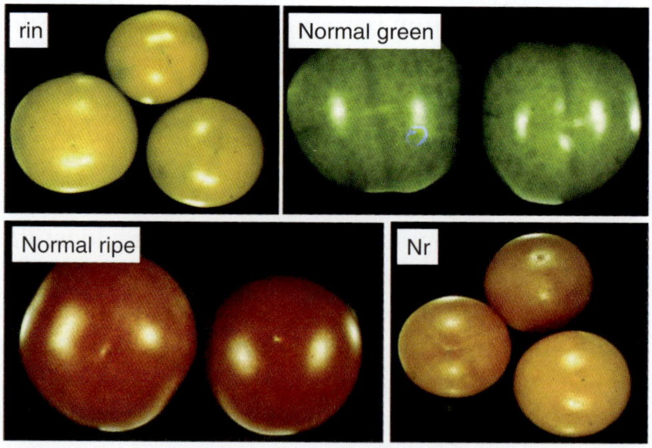

Figure 3.1 Mutants of ethylene action and ripening. Mature green and ripe Ailsa Craig tomato fruits, photographed at approximately 40 and 50 days respectively, together with fruit from near-isogenic lines of the *ripening inhibitor* (*rin*) *and Neverripe* (*Nr*) mutants, photographed at around 6 months old. As discussed in the text, the Nr mutation bred into the Pearson tomato cultivar (not shown) has a more severe phenotype and the fruits remain green.

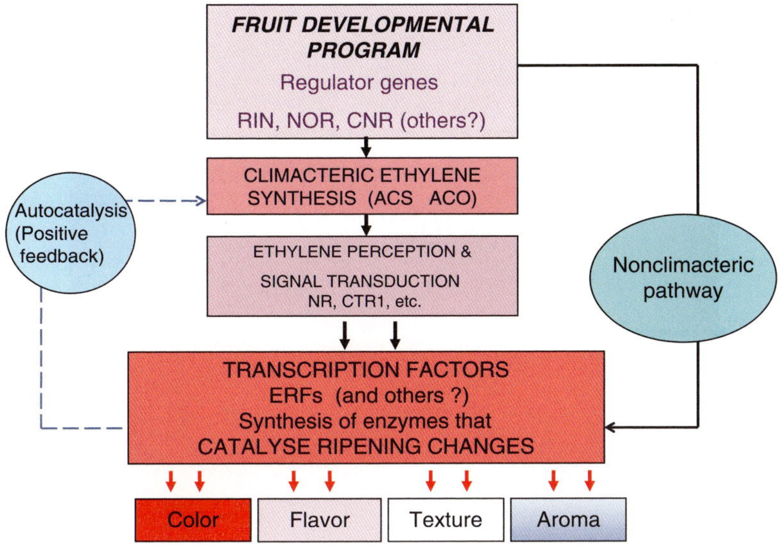

Figure 3.2 Control of ripening. The main features of the ripening process are indicated, based on studies in tomato. Note the regulator genes, the feedback loop promoting autocatalytic ethylene production, ethylene synthesis, perception and signaling mechanisms, and branches leading to separate responses controlled by transcription factors.

volatile production (a specific isoform of lipoxygenase (TOMLOXC)) could each be inhibited independently by gene silencing without affecting other ripening attributes. In the climacteric fruits (see later), ethylene has long been recognized as a ripening hormone. Thirty years before ethylene receptors were cloned and sequenced, Burg and Burg (1967) predicted the presence of a metal ion in the receptor based on the metal affinity of compounds that have ethylene-like or ethylene-antagonistic activities. For example, Ag(I), prevents ripening, by interfering with the ethylene-receptor interactions (Fig. 3.3). Three particularly intriguing mutants were *Nr, nonripening (nor),* and *rin*. The *Nr* mutation was shown to be due to a crucial amino acid change in a tomato ethylene receptor by Wilkinson et al. (1995), and the two other mutants had drastically reduced ethylene production, although ripening could not be restored by adding ethylene externally. The *rin* mutation was shown to be in a MADS-box transcription factor (now called LeMADS-RIN, Vrebalov et al., 2002) expressed in fruits, which when mutated results in a failure to synthesize or respond to ethylene. The *nor* mutation affects an NAC (*N*o *A*pical *M*eristem)-domain transcription factor (Giovannoni, 2004) and causes retardation of tomato fruit ripening with a phenotype which superficially looks similar to the *rin* mutant. A fourth regulatory mutant, *Colorless nonripening* (*Cnr),* was characterized more recently. This is an epigenetic change that alters the promoter methylation of a SQUAMOSA promoter-binding (SPB) protein, resulting in a pleiotropic-ripening inhibition phenotype and inhibited expression of ethylene-related genes *ACO1*, *E8*, and *NR*, and a range of ripening-related genes (Thompson et al., 1999). As discussed later, these transcription factors, and others, interact and regulate both ethylene-dependent and ethylene-independent aspects of ripening.

The concept of ripening as a highly coordinated series of genetically programmed events is consistent with the known effects of environmental factors, such as heat shock, on ripening. At

Figure 3.3 Ag^+ inhibits ethylene perception or action and consequently prevents ripening. Silver thiosulfate, which is a translocatable form of Ag +, was introduced to one side of the mature green fruit pedicel while it was still attached to the plant. A needle was used to insert a thread asymmetrically into the fruit stalk to act as a wick for the uptake of the solution. Ag + was only transported to half of the vascular tissue and delivered to one-half of the fruit, preventing those cells from responding to ethylene and they did not undergo ripening changes. See Davies et al. (1988), (1990).

temperatures above 35°C tomato ripening is inhibited, and this has been used as a postharvest treatment to delay ripening and prolong fruit storage. Picton and Grierson (1988) showed that in response to heat shock (HS) the fruit cells failed to accumulate carotenoids, or to soften or synthesize ethylene. Supplying ethylene to the fruit did not restore ripening, demonstrating that these responses were not caused by the lack of ethylene. The normal pattern of protein synthesis was disrupted and the fruit produced characteristic HS proteins instead. HS also led to the rapid disappearance of ripening-related mRNAs, and a cessation of synthesis of the specific proteins associated with normal ripening. This showed that HS exerts its effects by disrupting the genetic ripening program.

Ethylene and Climacteric and Nonclimacteric Fruits

All plants produce some ethylene during their life cycle. Often there is a low basal level, in tomato this is around 0.05 $nL.g^{-1}.h^{-1}$, which can increase 100-fold or more at particular stages of the life cycle, for example in response to wounding or pathogen attack, ripening, senescence, or abscission. Increased respiration (the climacteric) and a burst of ethylene biosynthesis are characteristics of the ripening process in many fleshy fruits, such as tomato, avocado, apple, melon, and banana, and these are called climacteric fruits (this term originally referred to the respiratory climacteric, but now is commonly used to discuss the onset of ethylene production and ripening). Other fruits, such as strawberry, grape, and citrus, show no increase in respiration and ethylene production during ripening (although they do respire) and they were originally called nonclimacteric fruits, in which ripening control was originally thought to be independent of ethylene (Fig. 3.2). It is becoming clear, however, that ethylene may be involved in the development and aspects of ripening of some nonclimacteric fruits, such as citrus, white grape, and strawberry, although it seems that only low levels of ethylene are required. External ethylene has long been known to stimulate de-greening of citrus, which is nonclimacteric, and there is recent evidence for small changes in ethylene biosynthesis genes, and ethylene production during ripening of strawberry (Trainotti et al., 2005). Fei et al. (2004) compared the ripening genes of climacteric tomato fruits and nonclimacteric grapes and showed that the two species shared a subset of common ripening transcription regulators, including members of the MADS-box, zinc finger, and bZIP transcription factor families, suggesting that these genes have been conserved for ripening control during evolution, and are active in both climacteric and nonclimacteric fruits. (See Wechter et al., 2008, for a study of expression of ethylene-related and other genes during ripening of nonclimacteric watermelon.)

Thus, it has gradually become clear that the distinction between climacteric and nonclimacteric fruits, and the view that ethylene is only involved in controlling ripening of climacteric fruits, is an oversimplification. In general, fruit with the highest respiration rate tend to ripen most rapidly. Ethylene hastens the ripening of climacteric fruits and the increase in ethylene production associated with the respiratory rise is autocatalytic. There is clear evidence that, at least in some fruits, ethylene causes the climacteric rise in respiration. Climacteric and non-climacteric types can be found in the same species, however, suggesting that small genetic differences underlie the climacteric trait, and in other cases different fruit parts, such as skin and pulp, may behave differently. It is likely, therefore, that as our molecular understanding increases, and the ripening behavior of a wider range of fruits is investigated at the genetic level, we may need to modify our classification of fruit types. The view that is emerging is of a basic underlying genetic program that controls climacteric and nonclimacteric ripening, which is modulated by a range of factors, including ethylene, particularly in the climacteric fruit types, but also involving other regulators (Fig. 3.2).

Many fruits can ripen on the plant or tree, but some, such as avocado, only ripen after they are picked. A "tree factor" that inhibits ripening has been postulated to explain this effect. Also, in strawberry it was recognized over 50 years ago that auxin, coming probably from the developing achenes, actually delayed the onset of ripening. Harvesting can also hasten the ripening of some fruits and in such cases wound- or dehydration stress-ethylene from the calyx or calyx scar may also stimulate ripening. It would be wrong to assume, however, that ethylene is the only hormone that affects ripening. Auxin inhibits the ripening of strawberry, which is nonclimacteric. Auxin is also involved in the regulation of some ethylene response factors (ERFs, transcription factors), discussed later, which are important in climacteric fruit ripening, and there is some evidence of interactions between auxin and ethylene signaling during ripening (Lin et al., 2009).

Many climacteric fruits can be harvested unripe and stimulated to ripen by supplying ethylene externally. This demonstrates that ripening is basically a conversion process, rather than one requiring import of new materials, although sometimes the flavor and other quality attributes may not be as well developed in detached fruits as in those allowed to ripen attached to the plant.

It has long been noted that fruits need to be mature before they can effectively respond to ethylene—that is, they develop a competence to respond to ethylene and to ripen. An intriguing observation, made many years ago is that "ethylene shortens the green life" (of banana). Put another way, when bananas are picked green they will take a certain number of days to start to ripen. If they are *transiently* exposed to ethylene after picking, the time that elapses before they eventually ripen is reduced. Similarly, picked green tomatoes do not respond to ethylene when they are very immature, and they take a progressively shorter time to respond to external ethylene when they are picked closer and closer to the time when they would naturally ripen. This tells us several things: firstly, ethylene synthesis is normally delayed until the fruit is ready to ripen, and secondly, either the ethylene response system is not developed, or is inhibited, until close to maturity. It is possible that other hormones play a role in these processes.

A Molecular Explanation for System-1 and System-2 Ethylene

Ethylene synthesis via the Yang cycle requires the action of two key enzymes, ACC synthase (ACS) and ACC oxidase (ACO) (Fig. 3.5). Direct evidence that ethylene synthesis is essential for climacteric fruit ripening was established by experiments showing that reduction in the expression of tomato *ACS2* and *ACO1* by expressing antisense genes in transgenic plants inhibited or delayed ripening and spoilage in tomato and melon (Oeller et al., 1991; Hamilton et al., 1990; Picton et al., 1993) (Figs. 3.4, 3.5, 3.6, and 3.7). Low ethylene fruits have a much longer shelf life, because the expression of some ripening genes that lead to over-ripening, infection, and decay (see later in this chapter) are reduced. Ethylene is also required for the

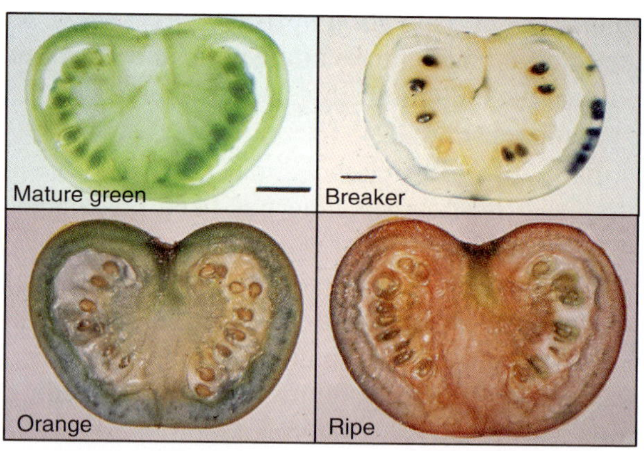

Figure 3.4 ACO1 promoter-Gus expression in transgenic tomatoes. The ACO1 gene promoter was fused to the GUS (β-glucuronidase) reporter gene, transformed into Ailsa Craig plants and tomatoes examined during fruit development and ripening. GUS activity, indicated by the blue staining, was evident around the vascular tissue at the breaker stage of ripening (first sign of color change) and was more uniformly distributed in orange fruits. Note that, although GUS staining is not really quantitative, the intensity of blue color corresponds approximately to the ethylene production curve during normal ripening (Fig. 3.5). See Blume and Grierson (1997).

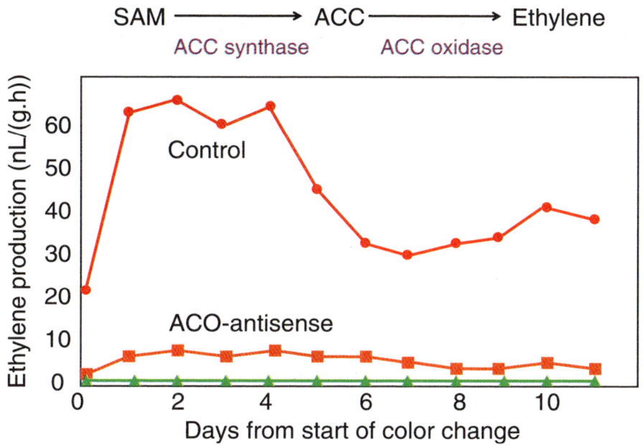

Figure 3.5 Gene silencing of ACO1 inhibits ethylene production in ripening fruit. Ethylene is synthesized from S-adenosyl methionine (SAM), to produce 1-amino-cyclopropane-1-carboxylic acid (ACC) by ACC synthase, which is then converted to ethylene by ACC oxidase. Before ACC oxidase was discovered, a possible candidate clone from tomato, TOM13, was used to construct an antisense gene under the control of the Cauliflower mosaic virus 35S promoter and transformed into tomato plants. Primary transformants were selected for those with the transgene and selfed, to produce progeny which were selected for 0, 1, or 2 transgenes and ethylene measured during fruit maturation and ripening. Control: red circles; one antisense gene: orange squares; two antisense genes: green triangles. The fruit with 2 antisense genes produced only 5% of the normal ethylene. They still ripened, but over-ripening and deterioration were strongly delayed. See Hamilton et al. (1990) and Figure 3.6. Expression of TOM13 in yeast confirmed that it encoded ACC oxidase (see Hamilton et al., 1991).

full development of flavor and aroma, however (Ayub et al., 1996; Flores et al., 2002), and biotechnological control of ethylene needs to strike a balance between enhancing shelf life while preserving flavor and aroma of fruits. This is complicated by the fact that there are often different threshold concentrations of ethylene that elicit different types of ethylene responses, which raises an interesting regulatory question about how this is brought about. We now know that there are at least 14 *ACS* and *ACO* genes in plants such as Arabidopsis and tomato, and

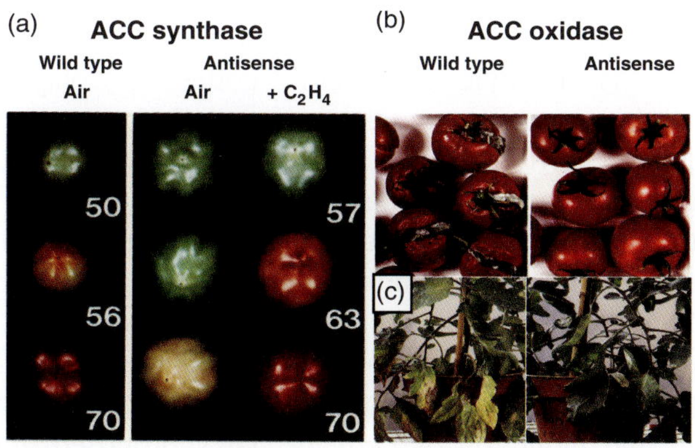

Figure 3.6 Antisense silencing of ACS (a) and ACO (b) inhibits ripening and senescence (c). Fruit expressing an ACS2 (ACC synthase 2) antisense gene, together with controls, are shown in (a). Note that in air the antisense fruit did not ripen, but ripening could be restored by adding external ethylene. The ACO1 (ACC oxidase) antisense gene only inhibited ethylene synthesis by approximately 95%: fruit ripened, but over-ripening and deterioration were greatly reduced (b). Also, leaf senescence was delayed (c). See Oeller et al. (1991); Picton et al. (1993), and Figure 3.5.

Figure 3.7 Inhibition of ripening of low ethylene melon by silencing the ACO1 gene. Transgenic melon expressing an ACO1 antisense gene produced extremely low levels of ethylene, and remained firm for much longer than the controls. In addition, the rind remained green and fruit remained attached to the plant for longer, because separation at the abscission zone was delayed, allowing fruit to accumulate more sugar. See Ayub et al. (1996).

these are expressed at different times in the life cycle and in response to different developmental and environmental cues. In addition, some ethylene receptors and signaling molecules show different expression patterns during development.

A characteristic feature of ethylene synthesis in climacteric fruits is that some parts of the plant, including unripe fruits, synthesize low levels of ethylene, which is regulated by feedback inhibition but, after the onset of ripening, unrestrained or autocatalytic ethylene production begins and a major burst of ethylene evolution occurs, and this stimulates ripening. McMurchie et al. (1972) explained this autocatalytic rise in ethylene in climacteric fruits by proposing that two systems (System-1 and System-2) were involved in ethylene biosynthesis (reviewed by Lelievre et al., 1998). System-1 functions during normal vegetative growth, is autoinhibited by ethylene, and is responsible for producing the basal levels of ethylene that are synthesized by all plant tissues. Nonclimacteric fruit may be considered to be locked in the System-1 stage and only produce low levels of ethylene. System-2 comes into play during the ripening of climacteric fruits, and during senescence, and requires the induction of new isoforms of ACS and ACO. These are regulated differently, give rise to much higher levels of ethylene, without autoinhibition (feedback inhibition) by ethylene.

In tomato, some *ACO* and *ACS* genes are expressed in unripe fruits while others show ripening-related expression, and the accumulation of their mRNAs can be either stimulated or inhibited by ethylene and developmental cues (Fig. 3.8). Barry et al. (2000) proposed that in tomato, System-1 ethylene is regulated by the expression of *LeACS1A* and *6*, and that during the transition from System-1 to System-2 the *RIN* gene is involved in causing increased expression of *LeACS1A* and induction of *LeACS4*. They proposed that the maintenance of system-2 ethylene production is due to the ethylene-dependent induction of *LeACS2* (Barry et al., 2000). A recent update of this scheme by Yokotani et al. (2009) depicts the relationship between developmental factors and individual *ACS* genes in regulating System-1 and System-2 ethylene during fruit development and ripening (Fig. 3.9). Note that the *ACO* genes also show

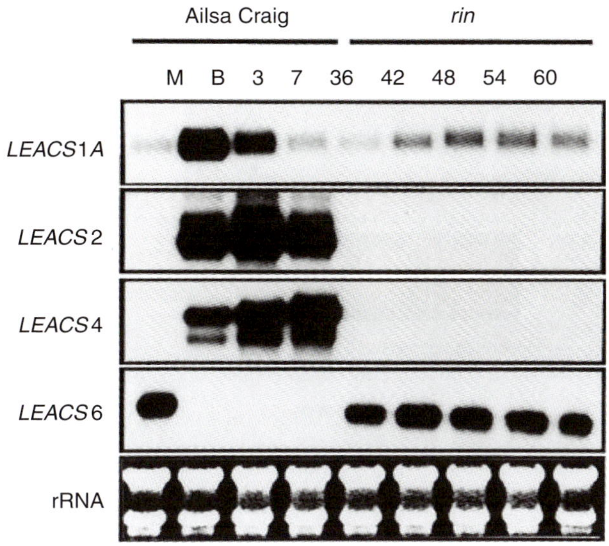

Figure 3.8 Expression of different ACS genes in developing tomato fruit. Gene-specific probes were hybridized with RNA from normal and *rin* fruit in the Pearson background at the mature green (M) and breaker (first sign of color change) stages, and at different days thereafter. In the case of *rin,* which does not ripen, fruits were studied for up to 60 days. Note that ACS2 and ACS4 are the most strongly expressed and ACS1A is transiently expressed during normal ripening, and these changes are greatly reduced or prevented in *rin* fruit. ACS6, on the other hand, is repressed during normal ripening but not in rin fruit. See Barry et al. (2000).

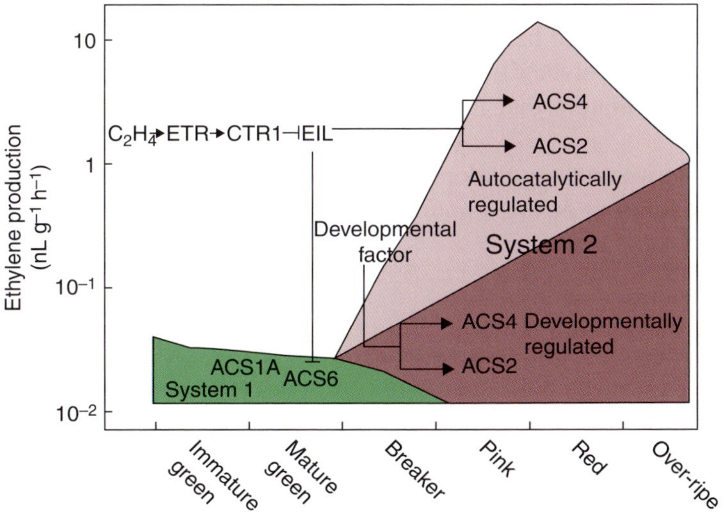

Figure 3.9 The involvement of ACS isoforms in System-1 and System-2 ethylene production during ripening of tomato. The contribution of each isoform to total ethylene evolution is shown at different stages of development and ripening. Note that EIL is shown promoting upregulation of System 2 (ACS4 and ACS2) and inhibiting System 1 (ACS1A and ACS6). Modified from Yokotani et al. (2009).

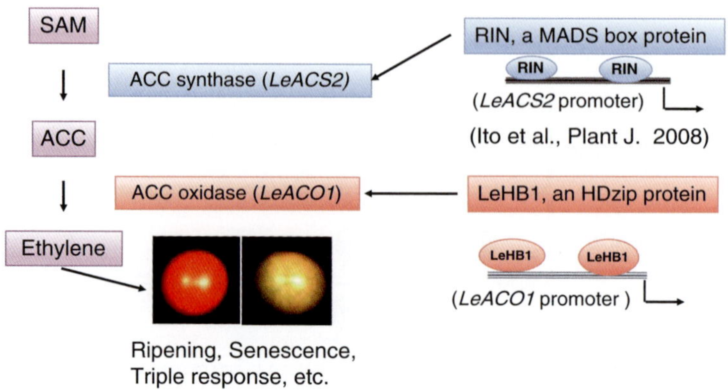

Figure 3.10 Transcriptional control of ethylene synthesis. The role of RIN and LeHB-1 transcription factors in controlling specific ethylene biosynthesis genes (ACS 2 and ACO1 respectively) during ripening. See Lin et al. (2008c) and Ito et al. (2008).

differential expression during development and ripening (Barry et al., 1996), but this has not received so much attention.

Using biochemical procedures, it has been demonstrated that the MADS-box protein RIN and the homeobox protein LeHB-1 control ethylene synthesis by binding to the promoters and activating transcription of two genes for ethylene biosynthetic enzymes: *LeACS2* (Ito et al., 2008) and *LeACO1* (Lin et al., 2008a; see later, Fig. 3.10). In addition, another regulatory gene, *TAGL1* (*TOMATO AGAMOUS-LIKE1*), has been shown to activate the *ACS2* promoter, and ERF2 from tobacco and tomato have been shown to regulate tomato *ACO3* (Itkin et al., 2009; Ito et al., 2008). As our understanding of the regulatory complexity improves we may yet find other transcriptional regulators that interact, perhaps in a combinatorial fashion, to regulate ethylene production.

In addition to transcriptional regulation, there is complex posttranslational control of ethylene synthesis by the modulation of the amount and activity of specific ACS isoforms. There are three classes of ACS:

Type 1 enzymes (e.g., Arabidopsis: AtACS1, AtACS2, AtACS6; Tomato: LeACS1A, LeACS1B, LeACS2, LeACS6) have multiple phosphorylation sites in an extended C-terminal region. These isoforms are active when phosphorylated by CDPK (calcium-dependent protein kinase) and MAPK (mitogen-activated protein kinase) and unphosphorylated proteins are degraded, probably by the ubiquitin 26S proteasome system.

Type 2 enzymes (e.g., Arabidopsis: AtACS9, AtACS8, ACS5, AtACS4; Tomato: LeACS8, LeACS7, LeACS3) are phosphorylated by CDPK only, and are known to interact with three proteins, ATO1, EOL1, and EOL2, involved in controlling their degradation by the ubiquitin 26S proteasome system.

Type 3 isoforms are not phosphorylated (e.g., Arabidopsis: AtACS11, AtACS7; Tomato: LeACS5, LeACS4).

Ethylene and Ripening Gene Networks in Flower and Fruit Development

Ethylene has been shown recently to influence the transition from the vegetative to the reproductive state in Arabidopsis (Achard et al., 2003, 2007), and there is very strong evidence from other plants that ethylene plays a role in the actual development of flowers as well as ripening of fruits. For example, Ishida et al. (1998) demonstrated that floral organ identity genes, such as *TAG1* (*TOMATO AGAMOUS1*, involved in flower development), were also involved in the ripening process. Sepals of the tomato cultivar VFNT Cherry, when cultured *in vitro* between 16° and 22°C, changed their developmental program to that of ripening fruits. This cool temperature-induced morphogenesis process (referred to here as sepal "ripening") involves several regulatory genes, in addition to *TAG1*, such as *TDR4*, an HD-Zip gene *VaHOX1*, and the ripening regulator *NOR*. Although *NOR* has been suggested to function independently of ethylene in the tomato-ripening process (Adams-Phillips et al., 2004; Giovannoni, 2004), Bartley and Ishida (2007) showed that the accumulation of *NOR* mRNA was affected by blocking ethylene perception with 1-MCP (1-methylcyclopropene, which inhibits ethylene responses at the receptor level) in cold-induced "ripening" sepals. Sepals treated with 1-MCP showed a 90% reduction in ethylene compared with untreated sepals and did not induce the ripening genes expression program. Transcripts of *TAG1* and *TDR4* were also greatly reduced in 1-MCP-treated sepals, compared with untreated controls, suggesting they may function downstream of ethylene (Bartley and Ishida, 2007). The expression of *VaHOX1*, however, was not suppressed by 1-MCP, suggesting that it is either upstream of, or independent from, ethylene. Further studies on a related protein, TAGL1 have shown that this MADS-box protein is involved in control of ripening of fruits. Firstly, the TAGL1 protein activated the ACS2 promoter in model experiments, and secondly, inhibiting its expression inhibited the development of several ripening attributes. Interestingly, *TAGL1* expression was enhanced in the presence of 1-MCP and appears not to be regulated by RIN (Itkin et al., 2009).

Another HD-Zip protein, LeHB-1, originally discovered during a search for transcription factors that bind the *LeACO1* promoter (Lin et al., 2008b), appears to play a particularly important role. LeHB-1 stimulates ethylene synthesis by activating transcription of *LeACO1* at the onset of ripening (Figs. 3.10 and 3.11) but it was also found that ectopic expression of LeHB-1, using a virus vector, modified tomato flower development, for example by causing the formation of multiple flowers within one sepal whorl and the conversion of sepals into fruit-like structures, which ripened (Fig. 3.12). These results indicate, at the very least, that ethylene plays an important role in flower and early fruit development, although it is not yet clear whether LeHB-1 normally plays a direct role, or whether some other regulator is also involved. Such a role for ethylene is not really surprising, even though it has not been discovered in Arabidopsis, since ethylene is used to promote synchronous flowering in pineapples, is involved in sex determination in melons and cucumbers, and can cause male sterility when applied to developing cereals (see Lin and Grierson, 2010).

There are also putative LeHB-1-binding sites in the promoters of *ACO2*, *RIN*, *PG* (PG, see later) (Lin et al., 2008c) and a number of other ripening genes. RIN itself activates *LeACS2*, so it and LeHB-1 appear crucial to switching on System-2 ethylene and ripening control. It therefore seems that a genetic network involving ethylene and homeotic regulators control aspects of

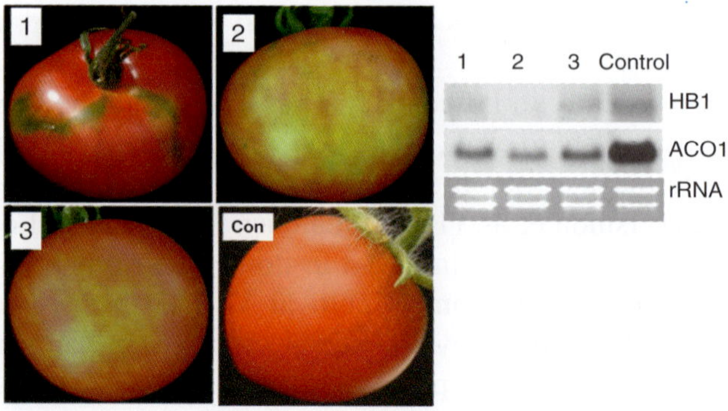

Figure 3.11 Inhibiting LeHB1 expression by VIGS inhibits ACO1 accumulation and delays ripening. Expression of LeHB-1 was inhibited in tomato fruit using a virus-induced gene silencing (VIGS) gene construct. This led to a reduction in LeHB-1 mRNA, a downregulation of ACO1 mRNA and an inhibition of ripening, compared to the control, revealed as streaks or blotchy sectors on the fruit (1–3), the extent of which depended on which cells the VIGS construct reached. See Lin et al. (2008c).

flower development, sex determination, fruit development, and ripening (see Fig. 3.16). It is not yet clear, however, whether this is confined to only some plants, such as melon and tomato, or also occurs in Arabidopsis, but has so far escaped attention.

Ethylene Perception and Signaling

An outline of the ethylene perception and signaling process is shown in Figure 3.13. This scheme was initially developed following the characterization of Arabidopsis mutants of ethylene

Figure 3.12 Ectopic expression of *LeHB-1* converts sepals into fruits. When applying VIGS constructs to whole plants it is difficult to control where the virus goes. Sometimes it is translocated away from the site of application and it can lead to overexpression rather than silencing. These photographs show swelling of green sepals (left) and the formation of fruitlets on the shoulders of a tomato fruit (right). That this was due to overexpression, rather than silencing, was shown by the fact that it was entirely prevented by using a construct with a stop codon in the LeHB-1 coding sequence, which would be expected to prevent the formation of a full-length LeBB-1 protein, but would not affect silencing at the level of mRNA production. See Lin et al. (2008c).

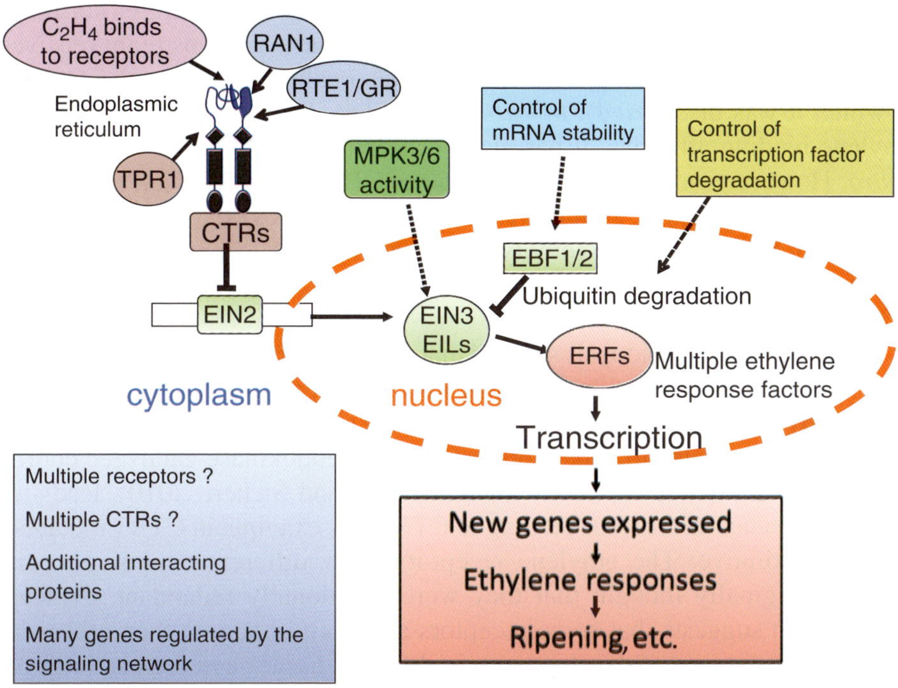

Figure 3.13 Ethylene signaling components. The main features of the ethylene perception and signaling mechanisms in plants are shown. See Lin et al. (2009). The roles of individual components are discussed in the text.

perception and signaling, and subsequently extended to and confirmed in other plants, including ripening fruits, where additional nuances and variations have been discovered.

The Receptors

Ethylene is perceived by a family of copper-binding membrane-associated Type I or Type II receptors. The dedicated work of Sisler led to the identification of a range of organic chemical inhibitors of ethylene action, in addition to Ag^+ (Fig. 3.2), culminating in the discovery of the volatile compound 1-MCP, which has been extensively used commercially in horticulture and floriculture to block ethylene action, thereby slowing down or preventing ethylene-stimulated processes such as ripening, abscission, and senescence (reviewed by Reid and Staby, 2008).

There are five ethylene receptors in Arabidopsis, and six in tomato with amino acid sequence similarity to the bacterial two-component histidine kinases. They have been classified into two subfamilies: subfamily I receptors in Arabidopsis are ETR1 and ERS1 (LeETR1, LeETR2, and NR in tomato) (most recently discovered tomato genes and proteins are prefixed with either Le, for *Lycopersicon esculentum*, or Sl, for *Solanum lycopersicon*, the new Latin binomial name for tomato). These receptors contain conserved histidine kinase motifs in a C-terminal domain, while subfamily II receptors (Arabidopsis ETR2, ERS2, and EIN4; tomato LeETR4, leETR5,

Le ETR6) have a more diverged histidine kinase region lacking some features normally required for activity (Schaller and Kieber, 2002). Some receptors have a downstream receiver domain, but this is lacking in ERS1, ERS2, and LeETR3 (also called NR, after the tomato-ripening mutant in which it was discovered). *In vitro* experiments have shown that all receptors have similar ethylene-binding affinities. It has also been demonstrated that they can autophosphorylate *in vitro*. However, only subfamily I receptors ETR1 and ERS1 can phosphorylate on a histidine residue, while the three subfamily II receptors and ERS1 have serine/threonine kinase activity rather than histidine kinase activity (Moussatche and Klee, 2004). Canonical histidine kinase activity in the transmitter domain of ETR1, however, is not essential for activity (Wang et al., 2002).

Genetic experiments in Arabidopsis showed that the receptors function as negative regulators; they interact with the Raf kinase-like protein CTR1 and repress ethylene responses. When the receptors bind ethylene, this inhibition is relieved and a signaling process, which may involve a conformational change related to phosphotransfer, or a pseudokinase-catalyzed conformational change (see Rajakulendran et al., 2009; Rajakulendran and Sicheri, 2010), leads to ethylene responses. In the absence of receptor binding to CTR1 (for example in CTR1 mutants) ethylene responses are constitutive. The ethylene receptors show different expression patterns and, although it was originally thought that they were functionally redundant, evidence is now accumulating which suggests that some receptors are more important than others in ethylene signaling at particular stages of development, in different tissues and organs, or in response to different cues (see Tieman and Klee, 1999; Whitelaw et al., 2002; Binder at al., 2006). The receptors form dimers by association of their N-termini, which are inserted into a membrane. Each protein dimer binds a single Cu(I), which is required for ethylene binding. Insertion of copper into the receptors is thought to involve the RAN1 protein, which is similar to human and yeast proteins known to be involved in copper movement. It appears that receptors can form homo- and heterodimers and can also combine as higher aggregates (Gao et al., 2008), but the functional significance of this is unclear.

The Arabidopsis ETR1 ethylene receptor was shown to be in the endoplasmic reticulum (ER) membrane and later was also observed in the Golgi apparatus in roots where it was co-localized with the protein RTE1 (discussed in Lin and Grierson, 2010). The tobacco ethylene receptor NTHK1, in contrast, was found in the plasma membrane while melon CmERS1 and tomato NR (LeETR3), which are ERS1-type receptors lacking a receiver domain, have been found at the ER and the plasma membranes (Ma et al., 2006; Zhong et al., 2008) and nuclear membranes (Lin et al., 2008b). These reports of multiple subcellular locations for ethylene receptors raise the possibility of functional specialization and/or receptor movement in ethylene perception and signaling, although further studies are needed to test these possibilities.

The Nr mutant phenotype of tomato (See Fig. 3.1) was shown by Klee and co-workers to be caused by a dominant mutation causing an amino acid change in the N-terminus of the NR ethylene receptor. This abolishes ethylene binding to the receptor and results in tomato plants that are insensitive to ethylene and produce nonripening fruit (Wilkinson et al., 1995). Because the mutant NR receptor cannot bind ethylene, it constitutively suppresses ethylene responses through its interaction with the tomato CTR proteins, even in the presence of ethylene. Antisense inhibition of production of the mutant mRNA in the Nr mutant resulted in failure to synthesize

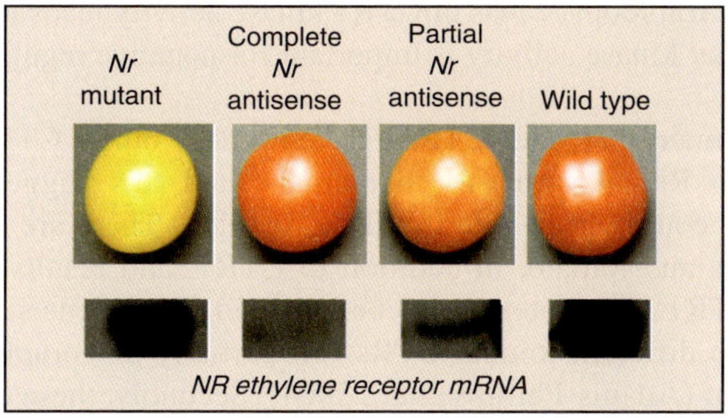

Figure 3.14 The mutant NR ethylene receptor inhibits ripening, even in the presence of ethylene. Silencing its expression (*Nr* antisense) restores ripening. See Hackett et al. (2000).

the mutant receptor protein, and partially or completely restored ripening (Hackett et al., 2000). This confirmed that the receptor inhibition model of ethylene signaling also applies to ripening (Fig. 3.14). Interestingly, the *Nr* mutation has a more severe effect on ripening when bred into the Pearson cultivar of tomatoes, when the fruits stay green, compared to the Ailsa Craig variety, where the fruits turn orange after several months (Fig. 3.1). Investigation of this variation might well provide more information about the perception and signaling apparatus and the modulation of tomato fruit ripening.

Since ethylene receptors are negative regulators, ethylene sensitivity would be expected to increase if the receptors were degraded, and this suggestion was supported by the finding that transgenic tomatoes with reduced ethylene receptor levels ripened early, and that this was due to increased ethylene sensitivity. In addition, reduction of the expression of the tomato ethylene receptors *LeETR4* or *LeETR6* elevated ethylene production and promoted and accelerated ripening (Tieman and Klee, 1999, 2000; Kevany et al., 2007, 2008). Klee and his co-workers went on to demonstrate that receptor degradation requires ethylene binding, since pretreatment with the competitive inhibitor 1-MCP stabilized the receptors. Receptor degradation can be blocked by MG132, a peptide aldehyde inhibitor of the protease activity of the 26S proteasome (Kevany et al., 2007), indicating that this pathway is involved in receptor degradation.

Interaction of Receptors with CTRs

The C-terminal domains of the Arabidopsis ethylene receptors ETR1 and ERS1 have been shown to have direct protein–protein interaction with the N-terminus of the serine/threonine protein kinase CTR1, which is co-localized with the receptors to the ER membrane (Clark et al., 1998; Gao et al., 2003). This interaction between the receptors and CTR1, and possibly the consequent recruitment of CTR1 to the ER membrane, is essential for CTR1 function in repressing the downstream ethylene response. Mutation of the conserved N-terminal CN motif in both the Arabidopsis CTR1 (Gao et al., 2003) and tomato CTR3 (Zhong et al., 2008) can disrupt their interaction with the receptors, and free CTR proteins then accumulate in the cytosol. In addition, in Arabidopsis, loss of CTR1 kinase activity leads to constitutive ethylene responses, indicating that kinase activity is important for negative regulation in the absence of ethylene.

The C-terminal domains of the CTR1 protein, and other CTR-like proteins, have sequence homology to Raf-like kinases in animals, and this suggests that phosphorylation/dephosphorylation could play a role in CTR1 function. There are three other CTR1-like proteins in Arabidopsis and tomato, in addition to CTR1, and results from tomato indicate that several of these CTR1-like proteins may be involved in ethylene signaling. The evidence for this is, firstly, three different tomato CTRs can partially or completely complement the Arabidopsis *ctr1* mutant (Adams-Phillips et al., 2004). Secondly, these different tomato CTRs can each bind one or more of the tomato ethylene receptors in model experiments (Zhong et al., 2008) Thirdly, a fourth tomato CTR-like protein (LeCTR2) also binds to some of the ethylene receptors. Transgenic plant experiments indicate that it is involved in disease response signaling, which involves ethylene (Lin et al., 2009a).

Mammalian Raf-1 can bind phosphatidic acid and this leads to its translocation from the cytosol to the plasma membrane. CTR1, like mammalian Raf-1, can bind to PA (Testerink et al., 2007) and affect the CTR1–receptor interactions *in vitro*, raising the question whether plant CTR proteins are regulated in a similar manner. Another unresolved question is whether different CTRs can interact with each other, or with different receptors or other signaling molecules. Some kinases can activate others in a "side-to-side" reaction and this might be important for regulation, signaling, or cross-talk with other pathways.

There is some evidence that all the receptors are capable of activating CTR1 to suppress downstream signaling, and if this is substantiated it indicates that a conserved histidine kinase motif and receiver domain is not required for ethylene signaling (Hua and Meyerowitz, 1998; Wang et al., 2003). However, the histidine kinase domains have been shown to be important for the actual association of the receptors with CTR1 (or CTR1-like proteins) from both Arabidopsis and tomato (Clark et al., 1998; Gao et al., 2003; Qu and Schaller, 2004). It is possible, as mentioned earlier, that a conformational change could be induced by pseudokinase activity, that is, without phosphotransfer.

Other Receptor-interacting Proteins

Several other ethylene-receptor-interacting proteins have been reported, including REVERSION-TO-ETHYLENE SENSITIVITY1 (RTE1) in Arabidopsis and its homolog GREEN-RIPE (GR) in tomato (Barry and Giovannoni, 2006; Resnick et al., 2006) (Fig. 3.13). The loss-of-function mutant *rte1* is hypersensitive to ethylene and overexpression of the functional RTE1 reduced ethylene sensitivity. Similarly, ectopic expression of GR inhibited fruit ripening. These results indicate that RTE1 and GR are negative regulators of ethylene signaling. It has been proposed that RTE1 regulates a conformational switch of the receptors upon binding ethylene. Interestingly, overexpression of tomato GR under control of the cauliflower mosaic virus (CaMV) 35S promoter led mainly to reduced ethylene responses in fruit and floral tissues, but not other parts of the plant, suggesting that GR could function in ethylene signaling in a tissue-specific manner, or, alternatively, GR may be involved in modulating the signal output of specific ethylene receptors, which have tissue-specific expression or function (Barry and Giovannoni, 2006). The receptor tissue-specific model is supported by the fact that the tomato ethylene receptors NR, LeETR4, and LeETR6 are preferentially expressed in fruits and have been suggested to have unique roles during ripening (Tieman et al., 2000; Kevany et al., 2007).

A tomato TPR (tetra*tr*ico*p*eptide *r*epeat) protein called SlTPR1 has recently been identified and shown to interact with the ethylene receptors NR and LeETR1 in yeast and *in vitro* and a related protein (TRP1) has been found in Arabidopsis (Lin et al., 2008a, 2009b) (Fig. 3.13). Overexpression of the tomato TPR1 protein causes a variety of phenotypes suggesting altered hormone responses, some related to ethylene and others reminiscent of auxin effects. SlTPR1 has over 50% sequence homology with the mammalian protein TCC1, which interacts with Ras and competes with Raf-1 for Ras binding (Lin et al., 2008b, 2009b). This is particularly interesting, as CTR1 (and the other CTR-like proteins) have similarity to Raf kinase. If SlTPR1 acts in the same fashion as the TTC1 protein to compete with LeCTRs for binding to the ethylene receptors, this could lead to unbound CTRs remaining in a nonactivated state, resulting in constitutive ethylene responses. SlTPR1 might also function as an adaptor for receptor degradation and affect ethylene responses in that way. The main effects of *TPR1* overexpression were on plant height, fertility (related to ethylene), production of parthenocarpic fruit (usually related to auxin), and abscission (usually ethylene), with no striking effect on ripening itself. This suggests that TPR1 can modulate some, but not all, ethylene responses (Lin et al., 2008b), and is again consistent with a signaling system with specific developmental components.

Signaling Downstream of CTR

EIN2 (ETHYLENE INSENSITIVE2) is a membrane protein that functions downstream of CTR1. The N-terminus of EIN2 shares sequence homology with NRAMP ion transporters; exactly how it functions in ethylene signaling is not clear, but it may be related to Ca^{2+} signaling. Inside the nucleus, responses to ethylene are amplified by a transcription factor cascade that includes ETHYLENE INSENSITIVE3 (EIN3), ETHYLENE RESPONSE FACTOR1 (ERF1), and their homologs (Solano et al., 1998). Counterparts of these ethylene perception and signal transduction components have now been identified in many plants and their fruits, but the complexity and subtlety of signaling mechanisms are still being established. When ethylene binds the receptors the inhibitory signal from CTR1 is switched off. EIN2 then activates the ethylene response through downstream transcription factors such as EIN3 and other EIN3-like proteins (EILs), which subsequently regulate other ethylene-responsive genes in the transcription cascade. There are four tomato EILs and it has been suggested that are functionally redundant (Tieman et al., 2001), but Chen et al. (2004a) showed that overexpressing LeEIL1 in the ethylene-insensitive ripening mutant Nr only restored expression of some ethylene-regulated genes, raising the possibility of distinct roles for specific EILs. Further work is needed to test this suggestion.

The EILs and EIN3 proteins are post-transcriptionally regulated in response to the ethylene signal (reviewed by Yoo et al., 2009) and are controlled by two F-box proteins (EBF1/2), which are Skp-Cullin-F-box (SCF) E3 ligases targeting EIN3 and EILs for 26S proteasome degradation (Guo and Ecker, 2003) (Fig. 3.13). Yang et al. (2010) showed that if accumulation of EBF1 and EBF2 mRNAs was inhibited by RNAi (gene silencing) this caused constitutive ethylene phenotypes, including accelerated ripening. In addition, EIN3 protein stability can be further regulated by MAPK phosphorylation, and Yoo et al. (2008) have suggested that the simultaneous activation of the MAPKK9 cascade and the inhibition of the CTR1 pathway control EIN3 levels (see the review by Yoo et al., 2009).

Ethylene Response Factors

Members of a large superfamily of plant-specific transcription factors, called AP2/ERF, operate downstream of EIN3 and the EILs and some of these control responses to ethylene during ripening. The Arabidopsis AP2/ERF superfamily has been divided into five groups of factors: AP2 (Apetala2-like, involved in developmental processes, 17–18 members), RAV (6 members), CBF/DREB (56 members, linked to abiotic stresses, particularly dehydration), and ERF (ethylene response factor like). Nakano et al. (2006) originally recognized 65 Arabidopsis ERFs, although Goremykin and Moser (2009) have suggested 122 members for this group. In tomato, and several other fruit species, around 100 ERF genes have been identified so far. In addition to ripening, the ERFs and some RAV proteins are involved in a wide range of plant processes, including responding to wounding, biotic stress, salt stress, anaerobiosis, and have been associated with the brassinosteroids, ethylene, jasmonic acid, and salicylic acid signaling pathways. There is significant overlap and cross-talk between these various signaling pathways and some of the transcription factors operate at the nodes of these networks.

The ERFs show tissue-specific patterns of expression and can be regulated by auxin as well as ethylene. ERFs bind the GCC box, a conserved motif of the cis-acting element found in the promoters of ethylene-responsive genes (originally called ethylene-responsive element-binding proteins, Ohmetakagi and Shinshi, 1995; Solano et al., 1998), but some ERFs can also bind other types of cis elements. Tomato ERFs have been divided into eight subclasses based on their structures. When tested *in vitro* some promote transcription while others with an EAR (ERF-associated amphiphilic repression domain) motif repress the expression of specific genes inhibit, and others have no apparent effect on mRNA production. The extent, and possibly the specificity of binding, of the ERFs to gene promoters is influenced by the actual sequence context of the GCC box (Tournier et al., 2003), and this may explain why there is such a range of affinities of ERF family members toward different target genes.

Ethylene and Ripening Gene Expression

The cascade of ripening genes that changes color, flavor, texture, aroma, and taste has been best studied in climacteric fruit such as tomato (reviewed by Gray et al., 1994; Lelievre et al., 1998; Alexander and Grierson, 2002; Giovannoni, 2004) and melon, but it also occurs in nonclimacteric fruits, where often similar genes are involved in ripening although they may be regulated differently. Cloning of tomato-ripening genes began in several laboratories in the 1980s and within a few years cDNAs for several important enzymes involved in cell wall degradation and the synthesis of ethylene and pigments were identified, including PSY, involved in carotenoid production (Chapter 4), PG, which metabolizes cell wall components (Chapter 7), ACC synthase, and ACC oxidase, required for ethylene synthesis (see previous sections). Investigation of the genes responsible for flavor volatiles followed later. The availability of cloned genes, and the development of methods for genetic transformation of tomato, and later melon, led to a major surge in understanding the role and regulation of ripening genes. Over 20 years ago, experiments where tomato ripening was inhibited with Ag+, provided clear evidence that ethylene was required for the expression of several cloned ripening mRNAs (Davies et al., 1988, 1990).

Antisense inhibition of TOM13 inhibited ethylene synthesis (Fig. 3.6) and TOM13 was shown to encode ACC oxidase by expressing it in yeast (Hamilton et al., 1991). The finding that antisense suppression of TOM5 produced yellow tomatoes with little or no lycopene accumulation (Bird et al., 1991) and overexpression of TOM5 in the mutant *yellow flesh* tomato restored lycopene synthesis (Fray and Grierson, 1993) showed that it encodes PSY, which catalyzes the first committed step in carotenoid biosynthesis.

Accumulation of PG mRNA was prevented by Ag+, pointing to it being regulated by ethylene (Davies et al., 1988, 1990). In fact, many ripening genes are regulated, either partly or completely, by ethylene (see Table 3.1), and others are expressed independently of ethylene.

Color

The first visible sign of color development at the onset of ripening of many fruits is loss of chlorophyll, followed by pigment accumulation during ripening. Chlorophyll loss is often stimulated by ethylene. De-greening of citrus rind, for example, is prevented by ethylene antagonists, and expression of chlorophyllase, the enzyme catalyzing the first step in chlorophyll degradation, is stimulated by adding ethylene (Jacob-Wilk et al., 2002, even though citrus is nonclimacteric). Two main types of pigments contribute to fruit color, carotenoids, which accumulate in chloroplasts as they are transformed to chromoplasts (Chapter 4), and anthocyanins, which are deposited in the vacuoles (Chapter 5). A few fruits, however, appear blue by a mechanism involving thin-film interference of light by parallel sheets of cell membranes, although this is rare. Some fruits naturally produce both anthocyanins and carotenoids, but a particularly spectacular synthetic example is the deep purple tomatoes engineered to accumulate their natural carotenoids together with anthocyanins (Butelli et al., 2008). A detailed review of chromoplast development in fruit may be found in Egea et al. (2010), and carotenoid biosynthesis, regulation, and metabolic engineering in tomato are discussed by Fraser et al. (2009) and in Chapter 4.

The two main carotenoids that accumulate in ripening tomatoes are lycopene and β-carotene, which are synthesized from phytoene. The condensation of two molecules of geranylgeranyl pyrophosphate (GGPP) to produce phytoene is catalyzed by a ripening-specific isoform of the

Table 3.1 Examples of enzymes and regulatory proteins induced by ethylene in ripening fruit.

Examples are taken mainly from studies on tomato, melon, peach, and kiwi fruit. Further information about the species, enzyme isoforms, or gene family members induced by ethylene can be obtained from the text and from Alexander and Grierson (2002), Cara and Giovannoni (2008), and Pech et al. *(2008), 2012.*

Ethylene synthesis, perception, and signaling	
	ACC synthase (ACS)
	ACC oxidase (ACO)
	E8 (Dioxygenase, related to ACO)
	Ethylene receptors (e.g., NR (LeETR3))
	F-box proteins SlEB1 and SlEB2 that regulate EIN2
	CTR1 negative regulator of ethylene signal transduction
Regulators of transcription and translation	
	Ethylene response factors (ERFs) (Both positive and negative ERFs exist. Some are regulated by auxin, rather than ethylene.)
	ER24 transcriptional co-activator MBF1 (Links EREBPs to TATA-box-binding protein)
	ER49 translation elongation factor EF-Ts
Color	
	Chlorophyllase
	Phytoene synthase 1
Flavor and Aroma	
	Alcohol acyl-transferase 1,2,3,4
	Lipoxygenase (TOMLOX C Tomato, LOX1 and five kiwi fruits)
Texture and Cell walls	
	Cellulase
	Expansin
	β-galactosidase
	Endo-(1,4)-β-mannanase
	Pectin methylesterase
	Polygalacturonase
	Rab GTPase involved in trafficking of cell-wall-modifying enzymes
	Xyloglucan endotransglucosidase/hydrolases
	β-D-Xylosidase
Others	
	Malate synthase (banana)
	E4 Methionine sulfoxide reductase

enzyme phytoene synthase (PSY1), which accumulates during ripening, and is distinct from PSPY2, which operates in green tissue (Fray and Grierson, 1993). *Psy1* mRNA accumulation is inhibited by Ag+ and in the *Nr* (ethylene-insensitive) mutant, which is consistent with *Psy1* being upregulated in response to ethylene (Davies et al., 1989, 1990). Alba et al., 2000 demonstrated phytochrome-mediated carotenoid biosynthesis, originally reported by Khudairi and Arboleda, 1971, by demonstrating that increased lycopene accumulation could be reversed by exposure to far red light. The number of plastids also has a bearing on the extent of pigment synthesis. The *high-pigment* tomato mutant (*hp*) has twice the normal level of carotenoids (Bramley, 1997) and an increased plastid number (Yen et al., 1997). An additional mutant, *hp-2*, is homologous to Arabidopsis DE-ETIOLATED 1 (Mustilli et al., 1999), also involved in light signaling, which indicates an important role for light in fruit pigment accumulation (Fig. 3.15). It remains to be established whether there is a direct signaling interaction between phytochrome and ethylene.

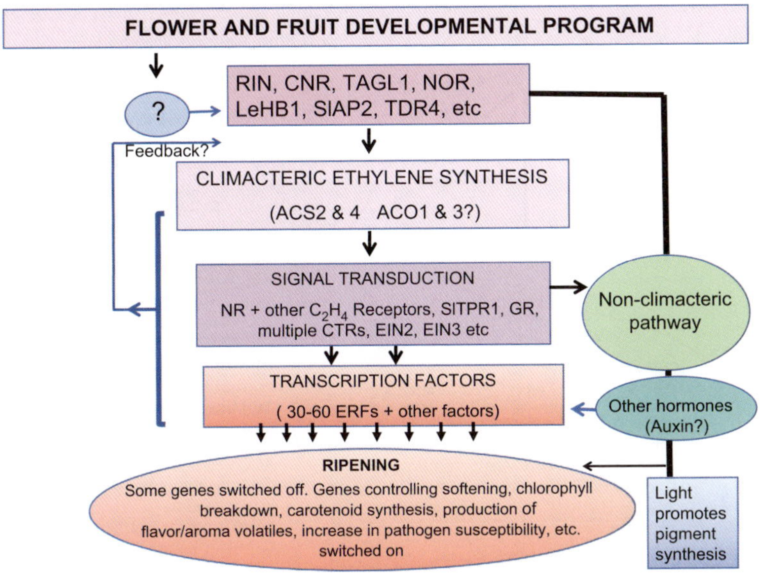

Figure 3.15 The main features of the genetic and hormonal control of ripening. The diagram is based on studies, mainly from tomato, discussed in the text. See also Matas et al. (2009). Compare with Figure 3.1. At the onset of ripening, some previously active genes are switched off, and a large number of new genes are switched on encoding proteins that catalyze ripening changes. The main uncertainties concern which proteins are involved in the positive- and negative-acting feedback mechanisms and how all the components interact in networks. It is known that ERF2 is involved in the feedback loop that stimulates autocatalytic ethylene synthesis in System 2. The role of individual AP2/ERF transcription factors in the control of expression of specific ripening proteins also needs clarification.

Texture

During ripening, the cell wall becomes increasingly hydrated and the pectin-rich middle lamella is partially hydrolyzed and other enzymatic modifications also take place. This makes it easier for the cells to separate from each another and plays a role in cell wall swelling, pectin solubilization, and texture changes (see Chapter 7). Tomato fruit PG is a major cell wall polyuronide-degrading enzyme that is synthesized *de novo* during the onset of ripening (Tucker and Grierson, 1982). The PG mRNA is very abundant in ripening fruits and a cDNA was cloned (Grierson et al., 1986; Dellapenna et al., 1986) and sequenced in 1986 (Grierson et al., 1986) and the PG gene was shown to be transcriptionally activated during ripening (DellaPenna et al., 1989). There has been disagreement as to whether PG is, or is not, induced by ethylene in ripening tomatoes. Early work had shown that treating tomatoes with inhibitors of ethylene action inhibited PG mRNA (Lincoln et al., 1987; Davies et al., 1988, 1990) and antisense inhibition of ACC oxidase certainly reduced PG mRNA accumulation (Picton et al., 1993), whereas inhibiting ethylene synthesis by antisense inhibition of tomato ACS2 surprisingly did not (Oeller et al., 1991). An extensive analysis of the *PG* gene promoter by Nicholass et al. (1995) defined elements required for reproducibly high levels of ripening-specific gene expression, using a reporter gene, and these have been used in several studies (Fernandez et al.,

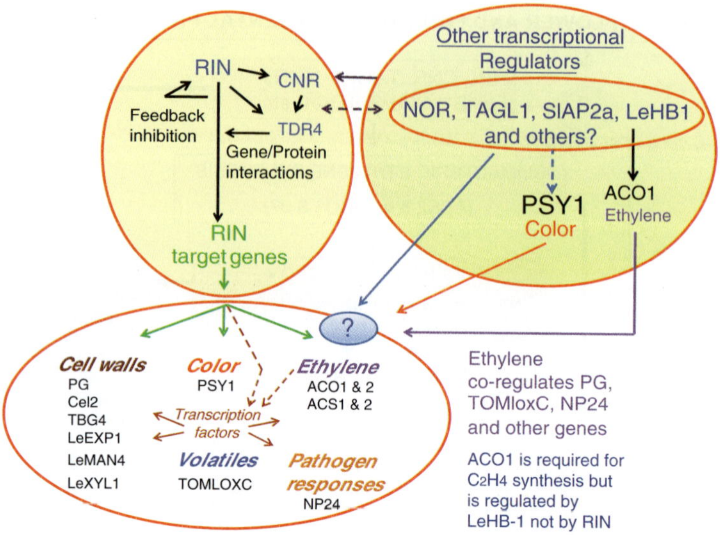

Figure 3.16 The roles of ethylene, RIN, and other ripening regulators in controlling the expression of tomato-ripening genes. This scheme is based on findings discussed in the text (See also Ito et al., 2008; Fujisawa et al., 2011a, 2011b; Karlova et al., 2011; Osorio et al., 2011).

2009). The observation that the *PG* promoter sequence contains ethylene-dependent ripening-specific control elements (Nicholass et al., 1995) indicated that *PG* was under ethylene control, but the ACS2 antisense results were inconsistent with this. Analysis of low-ethylene transgenic tomato plants showed that induction of *PG* mRNA occurs at very low ethylene levels (Sitrit and Bennett, 1998). This, taken together with the finding that the *PG* promoter contains a putative binding site for the HD-Zip protein LeHB-1, indicates that *PG* expression is probably under both ethylene control and developmental control (Figs. 3.15 and 3.16). This may also relate to the fact that different ethylene responses appear to have different threshold levels of ethylene. The situation is more complicated in melon (see Pech et al., 2012), where three PG genes (*CmPG1-3*) are expressed. *CmPG1* expression is totally dependent on ethylene, whereas regulation of *CmPG2* expression is ethylene independent and expression of *CmPG3* is regulated by both ethylene-dependent and ethylene-independent factors.

Despite the fact that the first cell-wall-modifying cDNA (PG) was cloned 25 years ago, and a wide range of other cell-wall-modifying enzymes has been studied (Table 3.1), we still have an imperfect understanding of how softening occurs and how it is regulated (see Chapter 7). One fact is clear; however, without ethylene, normal softening in climacteric fruits does not occur.

Other cell wall changes during ripening involve hydrolysis of polymeric galactose to free galactose. This is catalyzed by a family of at least seven β-galactosidases, one of which may be regulated by ethylene. Three *β-galactosidase* (*Gal1-3*) genes are also expressed in a melon fruit. 1-MCP completely inhibited *CmGal1* mRNA accumulation indicating it is regulated by ethylene, while *CmGal2* is ethylene independent (see Pech et al., 2008, 2012). Expansins, cell-wall-localized enzymes that are thought to cause wall loosening by reversibly disrupting the hydrogen bonds between cellulose microfibrils and matrix polysaccharides (Cosgrove, 2000), are expressed during tomato fruit development. Expansin EXP1 is fruit specific and is regulated by ethylene with mRNA transcripts accumulating either just before or at breaker stage (color change) of ripening (Rose et al., 1997). Endotransglycosylases/hydrolases (XTHs) and expansins have been suggested to play a synergistic role in the restructuration of the cellulose–xyloglucan network. Two melon *XTH* genes (*CmXTH1* and *CmXTH3*) and an *expansin* gene (*CmExp1*) showed ripening-related patterns of expression but were only partially ethylene dependent (Pech et al., 2008, 2012).

Flavor and Aroma Volatiles

The taste of fruits is strongly influenced by organic acids and sugars that accumulate during fruit development and ripening, but aroma volatiles such as alcohols, aldehydes, and esters produced during ripening and/or during maceration often contribute to the unique sensory notes specific to different fruit types. Thus different fruits produce different types of volatiles. Over 400 volatile compounds have been detected in tomatoes (See Chapter 6). Among the most important are hexanal, hexenal, hexenol, 3-methylbutanal, 3-methylbutanol, ethylnitrobutane, and isobutylthiazole. 3-methylbutanal and 3-methylbutanol are formed by the deamination and decarboxylation of amino acids whereas hexanal, hexenal, and hexenol are formed by lipid oxidation of unsaturated fatty acids during the maceration of fruits. The action of lipoxygenases (LOX) catalyzes the hydroperoxidation of polyunsaturated fatty acids containing a *cis,cis*-1,4 pentadiene structure, forming unsaturated fatty acid hydroperoxides (HPOs), which give rise to the C6 aldehydes through the action of hydroperoxide lyases (HPO-lyase). In tomato fruit, linoleic and linolenic acids are the main LOX substrates. Hexanal and hexenal are converted to hexenol and hexanol by the action of alcohol dehydrogenase (ADH).

LOX has a dual role in the production of volatiles for aroma and also defense signaling (Schwab et al., 2008) and tomato LOX is encoded by a family of at least five genes, *TOMLOXA,-B,-C,-D* and *TOMLOXE*. Levels of *TOMLOXA* mRNA decrease as ripening progresses and this is delayed in the mutants, *Nr* and *rin* as well as in transgenic low ethylene (ACO1-reduced by antisense) tomato fruit, indicating that this decrease is regulated by both ethylene and developmental factors, whereas levels of *TOMLOXB* increase during ripening (Griffiths et al., 1999a). *TOMLOXC* and the mainly leaf-expressed *TOMLOXD* differ from the other LOX enzymes in that they are chloroplast-targeted (Heitz et al., 1997). TOMLOXD probably functions in the octadecanoid defense signaling pathway which is activated in response to herbivore and pathogen attack (Heitz et al., 1997). *TOMLOXC* transcripts appear to be regulated by a developmental pathway that initiates expression and an ethylene component that enhances mRNA accumulation (Griffiths et al., 1999a). Silencing of TomloxA and TomloxB, by antisense gene knockout, failed to reduce flavor volatiles in ripening fruits (Griffiths et al., 1999b). When production of TOMLOXC was specifically reduced in transgenic tomatoes the levels of known flavor volatiles, including hexanal, hexenal, and hexenol, were reduced to as little as 1.5% of wild-type control levels following maceration of ripening fruits (Chen et al., 2004b).

Confocal imaging showed that a TOMLOXC-GFP fusion was localized in the chloroplasts, which are converted to chromoplasts during ripening. It is possible that TOMLOXC utilizes the polyunsaturated fatty acids from the redundant chloroplast thylakoids as substrates for the production of aroma volatiles hexanal and hexanol. Two ADHs (alcohol dehydrogenases) have been identified in tomato, ADH1 which is found only in pollen, seeds, and young seedlings and ADH2 which accumulates during the later stages of ripening. Genetic manipulation of ADH2 levels in ripening tomato fruit has been shown to affect the balance of some flavor aldehydes and alcohols and fruits with increased ADH2 levels had a more intense ripe fruit flavor (Speirs et al., 1998). Tomato ADH2 is induced by low oxygen stress and the increase in ADH2 activity during ripening may be due to decreasing oxygen concentration within ripening fruits (Speirs et al., 2002).

Ethylene has been shown to be extremely important in the production of aroma volatiles in cantaloupe charentais melon fruits (Pech et al., 2008, 2011), which generate a range of volatile aroma compounds, particularly esters synthesized from alcohols and acyl-CoAs, through the action of alcohol acyl-transferases (AATs). There are two different ADH enzymes in melon with a strong preference for aliphatic aldehydes. Both genes are specifically expressed in fruits, particularly during ripening, and at least three ATTs are known *(CmATT1*, *CmAAT3,* and *CmAAT4*), with different substrate preferences. All three can produce thiol esters but CmATT1 has a strong preference for producing *E*-2-hexenyl acetate and hexyl hexanoate. The role of ethylene became clear when the analysis of *Aco1*-antisense (low ethylene, Fig. 3.7) melons showed strong inhibition of aroma volatile production (Ayub et al., 1996; Flores et al., 2002). The levels of all three ATTs and the two ADHs were reduced in *ACO1*-antisense melons (Fig. 3.7), and in fruits treated with 1-MCP, indicating their expression is strongly influenced by ethylene (Pech et al., 2012). An AAT gene that plays a crucial role in flavor biogenesis has also been cloned from strawberry (Aharoni et al., 2000), emphasizing that similar types of ripening genes can be expressed in climacteric and nonclimacteric fruits.

Susceptibility to Pathogens

One consequence of ripening is that fruits become susceptible to invasion by pathogens. Several factors contribute to this. First, toxic compounds, such as alpha-tomatine in tomato, may be degraded during the ripening process. Second, spores of some pathogens, such as anthracnose (*Colletotrichum gloeosporioides*) remain quiescent, attached to the surface of unripe fruits. As ripening commences the spores sense the ethylene produced, they germinate, and invade the fruit tissue. The black spots that sometimes develop on the skin of ripening bananas provide an illustration of this. Third, the partial cell wall solubilization and degradation that occurs as fruit soften makes infection easier for some pathogens. The situation is more complicated than this, however, since ethylene is involved in resistance to some pathogens but actually stimulates symptom development in response to other diseases. This may be related to the role of ethylene in cross-talk with the SA (salicylic acid) and JA (jasmonic acid) signaling pathways (see Van der Ent and Pieterse, 2012). Fruit of the (ethylene-insensitive) *Nr* mutant showed less severe symptoms than wild type when infected by the fungus *Fusarium oxysporum* f.sp. *lycopersici* or by the bacteria *Xanthomonas campestris* pv. *vesicatoria* and *Pseudomonas syringae* pv. *Tomato* (Lund et al., 1998; Ciardi et al., 2000). Infection of ACO-antisense tomatoes by *Colletotrichum gloeosporioides* progressed more slowly than in the controls. Furthermore, infection of *rin* fruit was greatly reduced compared to controls, but could be stimulated by supplying ethylene (Cooper et al., 1998). Specific regulatory genes can also make a difference, however, since Cantu et al. (2009) showed that susceptibility of tomato fruits to *Botrytis cinerea* is dependent on the NOR regulator, but not on RIN or ethylene, and that the fungus could actually switch on some ripening genes in green fruits. The signaling mechanisms involved in these responses are not clear. Lin et al., found that LeCTR2 (a CTR-like protein in tomato), which has similarity to Arabidopsis EDR1, involved in disease resistance, could bind an ethylene receptor in tomato (Lin et al., 2008c) and specific members of the AP2/ERF (Apetala2/Ethylene Response Factor) superfamily of transcription factors have been implicated in ripening, JA signaling and ethylene signaling (Van der Ent and Pieterse, 2012). Since rotting of old fruit is widespread, perhaps it serves as "escape plan B" for the seeds, so that if the fruit is not eaten the seeds germinate and take root where the fruit fall.

Conclusions

Current ideas about how the network of ripening regulatory gene functions are discussed in more detail in the concluding Chapter (Chapter 8). The interplay between hormones and developmental regulator genes controls the synthesis of new proteins that catalyze the alterations in color, flavor, texture, and aroma that are essential features of ripening. Ethylene is an important regulator for climacteric fruits and also plays a role in nonclimacteric fruit ripening, but other hormones, particularly auxin, also appear to play a role. Some gene changes are ethylene dependent, while others are developmentally regulated and ethylene independent, and there are several examples where developmental factors and ethylene combine to regulate specific genes. Some of the transcription factors involved in this process control hormone production or action, whereas others are modulated by the same hormones. These interactions occur in a web of signaling networks, operating at the DNA–protein, RNA–protein, or protein–protein level. Acting as either negative or positive regulators, and involving multiple interactions and feedback loops, they switch developmental programs off and on and modulate expression of downstream ripening genes. Some regulators appear to act at several levels and current research is aimed at defining all the members of these networks and establishing how, where, and when they interact. Transcriptomics and bioinformatics approaches will be key to revealing these complex regulatory networks and establishing their interactions (Chapter 8).

There is good evidence that ethylene influences the timing of leaf senescence and fruit ripening and it would be interesting to know whether there is an interaction between ethylene and biological clock genes in measuring developmental time. In addition, ethylene is intimately involved in plant responses to biotic and abiotic stresses and increased knowledge should help devise strategies for improving crop protection and breeding.

Acknowledgments

Research in the author's laboratory was funded by the Biotechnology and Biological Sciences Research Council, its forerunners the Science and Engineering Research Council and the Agricultural and Food Research Council, the European Commission, and the University of Nottingham. I would like to express my appreciation to all colleagues who contributed to research discussed in this article.

References

Achard, P., Baghour, M., Chapple, A., Hedden, P., Van der Straeten, D., Genschik, P., Moretz, T., and Harberd, N.P. (2007) The plant stress hormone ethylene controls floral transition via DELLA dependent regulation of floral meristem-identity genes. *Proceedings of the National Academy of Sciences of the United States of America*, 104, 6484–6489.

Achard, P., Vriezen, W.H., Van Der Straeten, D., and Harberd, N.P. (2003) Ethylene regulates arabidopsis development via the modulation of DELLA protein growth repressor function. *The Plant Cell*, 15, 2816–2825.

Adams-Phillips, L., Barry, C., Kannanz, P., Leclercq, J., Bouzayen, M., and Giovannoni, J. (2004) Evidence that CTR1-mediated ethylene signal transduction in tomato is encoded by a multigene family whose members display distinct regulatory features. *Plant Molecular Biology*, 54, 387–404.

Aharoni, A., Keizer, L.C.P., Bouwmeester, H.J., Sun, Z., Alvarez-Huerta, M., Verhoeven, H.A., Blaas, J., van Houwelingen, A.M.M.L., De Vos, R.C.H., van der Voet, H., Jansen, R.C., Guis, M., Mol, J., Dawis, R.W., Schena, M., van Tunen, A.J., and O'Connell, A.P. (2000) Identification of the SAAT gene involved in strawberry flavor biogenesis by use of DNA microarrays. *The Plant Cell*, 12, 647–661.

Alba, R., Cordonnier-Pratt, M.M., and Pratt, L.H. (2000) Fruit-localized phytochromes regulate lycopene accumulation independently of ethylene production in tomato. *Plant Physiology*, 123, 363–370.

Alexander, L., and Grierson, D. (2002) Ethylene biosynthesis and action in tomato: a model for climacteric fruit ripening. *Journal of Experimental Botany*, 53, 2039–2055.

Ayub, R., Guis, M., Ben Amor, M., Gillot, L., Roustan, J.P., Latché, A., Bouzayen, M. and Pech, J.C. (1996) Expression of ACC oxidase antisense gene inhibits ripening of cantaloupe melon fruits. *Nature Biotechnology*, 14, 862–866.

Barry, C.S., Blume, B., Bouzayen, M., Cooper, W., Hamilton, A.J. and Grierson, D. (1996) Differential expression of the 1-aminocyclopropane-1-carboxylate oxidase gene family of tomato. *Plant Journal*, 9, 525–535.

Barry, C.S., and Giovannoni, J.J. (2006) Ripening in the tomato Greenripe mutant is inhibited by ectopic expression of a protein that disrupts ethylene signaling. *Proceedings of the National Academy of Sciences of the United States of America*, 103, 7923–7928.

Barry, C.S., Llop-Tous, I., and Grierson, D. (2000) The regulation of 1-aminocyclopropane-1-carboxylic acid synthase gene expression during the transition from system-1 to system-2 ethylene synthesis in tomato. *Plant Physiology*, 123, 979–986.

Bartley, G.E., and Ishida, B.K. (2007) Ethylene-sensitive and insensitive regulation of transcription factor expression during in vitro tomato sepal ripening. *Journal of Experimental Botany*, 58, 2043–2051.

Binder, B.M., O'Malley, R.C., Wang, W., Zutz, T.C., and Bleecker, A.B. (2006) Ethylene stimulates nutations that are dependent on the ETR1 receptor. *Plant Physiology*, 142, 1690–700.

Bird, C.R., Ray, J.A., Fletcher, J.D., Boniwell, J.M., Bird, A.S, Teulieres, C., Blain, I., Bramley, P.M., and Schuch, W. (1991) Using antisense RNA to study gene function: inhibition of carotenoid biosynthesis in transgenic tomatoes. *Biotechnology*, 9, 635–639.

Blume, B., and Grierson, D. (1997) Expression of ACC oxidase promoter-GUS fusions in tomato and *Nicotiana plumbaginifolia* regulated by developmental and environmental stimuli. *The Plant Journal*, 12, 731–746.

Bramley, P. (1997) The regulation and genetic manipulation of carotenoid biosynthesis in tomato fruit. *Pure and Applied Chemistry*, 69, 2159–2162.

Burg, S.P., and Burg, E.A. (1967) Molecular requirements for the biological activity of ethylene. *Plant Physiology*, 42, 144–152.

Butelli, E., Titta, L., Giorgio, M., Mock, H-P., Matros, A., Peterek, S., Schijlen, E.G.W.M., Hall, R.D., Bovy, A.G., Luo, J., and Martin, C. (2008) Enrichment of tomato fruit with health-promoting anthocyanins by expression of select transcription factors. *Nature Biotechnology*, 26, 1301–1308.

Cantu, D., Blanco-Ulate, B., Yang, L., Labavitch, J.M., Bennett, A.B., and Powell, A.L.T. (2009) Ripening-regulated susceptibility of tomato fruit to *Botrytis cinerea* requires *NOR* But Not *RIN* or ethylene. *Plant Physiology*, 150, 1434–1449.

Cara, B., and Giovannoni, J.J. (2008) Molecular Biology of ethylene during tomato fruit development and maturation. *Plant Science*, 175, 106–113.

Chen, G., Alexander, L., and Grierson, D. (2004a) Constitutive expression of EIL-like transcription factor partially restores ripening in the ethylene insensitive Nr tomato mutant. *Journal of Experimental Botany*, 55, 1491–1497.

Chen, G., Hackett, R., Walker, D., Taylor, A., Lin, Z., and Grierson, D. (2004b) Identification of a specific isoform of tomato lipoxygenase (TomLoxC) involved in the generation of fatty acid-derived flavor compounds. *Plant Physiology*, 136, 2641–2651.

Ciardi, J.A., Tieman, D.M., Lund, S.T., Jones, J.B., Stall, R.E., and Klee, H.J. (2000) Response to *Xanthomonas campestris* pv. *vesicatoria* in tomato involves regulation of ethylene receptor gene expression. *Plant Physiology*, 123, 81–92.

Clark, K.L., Larsen, P.B., Wang, X., Chang, C. (1998) Association of the Arabidopsis CTR1 Raf-like kinase with the ETR1 and ERS ethylene receptors. *Proceedings of the National Academy of Sciences of the United States of America*, 95, 5401–5406.

Cooper, W., Bouzayen, M., Hamilton, A., Barry, C., Rossall, S., and Grierson, D. (1998) Use of transgenic plants to study the role of ethylene and polygalacturonase during infection of tomato fruit by *Colletotrichum gloeosporioides*. *Plant Pathology*, 47, 308–316.

Cosgrove, D.J. (2000) Loosening of plant cell walls by expansins. *Nature*, 407, 321–326.

Davies, K.M., and Grierson, D. (1989) Identification of cDNA clones for tomato (Lycopersicon esculentum) mRNAs that accumulate during fruit ripening and leaf senescence in response to ethylene. *Planta*, 179, 73–80.

Davies, K.M., Hobson, G.E., and Grierson, D. (1988) Silver ions inhibit the ethylene-stimulated production of ripening-related mRNAs in tomato. *Plant, Cell and Environment*, 11, 729–738.

Davies, K.M., Hobson, G.E., and Grierson, D. (1990) Differential effect of silver ions on the accumulation of ripening related mRNAs in tomato. *Journal of Plant Physiology*, 135, 708–713.

DellaPenna, D., Alexander, D., and Bennett, A.B. (1986) Molecular cloning of tomato fruit polygalacturonase: Analysis of polygalacturonase mRNA levels during ripening. *Proceedings of the National Academy of Sciences of the United States of America*, 83, 6420–6424.

DellaPenna, D., Lincoln, J.E., Fischer, R.L., and Bennett, A.B. (1989)Transcriptional analysis of polygalacturonase and other ripening associated genes in Rutgers, rin, nor, and Nr tomato fruit. *Plant Physiology*, 90, 1372–1377.

Eak, I.L. (1972) Treatment of fruit with propylene gives information about the biogenesis of ethylene. *Nature*, 237, 235–236.

Egea, I., Barsan, C., Bian, W., Purgatto, E., Latché, A., Chervin, C., Bouzayen, M., and Pech, J-C. (2010) Chromoplast differentiation: current status and perspectives. *Plant Cell Physiology*, 51, 1601–1611.

Fei, Z., Tang, X., Alba, R., White, R., Ronning, C., Martin, G., Tanksley, S., Giovannoni, J. (2004) Comprehensive EST analysis of tomato and comparative genomics of fruit ripening. *The Plant Journal*, 40, 47–59.

Fernandez, A.I., Viron, N., Alhagdow, M., Karimi, M., Jones, M., Amsellem, Z., Sicard, A, Czerednik, A., Angenent, G., Grierson, D., May, S., Seymour, G., Eshed, Y., Lemaire-Chamley, M., Rothan, C., and Pierre, Hilson (2009) Flexible tools for gene expression and silencing in tomato. *Plant Physiology*, 151, 1729–1740.

Fraser, P.D., and Enfissi, E.M.A., and Bramley, P.M. (2009) Genetic engineering of carotenoid formation in tomato fruit and the potential application of systems and synthetic biology approaches. *Archives of Biochemistry and Biophysics*, 483, 196–204

Fray, R.G., and Grierson, D. (1993) Identification and genetic analysis of normal and mutant phytoene synthase genes of tomato by sequencing, complementation and co-suppression. *Plant Molecular Biology*, 22, 589–602.

Flores, F., Yahyyaoui, F.E., de Billerbeck, G., Romojaro, F., latche, A., Bouzayen, M., Pech, J.-C., and Ambid, C. (2002) Role of ethylene in the biosynthetic pathway of aliphatic ester aroma volatiles in Charentais Cantaloupe melons. *Journal of Experimental Botany*, 53, 201–206.

Fujisawa, M., Shima, Y., Higuchi, N., Nakano, T., Kasumi, T., and Ito, Y. (2011a) Identification of direct targets of the tomato ripening regulator RIN by transcriptome and chromatin immunoprecipitation. pp. 13–108, SOL & ICuGI Conference, Book of Abstracts, 2011 (28th Nov–2nd Dec. 2011).

Fujisawa, M., Nakano, T., and Ito, Y. (2011b) Identification of potential target genes for the tomato fruit-ripening regulator RIN by chromatin immunoprecipitation. *BMC Plant Biology*, 11, 26.

Gao, Z., Chen, Y.F., Randlett, M.D., Zhao, X.C., Findell, J.L., Kieber, J.J., and Schaller, G.E. (2003) Localization of the Raf-like kinase CTR1 to the endoplasmic reticulum of Arabidopsis through participation in ethylene receptor signaling complexes. *Journal of Biological Chemistry*, 278, 34725–34732.

Gao, Z., Wen, C., Binder, B.M., Chen, Y., Chang, J., Chiang, Y., Kerris, R.J., Chang, C., and Schaller, G.E. (2008) Heteromeric interactions among ethylene receptors mediate signalling in Arabidopsis. *Journal of Biological Chemistry*, 283, 23801–23810.

Giovannoni, J.J. (2004) Genetic regulation of fruit development and ripening. *The Plant Cell*, 16, S170–S180.

Giovannoni, J.J. (2007) Fruit ripening mutants yield insights into ripening control. *Current Opinion in Plant Biology*, 10, 283–289.

Goremykin, V., and Moser, C. (2009) Classification of the Arabidopsis *ERF* gene family based on Bayesian Inference. *Molecular Biology*, 43, 729–734

Gray, J.E., Picton, S., Giovannoni, J.J., and Grierson, D. (1994) The use of transgenic and naturally occurring mutants to understand and manipulate tomato ripening. *Plant, Cell and Environment*, 17, 557–571.

Grierson, D., Maunders, M.J., Slater, A., Ray, J., Bird, C.R., Schuch, W., Holdsworth, M.J., Tucker, G.A., and Knapp, J.E. (1986) Gene expression during tomato ripening. *Philosophical Transactions of the Royal Society of London*, B 314, 399–41020.

Griffiths, A., Barry, C., Alpuche-Solis, A.G., and Grierson, D. (1999a) Ethylene and developmental signals regulate expression of lipoxygenase genes during tomato fruit ripening. *Journal of Experimental Botany*, 50, 793–798.

Griffiths, A., Prestage, S., Linforth, R., Zhang, J., Taylor, A., and Grierson, D. (1999b) Fruit-specific lipoxygenase suppression in antisense-transgenic tomatoes. *Postharvest Biology and Technology*, 17, 163–173.

Guo, H., and Ecker, J. (2003) Plant responses to ethylene gas are mediated by SCFEBF1/EBF2-dependent proteolysis of EIN3 transcription factor. *Cell*, 115, 667–677.

Hackett, R.M., Ho, C.-W., Lin, Z., Foote, H.C.C., Fray, R.G., and Grierson, D. (2000) Antisense inhibition of the Nr gene restores normal ripening to the tomato never-ripe mutant, consistent with the ethylene receptor-inhibition model. *Plant Physiology*, 124, 1079–1085.

Hamilton, A.J., Bouzayen, N., and Grierson, D. (1991) Identification of a tomato gene for the ethylene forming enzyme by expression in yeast. *Proceedings of the National Academy of Sciences of the United States of America*, 88, 7434–7437.

Hamilton, A, Lycett, G.W., and Grierson, D. (1990) Antisense gene that inhibits synthesis of the hormone ethylene in transgenic plants. *Nature*, 346, 284–287.

Heitz, T., Bergey, D.R., and Ryan, C.A. (1997) A gene encoding a chloroplast-targeted lipoxygenase in tomato leaves is transiently induced by wounding, systemin, and methyl jasmonate. *Plant Physiology*, 114, 1085–1093.

Hua, J., and Meyerowitz, E.M. (1998) Ethylene responses are negatively regulated by a receptor gene family in Arabidopsis thaliana. *Cell*, 94, 261–271.

Ishida, B.K., Jenkins, S.M., and Say, B. (1998) Induction of AGAMOUS gene expression plays a key role in ripening of tomato sepals in vitro. *Plant Molecular Biology*, 36, 733–739.

Itkin, M., Seybold, H., Breitel, D., Rogachev, I., Meir, S., and Aharoni, A. (2009) TOMATO AGAMOUS-LIKE 1 is a component of the fruit ripening regulatory network. *The Plant Journal*, 60, 1081–1095.

Ito, Y., Kitagawa, M., Ihashi, H., Yabe, K., Kimbara, J., Yasuda, J., Ito, H., Inakuma, T., Hiroi, S., and Kasumi, T. (2008) DNA-binding specificity, transcriptional activation potential, and the rin mutation effect for the tomato fruit-ripening regulator RIN. *The Plant Journal*, 55, 212–223.

Jacob-Wilk, D., Holland, D., Goldschmidt, E.E., Riov, J., and Eyal, J.Y. (2002) Chlorophyll breakdown by chlorophyllase: isolation and functional expression of the Chlase1 gene from ethylene-treated Citrus fruit and its regulation during development. *The Plant Journal*, 20, 653–661.

Karlova, R., Rosin, F.M., Busscher-Lange, J., Parapunova, V., Do, P.T., Fernie, A.R., Fraser, P.D., Baxter, C., Angenent, G.C., and de Maagd, R.A. (2011) Transcriptome and metabolite profiling show that APETALA2a is a major regulator of tomato fruit ripening. *The Plant Cell March*, 23, 923–941.

Kevany, B.M., Tieman, D.M., Taylor, M.G., Dal Cin, V., and Klee, H.J. (2007) Ethylene receptor degradation controls the timing of ripening in tomato fruit. *The Plant Journal*, 51, 458–467.

Kevany, B.M., Taylor, M.G., and Klee, H.J. (2008) Fruit-specific suppression of the ethylene receptor LeETR4 results in early-ripening tomato fruit. *Plant Biotechnology Journal*, 6, 295–300.

Khudairi, A.K., and Arboleda, O.P. (1971) Phytochrome-mediated carotenoid biosynthesis and its influence by plant hormones. *Physiologia Plantarum*, 24, 18–22.

Lelievre, J.M., Latche, A., Jones, B., Bouzayen, M., and Pech, J.C. (1998) Ethylene and fruit ripening. *Physiologia Plantarum*, 101, 727–739.

Lin, Z., Alexander, L., Hackett, R., and Grierson, D. (2008a) LeCTR2, a CTR1-like protein kinase from tomato, plays a role in ethylene signalling, development and defence. *The Plant Journal*, 54, 1083–1093.

Lin, Z., Arciga-Reyes, L., Zhong, S., Alexander, L., Hackett, R., Wilson, I., and Grierson, D. (2008b) SlTPR1, a tomato tetratricopeptide repeat protein, interacts with the ethylene receptors NR and LeETR1, modulating ethylene and auxin responses and development. *Journal of Experimental Botany*, 59, 4271–4287.

Lin, Z., and Grierson, D. (2010) New perspectives in ethylene signaling. *Plant Signaling & Behavior*, 5, 1–5.

Lin, Z., Zhong, S., and Grierson, D. (2009) Recent advances in ethylene research. *Journal of Experimental Botany*, 60, 3311–3336.

Lin, Z., Ho, C.W., and Grierson, D. (2009b) AtTRP1 encodes a novel TPR protein that interacts with the ethylene receptor ERS1 and modulates development in Arabidopsis. *Journal of Experimental Botany*, 60, 3697–3714.

Lin, Z., Hong, Y., Yin, M., Li, C., Zhang, K., and Grierson, D. (2008c) A tomato HB-zip homeobox protein, LeHB-1, plays an important role in floral organogenesis and ripening. *The Plant Journal*, 55, 301–310.

Lincoln, J.E., Cordes, S., Read, E., and Fischer, R.L. (1987) Regulation of gene expression by ethylene during Lycopersicon esculentum (tomato) fruit development. *Proceedings of the National Academy of Sciences of the United States of America*, 84, 2793–2797.

Lund, S.T., Stall, R.E., and Klee, H.J. (1998) Ethylene regulates the susceptible response to pathogen infection in tomato. *Plant Cell*, 10, 371–382.

Ma, B., Cui, M.L., Sun, H.J., Takada, K., Mori, H., Kamada, H., and Ezura, H. (2006) Subcellular localization and membrane topology of the melon ethylene receptor CmERS1. *Plant Physiology*, 141, 587–597.

Matas, A.J., Gapper, N.E., Chung, M-Y., Giovannoni, J.J., and Rose, J.K.C. (2009) Biology and genetic engineering of fruit maturation for enhanced quality and shelf-life. *Current opinion in Biotechnology*, 20, 197–203.

McMurchie, E.J., Mc Glasson, W.B., and Eak, I.L. (1972) Treatment of fruit with propylene gives information about the biogenesis of ethylene. *Nature*, 237, 235–236.

Moussatche, P., and Klee, H. (2004) Autophosphorylation activity of the Arabidopsis ethylene receptor multigene family. *Journal of Biological Chemistry*, 279, 48734–48741.

Mustilli, A.C., Fenzi, F., Ciliento, R., Alfano, F., and Bowler, C. (1999) Phenotype of the tomato high pigment-2 mutant is caused by a mutation in the tomato homolog of DEETIOLATED1. *Plant Cell*, 11, 145–158.

Nakano, T., Suzuki, K., Fujimura, T., and Shinshi, H. (2006) Genome-wide analysis of the ERF gene family in Arabidopsis and rice. *Plant Physiology*, 140, 411–432.

Nicholass, F.J., Smith, C.J.S., Schuch, W., Bird, C.R., and Grierson, D. (1995) High levels of reporter gene expression directed by tomato fruit polygalacturonase gene flanking regions. *Plant Molecular Biology*, 28, 423–435.

Oeller, P.W., Lu, M.W., Taylor, L.P., Pike, D.A., and Theologis, A. (1991) Reversible inhibition of tomato fruit senescence by antisence RNA. *Science*, 254, 437–439.

Ohmetakagi, M., and Shinshi, H. (1995) Ethylene-inducible DNA-binding proteins that interact with an ethylene-responsive element. *Plant Cell*, 7, 173–182.

Osorio, S., Alba, R., Damasceno, C.M.B., Lopez-Casado, G., Lohse, M., Zanor, M.I., Tohge, T., Usadel, B., Rose, J.K.C., Fei, Z., Giovannoni, J.G., and Fernie, A.R. (2011) Systems biology of tomato fruit development: combined transcript, protein, and metabolite analysis of tomato transcription factor (nor, rin) and ethylene receptor (Nr) mutants reveals novel regulatory interactions. *Plant Physiology*, 157, 405–425.

Pech, J.C., Bouzayen, M., and Latché, A. (2008) Climacteric fruit ripening: ethylene-dependent and independent regulation of ripening pathways in melon fruit. *Plant Science*, 175, 114–120

Pech, J.C., Purgatto, E., Bouzayen, M., and Latché, A. (2012) Ethylene and fruit ripening. In: Annual Plant Reviews, volume 44, *The Plant Hormone Ethylene* (ed. M.T. McManus) pp. 275–304, Wiley-Blackwell, Oxford, 2012.

Picton, S., Barton, S.L., Bouzayen, M., Hamilton, A.J., and Grierson, D. (1993) Altered fruit ripening and leaf senescence in tomatoes expressing an antisense ethylene-forming enzyme transgene. *The Plant Journal*, 3, 469–481.

Picton, S., and Grierson, D., (1988) Inhibition of expression of tomato-ripening genes at high temperature. *Plant, Cell and Environment*, 11, 265–272.

Qu, X., and Schaller, G.E. (2004) Requirement of the histidine kinase domain for signal transduction by the ethylene receptor ETR1. *Plant Physiology*, 136, 2961–2970.

Rajakulendran, T., Sahmi, S., Lefrancois, M., Sicheri, F., and Therrien, M. (2009) A dimerization-dependent mechanism drives RAF catalytic activation. *Nature*, 461, 542–546.

Rajakulendran, T., and Sicheri, F. (2010) Allosteric protein kinase regulation by pseudokinases: insights from STRAD. *Science Signaling*, 3, pe8.

Resnick, J.S., Wen, C.K., Shockey, J.A., and Chang, C. (2006) REVERSIONTO-ETHYLENE SENSITIVITY1, a conserved gene that regulates ethylene receptor function in Arabidopsis. *Proceedings of the National Academy of Sciences of the United States of America*, 103, 7917–7922.

Reid, M.S., and Staby, G.L. (2008) A brief history of 1-methylcyclopropene. *HortScience*, 43, 83–85.

Rose, J.K.C., Lee, H.H., and Bennet, A.B. (1997) Expression of a divergent expansin gene is fruit-specific and ripening-regulated. *Proceedings of the National Academy of Sciences of the United States of America*, 94, 5955–5960.

Schaller, G.E., and Kieber, J.J. (2002) Ethylene. In: *The Arabidopsis Book. American Society of Plant Biologists*, Rockville, MD. e0071. doi: 10.1199/tab.0071.

Schwab, W., Davidovich-Rikanati, R., and Lewinsohn, E. (2008) Biosynthesis of plant derived flavour compounds. *Plant Journal*, 54, 712–732.

Sitrit, Y., and Bennett, A.B. (1998) Regulation of tomato fruit polygalacturonase mRNA accumulation by ethylene: a re-examination. *Plant Physiology*, 116, 1145–1150.

Solano, R., Stepanova, A., Chao, Q., and Ecker, J.R. (1998) Nuclear events in ethylene signaling: a transcriptional cascade mediated by ETHYLENE-INSENSITIVE3 and ETHYLENE-RESPONSE-FACTOR1. *Genes and Development*, 12, 3703–3714

Speirs, J., Corel, R., and Cain, P. (2002) Relationship between ADH activity, ripeness and softness in six tomato cultivars. *Scientia Horticulturae*, 93, 137–142.

Speirs, J., Lee, E., Holt, K., Yong-Duk, K., Steele, N., Loveys, B., and Schuch, W. (1998) Genetic manipulation of alcohol dehydrogenase levels in ripening tomato fruit affects the balance of some flavor aldehydes and alcohols. *Plant Physiology*, 117, 1047–1058.

Testerink, C., Larsen, P.B., van der Does, D., van Himbergen, J.A., and Munnik, T. (2007) Phosphatidic acid binds to and inhibits the activity of Arabidopsis CTR1. *Journal of Experimental Botany*, 58, 3905–3914.

Thompson, A.J., Tor, M., Barry, C.S., Vrebalov, J., Orfila, C., Jarvis, M.C., Giovannoni, J.J., Grierson, D., and Seymour, G.B. (1999) Molecular and genetic characterization of a novel pleiotropic tomato-ripening mutant. *Plant Physiology*, 120, 383–389.

Tieman, D.M., Ciardi, J.A., Taylor, M.G., and Klee, H.J. (2001) Members of the tomato LeEIL (EIN3-like) gene family are functionally redundant and regulate ethylene responses throughout plant development. *The Plant Journal*, 26, 47–58.

Tieman, D.M., and Klee, H.J. (1999) Differential expression of two novel members of the tomato ethylene-receptor family. *Plant Physiology*, 120, 165–172.

Tieman, D.M., Taylor, M.G., Ciadi, J.A., and Klee, H.J. (2000) The tomato ethylene receptors NR and LeETR4 are negative regulators of ethylene response and exhibit functional compensation within a multigene family. *Proceedings of the National Academy of Sciences of the United States of America*, 97, 5663–5668.

Tournier, B., Sanchez-Ballesta, M.T., Jones, B. Pesquet, E., Regad, F., Latché, A., Pech, J.C., and Bouzayen, M. (2003) New members of the tomato ERF family show specific expression pattern and diverse DNA-binding capacity to the GCC box element. *Febs Letters*, 550, 149–154.

Trainotti, L., Pavanello, A., Casadoro, G. (2005) Different ethylene receptors show an increased expression during the ripening of strawberries: does such an increment imply a role for ethylene in the ripening of these non-climacteric fruits? *Journal of Experimental Botany*, 56, 2037–2046.

Tucker, G.A., and Grierson, D. (1982) Synthesis of polygalacturonase during tomato fruit ripening. *Planta*, 155, 64–67.

Van der Ent, S., and Pieterse, C.M.J. (2012) Ethylene: multitasker in plant-attacker interactions. In: Annual Plant Reviews, volume 44, *The Plant Hormone Ethylene* (ed. M.T. McManus) pp. 343–378, Wiley-Blackwell, Oxford, 2012.

Vrebalov, J., Ruezinsky, D., Padmanabhan, V., White, R., Medrano, D, Drake, R., Schuch, W., Giovannoni, J. (2002) A MADS-box gene necessary for fruit ripening at the tomato ripening-inhibitor (Rin) locus. *Science*, 296, 343–346.

Wang, K.L-C., Li, H., and Ecker, J.R. (2002) Ethylene biosynthesis and signalling networks. *The Plant Cell Supplement*, S131–S151.

Wang, W., Hall, A.E., O'Malley, R., and Bleecker, A.B. (2003) Canonical histidine kinase activity of the transmitter domain of the ETR1 ethylene receptor from Arabidopsis is not required for signal transmission. *Proceedings of the National Academy of Sciences of the United States of America*, 100, 352–357.

Whitelaw, C.A., Lyssenko, N.N., Chen, L., Zhou, D., Mattoo, A.K., and Tucker, M.L. (2002) Delayed abscission and shorter Internodes correlate with a reduction in the ethylene receptor LeETR1 transcript in transgenic tomato. *Plant Physiology*, 128, 978–87.

Wechter, W.P., Levi, A., Harris, R., Davis, A., Fei, Z., Katzir, N., Giovannoni, J.J., Saman-Minkov, A., Hernandez, A., Thimmapuran, J., Tadmor, Y., Portnoy, V., and Trebitsh, T. (2008) Gene expression in developing watermelon fruit. *BMC Genomics*, 9, 275. doi:10.1186/1471-2164-9-275.

Wilkinson, J.Q., Lanahan, M.B., Yen, H.C., Giovannoni, J.J., and Klee, H.J. (1995) An ethylene-inducible component of signal transduction encoded by Neverripe. *Science*, 270, 1807–1809.

Yang, Y., Wu, Y., Pirello, J., Regad, F., Bouzayen, M., Deng, W., Zengguo, L. (2010) Silencing Sl-EBF1 and Sl-EBF2 expression causes constitutive ethylene response phenotype, accelerated plant senescence, and fruit ripening in tomato. *Journal of Experimental Botany*, 61, 697–708.

Yen, H., Shelton, A., Howard, L., Vrebalov, J., and Giovannoni, J.J. (1997) The tomato *high pigment* (*hp*) locus maps to chromosme 2 and influences plastome copy number and fruit quality. *Theoretical and Applied Genetics*, 95, 1069–1079.

Yokotani, N., Nakano, R., Imanishi, S., Nagata, M., Inaba, A., and Kubo, Y. (2009) Ripening-associated ethylene biosynthesis in tomato fruit is utocatalytically and developmentally regulated. *Journal of Experimental Botany*, 60, 3433–3442.

Yoo, S-D., and Cho, Y., Sheen, J. (2009) Emerging connections in the ethylene signaling network. *Trends in Plant Science*, 14, 270–279.

Yoo, S., Cho, Y., Tena, G., Xiong, Y., and Sheen, J. (2008) Dual control of nuclear EIN3 by bifurcate MAPK cascades in C_2H_4 signaling. *Nature*, 451, 789–795.

Zhong, S., Lin, Z., and Grierson, D. (2008) Tomato ethylene receptor interaction: visualization of NEVER-RIPE interaction with multiple CTRs at the endoplasmic reticulum. *Journal of Experimental Botany*, 59, 965–972.

4 Carotenoid Biosynthesis and Chlorophyll Degradation

Peter M. Bramley

Introduction

The pigment content of fruits has been studied for many decades not only because of the esthetic appeal of ripe fruit colors, but also due to the scientific curiosity about the wide variety of compounds that are present and are responsible for the characteristic green, yellow, orange, red, blue, and purple colors we all know well. Plant breeders have continually focused upon varieties that show the greatest color intensity in ripe fruits not only because consumers associate this with quality and taste (Vogel et al., 2010), but also due to the increasing body of evidence that such pigments are beneficial to health, when consumed in sufficiently large amounts. It is believed that consuming them in fruits and vegetables is preferable to taking supplements. Their potential benefits to health have been documented in several recent reviews (Liu, 2003; Johnson and Krinsky, 2009; Robles-Sardin et al., 2009; Landrum, 2010). It is also reported that biological activities attributed to chlorophyll derivatives are linked to cancer prevention (reviewed by Ferruzzi and Blakeslee, 2007).

Fruits typically contain a mixture of pigments, including the green chlorophylls; yellow, orange, and red carotenoids; red, blue, and violet anthocyanins; and yellow flavonoids. This chapter focuses on the chlorophylls and carotenoids and in particular the changes that occur during ripening of fruits. Reviews on other pigments in fruits can be found elsewhere (Willson and Whelan, 1990; Shirley, 1996; Harborne and Williams, 2000; MacDougall, 2002).

Distribution of Carotenoids and Chlorophylls in Fruit

The distribution of carotenoids in fruit is extremely complex and variable, giving patterns that are characteristic of a particular species or variety. These vary from just a few carotenoids to over 50, as in citrus fruit. In addition, the content often varies between pulp and peel. Biosynthesis is autonomous in most fruits and continues after harvesting from the parent plants.

Green fruit contains the same chlorophylls and carotenoids as other photosynthetic tissues, within the chloroplast, as part of the pigment–protein complexes. The carotenoid

The Molecular Biology and Biochemistry of Fruit Ripening, First Edition.
Edited by Graham B. Seymour, Mervin Poole, James J. Giovannoni and Gregory A. Tucker.

content of chloroplasts is quite constant, with the main components being β-carotene (25–30% of the total), lutein (40–50%), violaxanthin (15%), and neoxanthin (15%). Small amounts of other carotenoids may be found, such as ζ-carotene, zeaxanthin, antheraxanthin, and lutein 5,6-epoxide. However, on the onset of ripening, the chloroplasts differentiate into chromoplasts and there is often, but not always, *de novo* synthesis of carotenoids as well as degradation of chlorophylls, as typified by the green-to-red transition of tomato fruit. Carotenoid formation during tomato fruit ripening has been studied extensively and has become the best model system for other chromoplast-containing tissues. During ripening, the concentration of carotenoids increases between 10- and 14-fold, due mainly to the accumulation of lycopene (Fraser et al., 1994). Some tomato species, however, ripen without changing their green color, for example, *Solanum pennellii* (http://www.nhm.ac.uk/research-curation/research/projects/solanaceaesource), as do some commercial varieties (see http://www.tomatogrowers.com/Tomatoes/departments/1). Among the 12 wild relatives of tomato, there is only one (*Solanum pimpinellifolium* B. Juss.) with a red berry and two with yellow, yellow-green, or orange fruits (*Solanum cheesmaniae* (L. Riley) Fosberg; *Solanum galapagense* S.C. Darwin and Peralta), whereas all the other species have green, yellow-green, dark green, or black fruits (Peralta et al., 2006; Peralta and Spooner, 2007).

According to Goodwin (1980), ripe fruits can be approximately divided into eight main groups according to their carotenoid contents:

(1) Insignificant amounts, for example, strawberry.
(2) Large amounts of chloroplast carotenoids, for example, lime, goji berry.
(3) Large quantities of lycopene and its derivatives, for example, tomato, gac (*Momordica cochinchinensis*), water melon.
(4) Large levels of β-carotene, for example, peach, mango, papaya.
(5) Large amounts of epoxides, for example, carambola.
(6) Unusual carotenoids, for example, capsanthin and capsorubin in pepper.
(7) Poly-*cis* carotenoids, for example, prolycopene in tangerine tomatoes.
(8) Apocarotenoids, for example, in the peel and flavedo of Citrus species.

In all cases, the carotenoids are a mixture of geometrical isomers, although the all-*trans* (E) configuration predominates. The structures of common carotenoids are shown in Figure 4.1. Xanthophylls and xanthophyll epoxides can be esterified to fatty acids such as lauric, stearic, and palmitic, in some fruit such as mango (de Jesus Ornelas-Paz et al., 2007). There are, however, significant differences in the qualitative distribution of carotenoids and their esters in different cultivars of all fruits and also when grown in different environmental conditions and across seasons. Comprehensive tables detailing the carotenoid content of many fruits can be found in Goodwin and Goad (1970), Goodwin (1980), Goss (1991), Khachik et al. (1991), and Matsumato et al. (2007).

Chlorophylls are the most widely distributed of the natural pigments. It is estimated that more than 109 tons of chlorophyll are biosynthesized and degraded every year on the earth (Brown et al., 1991). They are porphyrins containing a tetrapyrrole ring, coordinated with an Mg^{2+} ion and a phytol side chain, which creates its overall hydrophobicity (Fig. 4.2). The predominant

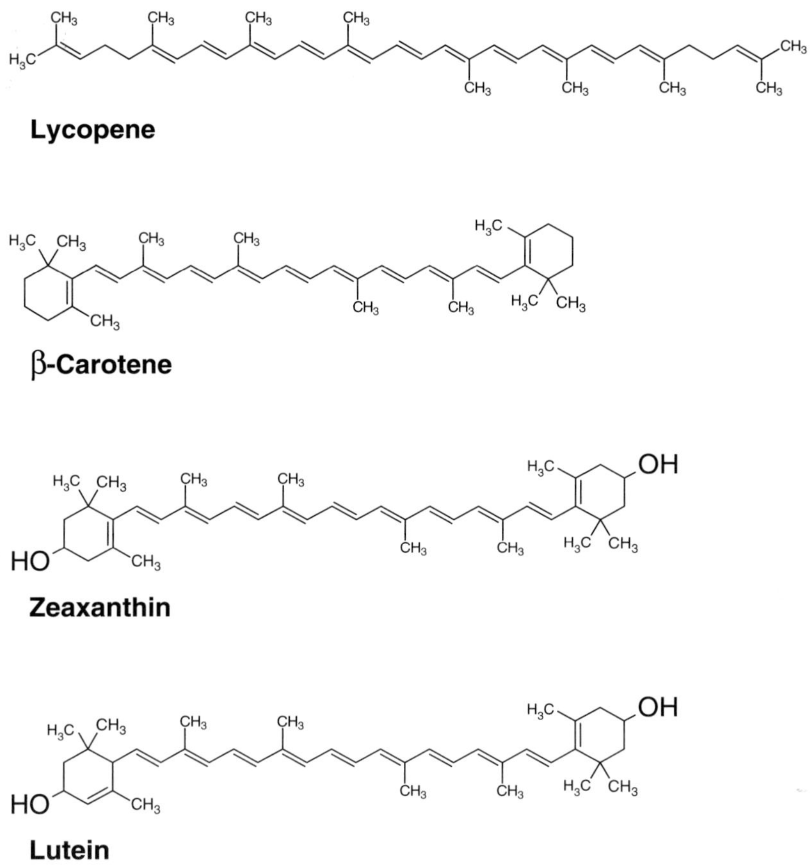

Figure 4.1 Structures of common plant carotenoids.

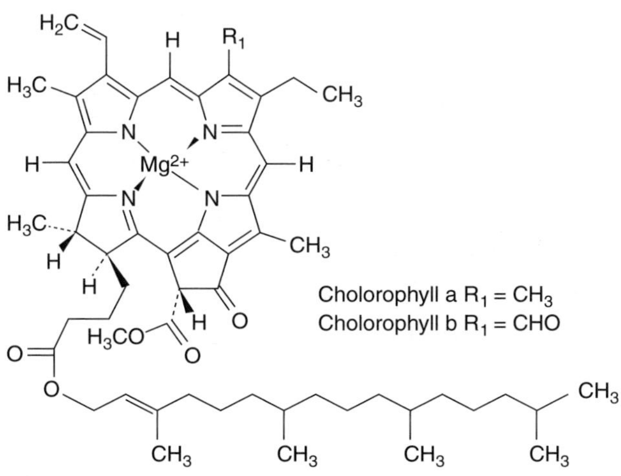

Figure 4.2 Structures of chlorophyll a and b.

forms are chlorophyll *a* and *b*, but *c*, *d* and, more recently, the *f* form has been characterized (Chen et al., 2010a). All are present in light-harvesting complexes and enable plants to absorb light in the red and blue regions of the spectrum. Chlorophylls do not occur as free pigments. They are bound, together with carotenoids, to protein constituents of photosystems I and II and arranged in the lipid bilayers of the thylakoid membranes, in order to reduce the risk of uncontrolled interaction of excited pigments with oxygen (Blankenship, 2002). Chlorophyll biosynthesis and its regulation have been studied intensively for many decades. The pathway has been established and shown to comprise at least 15 steps, within the chloroplast. As this is outside the topic of this chapter, details can be found in several reviews (von Wettstein et al., 1995; Tanaka and Tanaka, 2007; Reinbothe et al., 2010; Mochizuki et al., 2010).

Chlorophyll Degradation and Recycling

Chlorophyll Degradation During Senescence and Fruit Ripening

Chlorophylls are extremely labile, being sensitive to light, heat, oxygen, and chemical degradation that give rise to a series of derivatives such as the chlorophyllides, pheophytins, pheophorbides, and pyrochlorophylls (Goss, 1991). However, in contrast to chlorophyll biosynthesis, the mechanism for the degradation of chlorophylls has been the subject of less attention. Therefore, chlorophyll degradation which occurs during leaf senescence and fruit ripening, but also as a response to biotic and abiotic stresses, is less well understood. Considerable progress on the biochemistry of this process has been made recently, albeit mainly in leaves rather than fruit.

During senescence, both chlorophylls *a* and *b* are degraded to polar, colorless, and nonfluorescent chlorophyll catabolites, which accumulate in the vacuoles of senescent leaves. The initial step is either the loss of the magnesium or the loss of the phytol tail. The former reaction results in the formation of pheophytin, while the latter produces chlorophyllide. Further degradation of either the pheophytin or the chlorophyllide produces a pheophorbide: pheophytin is degraded by the loss of the phytol tail and a chlorophyllide loses its magnesium ion. In this pathway chlorophyll is converted to colorless nonfluorescent end products, known as NCCs. Species-specific modifications are found in the chemical structures of NCCs. The key reaction of this pathway is the conversion of pheophorbide a to a fluorescent catabolite (pFCC) in two steps, catalyzed by the joint action of pheophorbide a oxygenase (PAO) and RCC reductase (RCCR). This reaction causes the loss of the green chlorophyll color. Among the catabolic enzymes, PAO, a Rieske-type iron–sulfur-cluster-containing oxygenase, has been demonstrated to be exclusively active during senescence (Ye et al., 2006). The pathway is outlined in Figure 4.3 and a topographical model of chlorophyll breakdown in a senescent leaf cell is shown in Figure 4.4. Chlorophylls *a* and *b* are known to be interconvertible, in a process known as the chlorophyll cycle, which involves three enzymes: chlorophyllide *a* oxygenase, chlorophyll *b* reductase, and 7-hydroxymethylchlorophyll reductase (reviewed by Tanaka and Tanaka, 2011). The biochemistry of the chlorophyll degradation pathway has been reviewed comprehensively by Krautler (2008) and Hörtensteiner and Krautler (2011). Chloroplast proteomics have shown that the stroma and both thylakoid and envelope membranes are involved

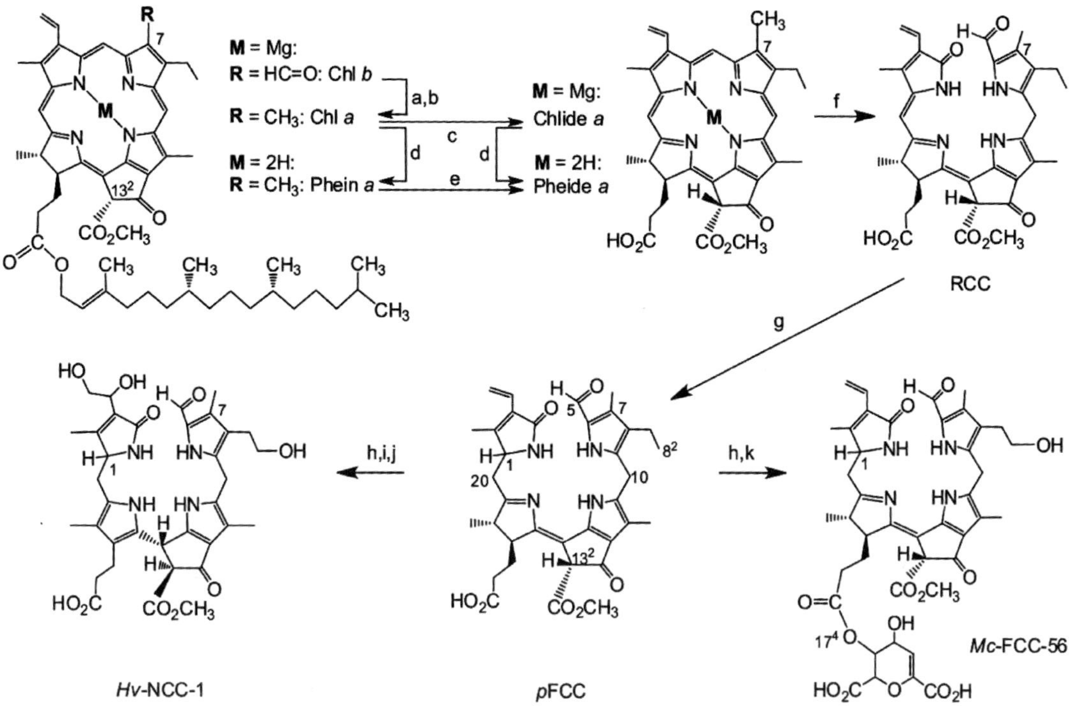

Figure 4.3 Chlorophyll degradation pathway. Representative structural outline of major catabolites delineating the main paths of chlorophyll breakdown in higher plants. Chlorophylls are degraded in the chloroplast byenzyme-catalyzed processes via pheophorbide (Pheide) *a* and the red chlorophyll catabolite (RCC) to give primary fluorescent chlorophyll catabolites (pFCC, or its C1-epimer, epipFCC). The relevant enzymes involved in this part are (a) Chl b reductase; (b) 7-hydroxymethyl Chl reductase; (c) chlorophyllase (CLH); (d) magnesium dechelatase; (e) pheophytinase (PPH); (f) pheide a oxygenase (PAO); (g) RCC reductase (RCCR). pFCCs are modified further by unidentified hydroxylating enzymes (h, i). When carrying a free propionic acid group, FCCs are transported into the vacuole, where they are suggested to isomerize by a spontaneous, acid catalyzed reaction (j) to the corresponding nonfluorescent chlorophyll catabolites (NCCs) such as Hv-NCC-1 (the main tetrapyrrolic catabolite found in senescent leaves of barley, *Hordeum vulgare*). Else, they are esterified by unknown enzymes at the propionic acid group (k) to give "persistent" hypermodified FCCs such as Mc-FCC-56 (the main FCC in peels of ripe bananas, *Musa acuminata*, cavendish cultivar). Relevant atom numbering is specified. Reprinted from Hörtensteiner and Kräutler (2011). Copyright 2011, with permission from Elsevier.

in chlorophyll catabolism, although the precise subcellular localization of these proteins awaits elucidation (reviewed by Joyard et al., 2009).

The same level of detail regarding chlorophyll degradation and its functions has yet to be elucidated in ripening fruit. Chlorophyll metabolites (NCCs) have been detected and structurally elucidated in ripening apples and pears. They are the same as those from senescing leaves, indicating that the pathway is probably common in both cases (Schelbert et al., 2009), although the range of FCCs and NCCs in ripening bananas suggests a different pathway beyond pFCC (Moser et al., 2008; 2009).

The *green flesh* (*gf*) mutant of tomato was found by Kerr (1956). In this mutant, chlorophyll degradation is defective and normal chloroplast–chromoplast transition is affected, giving rise to a rusty red fruit and dark green senescent leaf color (Cheung et al., 1993). *LeSGR1* has been

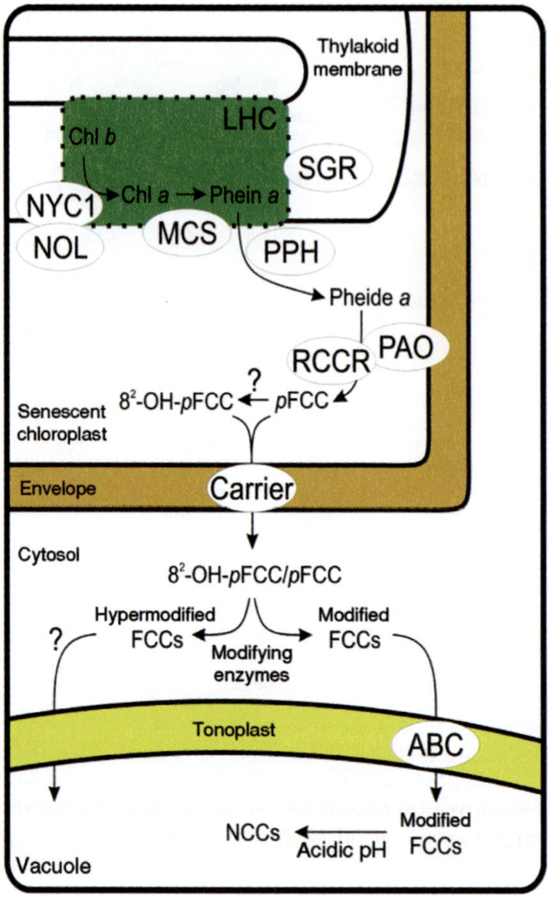

Figure 4.4 Topographical model of chlorophyll breakdown. Topographical model of chlorophyll breakdown in a senescent leaf cell. The model incorporates the current knowledge about chlorophyll catabolites and chlorophyll catabolic reactions and shows their presumed subcellular localizations. Putative steps are labeled with question marks. Note that for NYC1/NOL and PAO/RCCR physical interaction has been demonstrated. The pathway is split at the late reactions, thereby either producing hypermodified FCCs or (ultimately) NCCs. FCC, fluorescent chlorophyll catabolite; pFCC, primary FCC; LHC, light-harvesting complex; MCS, metal-chelating substance; NCC, nonfluorescent chlorophyll catabolites; NOL, NONYELLOW COLORING1 LIKE; NYC1, NONYELLOW COLORING1; PAO, pheophorbide a oxygenase; PPH, pheophytinase; RCCR, reduced chlorophyll catabolite reductase; SGR, stay green protein.

found to be crucial for chlorophyll degradation and a mutation of the SGR1 protein may be responsible for the properties of *gf* (Hu et al., 2011). Four mutant alleles of *gf* have been identified in tomato heirloom varieties (Barry and Pandy, 2009). A single recessive mutation designated *chlorophyll retainer* (*cl*) in pepper (Smith, 1950) blocks the breakdown of chlorophylls during the chloroplast to chromoplast differentiation in ripening fruits (Cheung et al., 1993; Akhtar et al., 1999; Roca and Minguez-Mosquera, 2006; reviewed by Bouvier et al. (2009)). The chlorophyll retainer (*cl*) mutation in pepper has been identified as an SNP in an SGR gene, which is induced during fruit ripening, but to a lesser extent in senescing leaves (Borovsky and

Paran, 2008). Chlorophyll breakdown in banana and plantain at high temperatures of ripening (20° and 30°C) has been shown to be reduced due to the reduction in Mg-dechelatase activity in the peel, but an increase in chlorophyllase activity occurs, causing uneven de-greening (Yang et al., 2009). PCCR and RAO have been found in the chromoplast membranes of tomato and pepper (Moser and Matile, 1997; Akhtar et al., 1999). Studies on citrus chlorophyllase at the breaker stage of ripening have shown that it is located in the plastid and its levels, analysed by immunoblotting, are negatively correlated with the chlorophyll content. The protein is initially a 35 kDa precursor that is N-terminally processed to 33 kDa (Azoulay Shemer et al., 2008).

Recycling of Chlorophyll Metabolites

The rationale for chlorophyll breakdown during senescence is not fully understood, although it is highly conserved in higher and lower plants (Hörtensteiner, 2006). However, since the prime function of leaf senescence is to recycle compounds accumulated during leaf growth and maturation into reusable and exportable nutrients to supply sink organs, it is likely that chlorophyll degradation products fall into this category.

The remobilization of nitrogen from the pyrrole ring was hypothesized several decades ago to be the main reason for chlorophyll degradation (Hendry et al., 1987), but the more recent discovery that the end products of catabolism are NCCs, which still contain nitrogen atoms (see Figure 4.4), makes this less likely. It may be that chlorophyll breakdown to nontoxic products allows cell viability to be maintained for the degradation of chlorophyll-binding proteins and the subsequent recycling of organic nitrogen from amino acids (Hörtensteiner, 2004). Furthermore, expression of many of the chlorophyll catabolic genes are upregulated following biotic and abiotic stresses (Zimmermann et al., 2004) and this may result in chlorophyll catabolites being involved in causing programmed cell death after pathogen infections as a defense mechanism (Mur et al., 2010). The identification of the so-called "stay green" mutants of crops (Thomas and Howarth, 2000; Gong et al., 2005; Borovsky and Paran, 2008), that are delayed in senescence, due to mutations in the *stay-green* gene (*SGR*) that controls protein factors involved in LHCII disassembly, has led to suggestions of increasing crop yields via a longer growing period (Hörtensteiner, 2009).

Phytol comprises some one-third of the mass of chlorophyll and is generally considered to be the most abundant acyclic isoprenoid in nature (Volkman and Maxwell, 1986). Free phytol is generated during chlorophyll catabolism by chlorophyllase (Fig. 4.3). While the fate of chlorophyllide *a* and its derived pigments has been characterized, less is known about the metabolic fate of phytol. It has been established that phytol derived from chlorophyll breakdown is re-esterified to acetic acid or other fatty acids (Peisker et al., 1989; Patterson et al., 1993), while a portion is subjected to photodegradation (Rontani et al., 1996). It has been suggested that chlorophyll-derived phytol may be a precursor for the biosynthesis of tocopherols, based on the temporal patterns of chlorophyll degradation and leaf and seed tocopherol accumulation during leaf senescence (Rise et al., 1989) and seed maturation (Goffman et al., 1999). Evidence to support this hypothesis has now been reported, at least for leaves and seeds. Studies on a low tocopherol *Arabidopsis thaliana* mutant, and subsequent cloning and expression in

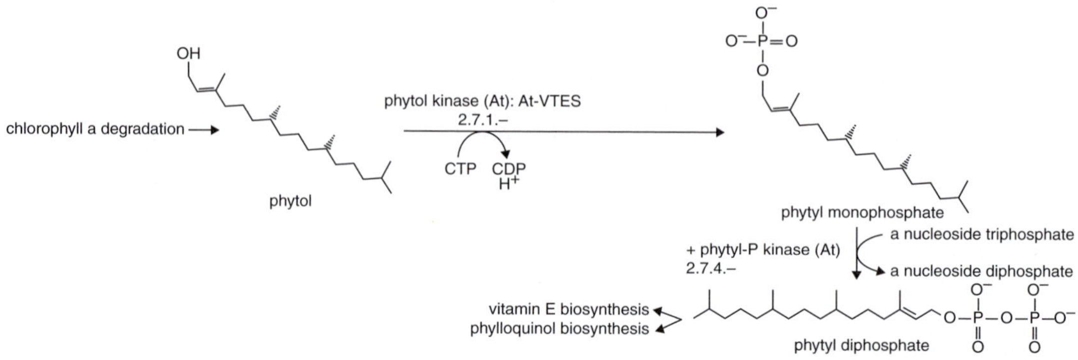

Figure 4.5 Pathway for recycling phytol into tocopherol. During leaf senescence and chlorophyll breakdown, the enzyme phytol kinase, located in the chloroplast, converts phytol into phytol monophosphate, followed by a further phosphorylation catalyzed by phytol phosphate kinase. Phytol diphosphate can then be recycled into vitamin E and phylloquinol formation.

Escherichia coli supplemented with free phytol, plus *in vitro* assays of the recombinant protein produced phytyl monophosphate, suggesting that *VTE5* encodes for a phytol kinase (Valentin et al., 2006). Homologs of this gene are found in other plants as well as in cyanobacteria. For example, *slr1652*, the *Synechocystis* spPCC 6803 homolog, is essential for normal tocopherol accumulation *in vivo* and At5g04490 and slr1652 proteins have phytol kinase activity (Valentin et al., 2006). The establishment of phytol kinase enzyme activity shows that a biosynthetic pathway for the activation and reutilization of phytol from chlorophyll degradation in plant metabolism, including tocopherol biosynthesis, is feasible.

Phytol can be incorporated into chlorophyll, tocopherol, and lipid esters by *Arabidopsis* seedlings. Phytol is phosphorylated to phytyl phosphate and phytyl diphosphate by two successive kinase activities associated with chloroplast envelope membranes of *Arabidopsis*. Although phytol kinase is CTP dependent, the second kinase reaction, phytyl phosphate kinase, shows broader specificity for CTP, GTP, UTP, and ATP. Therefore, in addition to *de novo* synthesis from geranylgeranyl diphosphate (GGDP), phosphorylation of free phytol represents an alternative route for phytyl diphosphate production as the precursor for chloroplast prenyl lipid synthesis (Ischebeck et al., 2006).

The pathway for recycling phytol into tocopherols in leaves is shown in Figure 4.5. Evidence for such recycling has yet to be reported for fruit, although it is tempting to speculate that it does occur, given that fruits are autonomous in ripening once detached from the parent plant and an economy of energy for the considerable metabolic changes that occur, would be beneficial.

Carotenoids and Carotenoid Metabolites

Structures and Functions in Higher Plants

Carotenoids

Carotenoids were first isolated from carrots by Wackenroder (1831). Structurally, they are isoprenoids, typically comprising 8 isoprene units joined together so that the linking of units is reversed at the center of the molecule, as typified by the carotenes, lycopene and β-carotene (Fig. 4.1). They are subdivided into hydrocarbon carotenes and oxygenated xanthophylls, for example, lutein, zeaxanthin, (Fig. 4.1). Their most obvious feature is the polyene chain, which can extend from 3 to 15 conjugated double bonds. The length of this chromophore chain determines the absorption spectrum of the particular carotenoid, which can be from the UV to the red regions of the spectrum (Davies, 1976; Britton, 1985, 1991). In their comprehensive Carotenoids' Handbook, Britton et al. (2004) reported some 750 structures, while about 20 new structures are identified annually (Maoka, 2009). Further details of their three-dimensional structures, chirality, and many geometric isomers can be found in Britton et al. (2008).

In photosynthetic tissues, carotenoids fulfil several vital functions. They are essential components of the photosynthetic apparatus, being components of every pigment–protein complex. They fulfil indispensable functions in light harvesting, protection against photo-oxidation, and regulation of Photosystem II efficiency (Demmig-Adams et al., 1996; Niyogi, 2000; Yamamoto, 2004; Lu and Li, 2008). Within thylakoid membranes, carotenoids are bound mostly to specific chlorophyll–carotenoid-binding protein complexes of PSI and PSII. PSI is enriched in β-carotene, while in PSII lutein predominates. Within PSII, the bulk carotenoid is present in the core complexes closely surrounding the reaction center. The xanthophylls prevail in the remaining, chlorophyll *a*- and *b*-binding, light-harvesting antennae. Colored carotenoids absorb light in the blue region of the spectrum (400–600 nm) and the energy absorbed is transferred to chlorophylls, which in themselves do not absorb such light energy. In addition, the carotenoids of the violaxanthin cycle, that is, violaxanthin, antheraxanthin, and zeaxanthin, are ubiquitous in the thylakoid membranes (Demmig-Adams et al., 1996). This transmembrane cycle catalyzes de-epoxidation on the lumen side and epoxidation on the stromal side of membranes (Yamamoto et al., 1999).

Carotenoids also play an essential role in protecting the cell against photo-oxidation via energy dissipation, free radical quenching, and limiting damage to membranes (Bartley and Scolnik, 1995; Pogson et al., 2005). They are able to quench triplet chlorophyll and singlet oxygen, due to their antioxidant properties, which are a result of the conjugated double bonds in the chromophore. Other studies have shown that carotenoids and some other isoprenoids, such as tocopherols, stabilize and photo protect the lipids of thylakoid membranes. When plants are exposed to strong light and/or elevated temperatures, violaxanthin and the products of its enzymic de-epoxidation, antheraxanthin and zeaxanthin (the violaxanthin cycle), partition between the light-harvesting complexes and the lipid phase of the thylakoid membranes. The interaction of the xanthophylls and the membrane lipids decreases membrane fluidity, increases membrane thermostability, and lowers susceptibility to lipid peroxidation (Havaux and Niyogi, 1999). The crystal structures of light-harvesting complexes show that carotenoids span the lipid bilayer and connect components of the complexes on both membrane surfaces and/or to mediate the interaction of transmembrane protein helices. The chlorophyll *a*/*b* light-harvesting complexes also assemble only when carotenoids are present, although, at least in *Arabidopsis,* the presence of lutein is not essential (Pogson et al., 1996). The molecular factors of carotenoid–protein complexes that affect light harvesting have been reviewed by Polivka and Frank (2010).

Carotenoids are also found in the chloroplast envelope membranes, being some 0.2% of the total membrane lipid. Violaxanthin is the major carotenoid in this respect (Jeffrey et al., 1974). In the envelope, violaxanthin undergoes a light-induced decrease without the concomitant increase in zeaxanthin, as the violaxanthin cycle is absent (Siefermann-Harms et al., 1978). Yamamoto (2006) has suggested that differential partitioning of violaxanthin between the envelope and thylakoids is due to relative solubility in these membranes and that the violaxanthin cycle links the thylakoids and envelope for signal transduction of light stress.

Carotenoids are also located in plastoglobuli (Lichtenthaler, 1968), typically esterified and more abundant during senescence (Devini and Steinmuller, 1985). More recent investigations, using three-dimensional tomography combined with immunolabeling techniques, have shown that plastoglobules function as both lipid biosynthesis and storage subcompartments of thylakoid membranes and that the lipid molecules contained in the plastoglobule cores (carotenoids, plastoquinone, and tocopherol) are in a dynamic equilibrium with those located in the thylakoid membranes. The role of these carotenoids is not clear and they may simply accumulate in these oil droplets as a metabolic, osmotically inert sink. Alternatively, it has been suggested that they act as antioxidants *in situ* (Bouvier et al., 1998a), as photoprotectors (Merzlyak and Gitelson, 1995; Merzlyak et al., 1999) or even as conduits of heat from light through absorption for transpiration and metabolism in autumn leaves (Schmidt, 1988; Biswell, 1995). Photoprotection in apple fruit is a consequence of the combined effects of carotenoids, anthocyanins, and flavonols (Merzlyak et al., 2008).

An additional function of carotenes, or perhaps a carotene metabolite, has been reported recently by Leon and coworkers. A new signal derived from linear carotenoids regulates early chloroplast and leaf development in *Arabidopsis* (Leon et al., 2011). Synthesis of this novel signal depends on ζ-carotene desaturase (ZDS) activity (see Section Desaturation, cyclization, and oxidation reactions from phytoene).

The evidence for functions of carotenoids in ripening, or ripe fruit, is less well established. It has long been suggested that their presence in ripe fruit is attractive to seed-dispersing animals, but this does not account for the variety of pigments and colors found in such tissues (see Section Distribution of carotenoids and chlorophylls in fruit). Studies on ripening apple fruit, using reflectance spectroscopy, have shown two pools of carotenoids, one closely associated with chlorophyll in chloroplasts and the other localized in plastoglobuli and appearing in chloroplasts that are being transformed to gerontoplasts during ripening. The latter pool is thought to provide protection from irradiation (Merzlyak and Solovchenko, 2002; Solovchenko et al., 2010).

Carotenoid Metabolites (Apocarotenoids)

Enzymatic cleavage of carotenoids produces various types of apocarotenoids, many of which have important biological functions in the growth and development of plants and animals ranging from being scent and aroma compounds (see Chapter 6), accessory light-harvesting pigments (Markwell et al., 1992), regulatory molecules in plant development (see later), while others are of unknown function. They include abscisic acid (ABA), flavor volatiles, and strigolactones. The oxidative cleavages are catalyzed by a family of carotenoid cleavage dioxygenases (CCDs; also referred to as CCOs, carotenoid cleavage oxygenases), and are described in Section Apocarotenoids. Reviews on the topic include those of Giuliano et al. (2003), Bouvier et al. (2005), Vallabhaneni et al. (2010), and Walter et al. (2010).

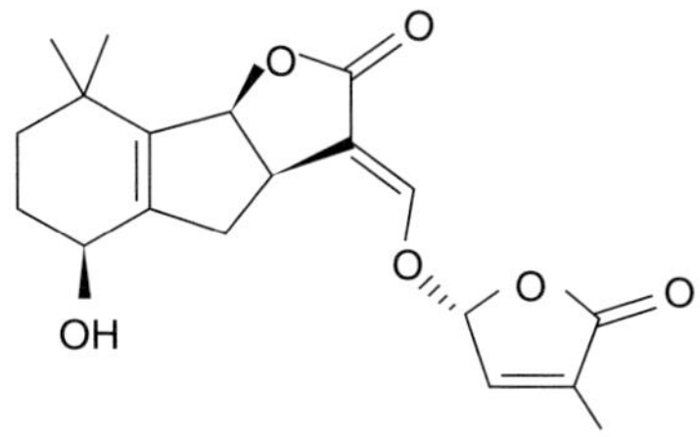

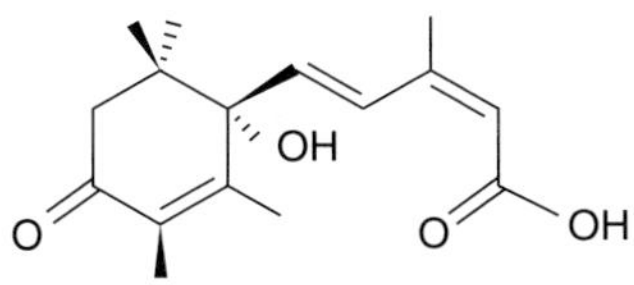

Figure 4.6 Structures of apocarotenoids. All apocarotenoids are derived from carotenoids, with strigol being a member of the strigolactone family, which act as *plant hormones* that stimulate the branching and growth of symbiotic *arbuscular mycorrhizal fungi*. Abscisic acid is also a plant hormone, and involved in many plant developmental processes (see Section Abscisic acid).

Abscisic Acid

The phytohormone ABA (Fig. 4.6) is probably the best known of the apocarotenoids, having been discovered in the 1960s (reviewed by Milborrow, 1974; Schwartz and Zeevart, 2010). It regulates numerous developmental processes and abiotic and biotic adaptive stress responses in plants, including embryo and seed development, stomatal opening and closure, vegetative development, and fruit ripening (reviewed by Nambara and Marion-Poll, 2005; Cutler et al., 2010). The multitude of signaling networks involving ABA has been recently documented (Hauser et al., 2011).

There are numerous reports on the functions of ABA in fruit ripening. It is believed to play an important role not only in the regulation of fruit ripening, but also to stimulate sugar accumulation, leading to optimized yields and flavor (Lu et al., 1999). The similarity of ABA changes in fruit with that of ethylene (see Chapter 3) suggests a role in fruit maturation and ripening (Giovannoni 2001, 2004). More recently, it has been concluded that ABA plays a different role in ethylene synthesis in different stages of tomato fruit ripening (Zhang et al., 2009a). The maximum ABA content precedes ethylene production in both the seeds and the flesh of tomato (Zhang et al., 2009b). It is also involved in tomato fruit set (Nitsch et al., 2009). Other studies have shown that tomato plants overexpressing a gene encoding 9-*cis*-epoxycarotenoid dioxygenase, the enzyme that catalyzes a key rate-limiting step in ABA biosynthesis, not only increased ABA production, but also had higher transpiration efficiency, resulting in plants that required less water for growth and development (Thompson et al., 2007; Tung et al., 2008). ABA also prevents blossom end rot in tomato fruit by increasing calcium uptake and improving water use efficiency (Tonetto de Freitas et al., 2011). A recent report has shown that ABA promotes strawberry (*Fragaria ananassa*) fruit ripening (Jia et al., 2011), reviewed by Li et al. (2011).

Flavor Volatiles

Several carotenoid metabolites, including β-ionone, geranyl acetone, and 6-methyl-5-hepten-2-one, are important components of flavor and aroma (organoleptic properties) and many fruits and vegetables and are produced by either nonenzymatic oxidative cleavage or through the action of CCDs.

Strigolactones

The strigolactones are the most recently discovered and chemically identified group of plant hormones, although their presence has been assumed for many decades. They are apocarotenoids, chemically based on strigol (Fig. 4.6) and responsible for the inhibition of shoot branching (Gomez-Roldan et al., 2008). They were originally discovered as exudates of plant roots that stimulated germination of plant parasitic weeds (Cook et al., 1966), but later found to be indispensable chemical signals for root colonization by symbiotic arbuscular mycorrhizal fungi. Further studies have established that they, or their metabolites, are involved in the control of shoot branching (Gomez-Roldan et al., 2008), rhizosphere communications (Akiyama et al., 2005), and root development in concert with auxin and ethylene (Kapulnik et al., 2011; Koltai, 2011). Further details can be found in reviews by Chen et al. (2009), Dun et al. (2009), Walter et al. (2010), and Xie et al. (2010). The structures of many strigolactones have now been elucidated (reviewed by Yoneyama et al., 2009; Xie et al., 2010).

With respect to fruit yield and quality, a reduction in strigolactone formation has been shown to reduce susceptibility to the root parasite *Orobanche* in the high pigment mutant of tomato (Lopez-Raez et al., 2008), and there have been suggestions that plant vigor can be improved through plant biotechnology approaches to manipulating strigolactone levels (Csukasi et al., 2009; Rameau et al., 2011).

Carotenoid and Apocarotenoid Biosynthesis

Carotenoids

Studies on the biosynthesis of carotenoids and regulation of the pathways have been undertaken for over 60 years. Experimentally, several approaches have been used:

- Pioneering studies using natural product chemistry and specific enzyme inhibitors, to elucidate structures and formulate pathways
- *In vitro* approaches, largely using radioactive substrates, to confirm pathways and elucidate the biochemical properties of the enzymes
- Purification of the enzymes from native and recombinant sources
- Gene cloning and expression in recombinant systems and transgenic plants
- The use of "omic" technologies to analyse gene structure, transcription, and metabolic fluxes

These strategies have used a wide variety of higher plants and microorganisms, including a multitude of mutants. Of the many reviews on this topic, those of Spurgeon and Porter (1983), Bramley (1985), Jones and Porter (1986), Britton (1990), Sandmann (1991), and Bramley et al. (1993) describe the earlier studies on the subject, while those of Bartley and Scolnik (1995), Cunningham and Gantt (1998), Hirschberg (2001), Cunningham, (2002), Fraser and Bramley (2004), Römer and Fraser (2005), Rodríguez-Concepción (2006), Lu and Li (2008),

Cazzonelli et al. (2010), Cazzonelli and Pogson (2010), and Cazzonelli (2011) discuss our current understanding of the pathways and their regulation. The potential of transgenic crop plants to increase carotenoids in the diet has also been reviewed extensively (Botella-Pavía and Rodríguez-Concepción, 2006; Sandmann et al., 2006; Zhu et al., 2007; Giuliano et al., 2008; Fraser and Bramley, 2009; Fraser et al., 2009; Rosati et al., 2009; Farré et al., 2010, 2011).

The Early Steps Carotenoids are formed from the C_5 building block, isopentenyl diphosphate (IPP), and its allylic isomer dimethylallyl diphosphate (DMAPP). For many years, it was assumed that IPP was synthesized from acetyl-CoA via mevalonic acid (MVA) and then isomerized to DMAPP in all organisms (Chappell, 1995), but it has now been firmly established that IPP and DMAPP used for carotenoid formation (and also dolichols, tocopherols, and chlorophyll side chains) are derived from the methylerythritol (MEP) pathway, which does not utilize MVA (Rohmer et al., 1993; Lichtenthaler, 1999; Hunter, 2007). The MEP pathway synthesizes IPP and DMAPP in plastids, whereas the MVA pathway produces cytosolic IPP which is a precursor of sterols, sesquiterpenes, and dolichols (Lange et al., 2000; Bouvier et al., 2005). However, there is evidence for the interchange of IPP between the two compartments (reviewed by Eisenreich et al., 2001) and MVA-derived precursors imported into the plastid can sustain a limited production of carotenoids (Rodríguez-Concepción et al., 2004). The mechanism regulating this exchange, however, remains unknown. A recent report has shown that it is possible to express the cytosolic MVA pathway in tobacco chloroplasts resulting in elevated levels of carotenoids (Kumar et al., 2011).

The MEP pathway comprises seven reactions, the first being a transketolase-like decarboxylation reaction catalyzed by 1-deoxy-D-xylulose 5-phosphate (DXP) synthase (DXS) using pyruvate and glyceraldehyde-3-phosphate (GAP) as substrates, to yield 1-deoxy-D-xylulose-5-phosphate (Lange et al., 2000; Rodríguez-Concepción and Boronat, 2002). The complete pathway is shown in Figure 4.7. Following the formation of IPP and DMAPP, prenyl transferases catalyze the sequential head to tail condensation of 2 × C_5 units to form geranyl diphosphate (GPP), C_{15} farnesyl diphosphate (FPP), and finally C_{20}, GGPP, the direct precursor of phytoene. Detailed reviews on the MEP pathway can be found elsewhere (Eisenreich et al., 1998; Rodríguez-Concepción and Boronat, 2002; Cordoba et al., 2009; Ganjewala et al., 2009).

Desaturation, Cyclization, and Oxidation Reactions from Phytoene The elucidation of the carotenoid biosynthetic pathway began over 60 years ago with the pioneering studies of Porter and his coworkers (Porter and Lincoln, 1950; Porter and Anderson, 1962). Analysis of color mutants of tomato led them to suggest a sequence of four dehydrogenation reactions from phytoene (colorless, 3 conjugated double bonds) to lycopene (red, 11 conjugated double bonds); the so-called "Porter–Lincoln series," after which they went on to show these conversions using *in vitro* cell extracts (Kushwaha et al., 1970; Qureshi et al., 1974). In the subsequent years, their pathway has been shown to be essentially correct, although it has been refined as our understanding of the *cis/trans* isomeric nature of the carotenes has been elucidated, a concept that was established by Zechmeister and coworkers in the early 1940s, with the identification of poly-*cis* lycopene (prolycopene; Zechmeister et al., 1941; LeRosen and Zechmeister, 1942) and confirmed with the structural elucidation of multiple *cis* isomers of carotenes in the tangerine

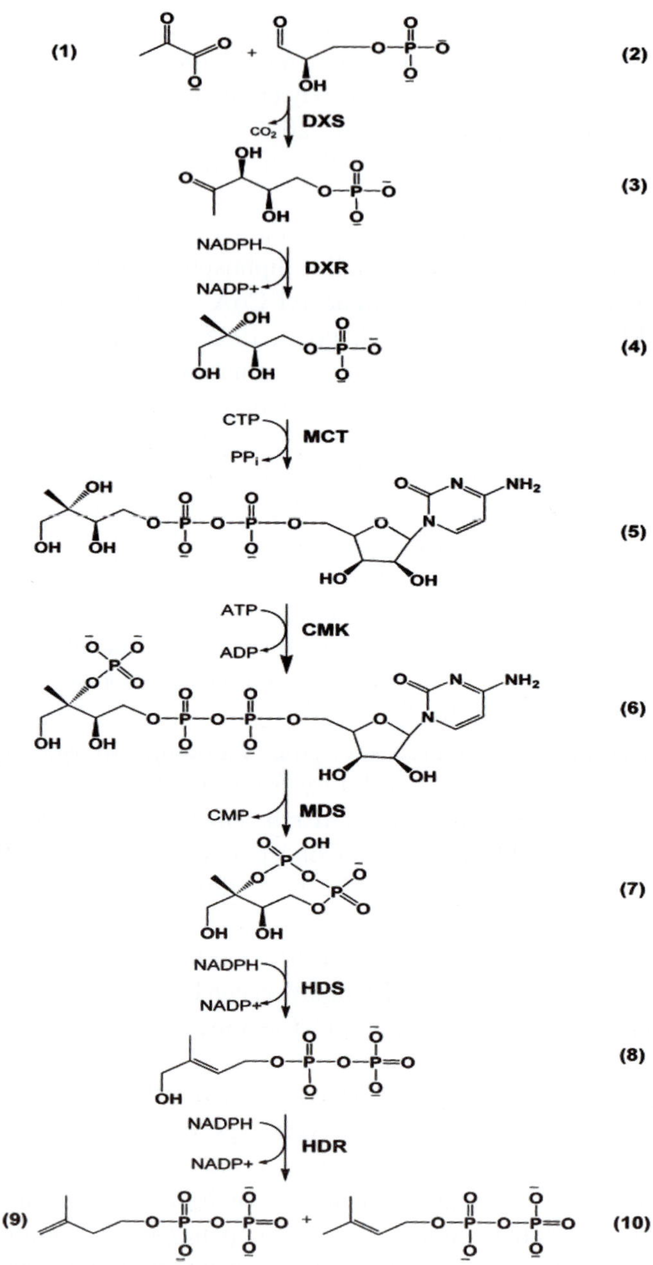

Figure 4.7 Methyl erythritol 5-phosphate (MEP) pathway. The intermediates involved in the MEP pathway to IPP (**9**) and DMAPP (**10**) formation are: (**1**), pyruvate; (**2**), D-glyceraldehyde 3-phosphate; (**3**), 1-deoxy-D-xylulose 5-phosphate (DXS); (**4**), 2-*C*-methyl-D-erythritol 4-phosphate (MEP); (**5**), 4-(cytidine 5′-diphospho)-2-C-methyl-D-erythritol; (**6**), 4-(cytidine 5′-diphospho)-2-C-methyl-D-erythritol phosphate; (**7**), 2-C-methyl-D-erythritol 2,4-cyclodiphosphate; (**8**), (E)-4-hydroxy-3-methyl but-2-enoyl diphosphate. The enzymes involved are DXS, 1-deoxy-D-xylulose 5-phosphate synthase; DXR, 1-deoxy-D-xylulose 5-phosphate reductoisomerase; MCT, 2-C-methyl-D-erythritol 4-phosphate cytidylyl transferase; CMK, 4-(cytidine 5-diphospho)-2-C-methyl-D-erythritol kinase; MDS, 2-C-methyl-D-eythritol 2,4-cyclodiphosphate synthase; HDS, (E)-4-hydroxy-3-methyl but-2-enoyl diphosphate synthase; HDR, (E)-4-hydroxy-3-methyl but-2-enoyl diphosphate reductase. Reprinted from Phillips et al. (2008). Copyright 2008, with permission from Elsevier.

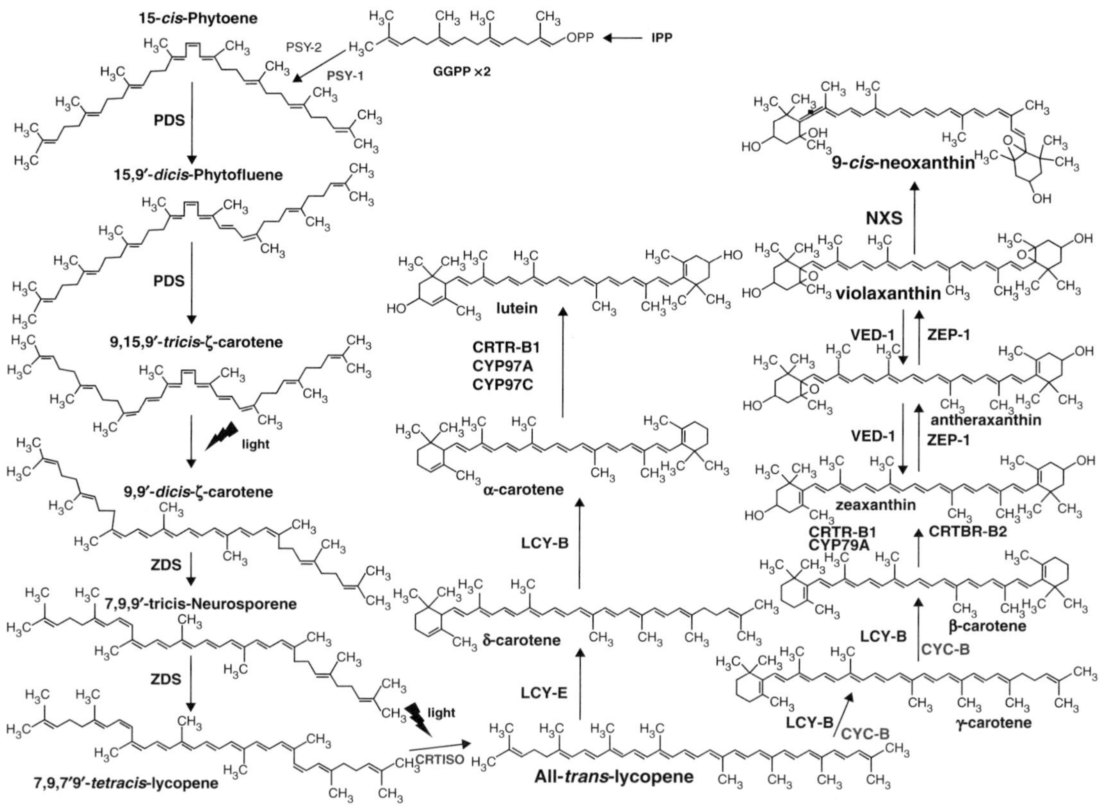

Figure 4.8 Carotenoid biosynthesis in higher plants. GGPP, geranylgeranyl diphosphate; PDS, phytoene destaurae; PSY1, phytoene synthase 1; ZDS, ζ-carotene desaturase; CRTISO, carotene isomerase; LCY-B, lycopene β-cyclase; LCY-E, lycopene ε-cyclase; CRTR-B1, β-carotene hydroxylase 1; CRTR-2, chromoplast-specific β-carotene hydroxylase-2; CYC-B, β-cyclase; CYP97A and C, cyt P450 carotene hydroxylases; CYP79A, carotene hydroxylase; ZEP-1, zeaxanthin epoxidase; VDE-1, violaxanthin de-epoxidase; NXS, neoxanthin synthase. For further details, see Section Carotenoid and apocarotenoid biosynthesis. Reprinted from Fraser et al., 2009. Copyright 2009, with permission from Elsevier.

mutant of tomato (Clough and Pattenden, 1979, 1983). Lycopene is a branch point to cyclic carotenes, which are subsequently oxidized to xanthophylls. Our current understanding of the biosynthetic pathway in plants is shown in Figure 4.8.

The formation of phytoene, from two molecules of GGPP through a head to tail condensation, is the first committed step in carotenoid biosynthesis. In higher plants, the 15-*cis* isomer is formed through the catalytic activity of phytoene synthase (PSY). Phytoene undergoes two desaturation reactions to form ζ-carotene, catalyzed by phytoene desaturase (PDS), which in turn is desaturated to neurosporene and finally lycopene, containing 11 conjugated double bonds, through the action of ZDS. As Figure 4.8 shows, the product of PDS is thought to be 9,15,9′-tri-*cis*-ζ-carotene, which is isomerized to the 9,9′-di-*cis* isomer by ζ-carotene isomerase, a transmembrane protein with similar structure to oxidoreductases (Z-ISO, Chen et al., 2010b), before conversion to 7,9,9′-tri-*cis*-neurosporene, 7,9,7′,9′-tetra-*cis*-lycopene, and finally all-trans lycopene, through the activity of another carotene isomerase, CRTISO (Isaacson et al., 2002, 2004; Breitenbach and Sandmann, 2005). It is reported that, in green tissues, the isomerizations occur nonenzymatically in the presence of light and chlorophyll, although whether such precise *cis–trans* isomerizations are feasible at a sufficient rate is open to debate. Indeed, the biological rationale for such a complex series of isomerizations has yet to be established, as has the thermodynamic stability of the *cis* isomers *in planta*, or definite identification of the geometric isomers by NMR. Intriguingly, the complete pathway from phytoene to lycopene is catalyzed by a single enzyme (CRTI) in prokaryotes (Giuliano et al., 1986; Misawa et al., 1990). Further discussion of *cis/trans* isomers of carotenoids can be found in Chasse et al. (2001) and Liaaen-Jensen and Lutnoes (2008). The desaturation reactions require cofactors from a redox pathway, probably using a plastid terminal oxidase (PTOX; Morstadt et al., 2002; Shahbazi et al., 2007). The involvement of a plastoquinol oxidase activity in carotenoid desaturation is in agreement with biochemical data showing that, in daffodil chromoplasts, the desaturation of phytoene employs oxygen as a terminal electron acceptor, and that O_2 can be replaced by artificial quinones (Mayer et al., 1990). These data were later confirmed by genetic data showing that plastoquinone-deficient mutants are disrupted in phytoene desaturation (Norris et al., 1995). The requirement for PTOX activity, and more generally for a functional redox chain for carotenoid biosynthesis, is probably due to a need for membrane energization in order to derive the desaturation reactions (Kuntz, 2004).

All-*trans* lycopene has two biosynthetic fates. It is either cyclized at both ends of the molecule consecutively by lycopene β-cyclase (LYCB), to form bicyclic carotenoids with two β-rings, such as β-carotene, or cyclized at one end by LYCB and at the other by lycopene ε-cyclase (LYCE) to bicyclic carotenoids with two different rings, such as α-carotene (Fig. 4.8). It is unusual for lycopene to accumulate in plant tissues (with the notable exceptions of tomato fruit and watermelon) and so the formation of lycopene is presumably the rate-limiting step in the formation of cyclic carotenes. The ratio of β- to ε-ring carotenoids in maize has been shown to be significantly influenced by the expression of the *LycE* gene (Harjes et al., 2008).

Two structurally distinct classes of mixed function oxygenases (MFOs) are responsible for the formation of hydroxy carotenoids (Fig. 4.8). One class catalyzes the hydroxylation of β-rings. It is a nonheme diiron, integral membrane mono-oxygenase, containing four iron-binding histidine residues (Tian and DellaPenna, 2004; Bertrand et al., 2005) and has been shown to catalyze this reaction in recombinant *E. coli* (Sun et al., 1996). This is often called a β-carotene hydroxylase (BCH). The second class comprises a family of cytochrome P450 enzymes, CYP97, capable of hydroxylating both β- and ε-rings (Tian et al., 2004; Kim and DellaPenna, 2006). Each enzyme has a P450 domain, a conserved oxygen-binding signature, and a heme thiolate (Tian and DellaPenna, 2004). The rice *CYP97* has been expressed in *E. coli*, showing that CYP97A4 hydroxylates the β-ring (Quinlan et al., 2007). More recently, the tomato P450s have been identified. CYP97A29 and CYP97C11 hydroxylate the β- and ε-rings, respectively (Stigliani et al., 2011).

Zeaxanthin epoxidase (ZEP) converts zeaxanthin into antheraxanthin and then violaxanthin, thus entering the violaxanthin cycle (Fig. 4.10). These reactions are reversed by the action of violaxanthin de-epoxidase (VDE). Under excess light, violaxanthin is rapidly and reversibly converted via antheraxanthin to zeaxanthin under the action of VDE (Demmig-Adams and Adams, 1996), in order to protect thylakoid membrane lipids against photo-oxidation (Havaux and Niyogi, 1999). Violaxanthin is converted into neoxanthin, the final product in green tissues,

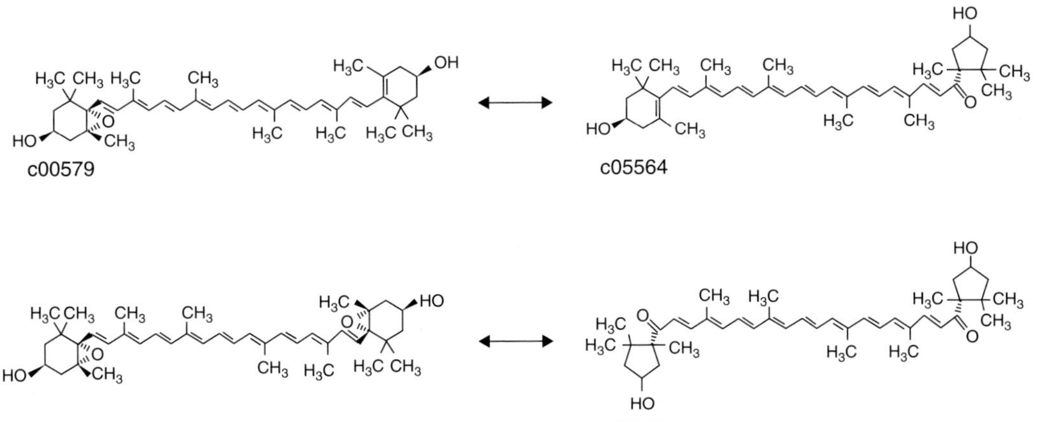

Figure 4.9 Formation of capsanthin and capsorubin. A, Antheraxanthin into capsanthin; B, violaxanthin into capsorubin. These ketocarotenoids are formed during pepper fruit ripening by the action of the enzyme capsanthin–capsorubin synthase. This converts antheraxanthin into capsanthin and violaxanthin into capsorubin. Zeaxanthin epoxidase converts capsanthin into capsorubin.

by neoxanthin synthase (NSY), a carotenoid cyclase paralog (Bouvier et al., 2000; Al-Babili et al., 2000). The ABA4 protein from *Arabidopsis* is reported to have a direct role in neoxanthin synthesis (North et al., 2007). Neoxanthin is the precursor of the apocarotenoid ABA (see Section ABA).

Ketocarotenoids are very rare in higher plants, with the notable exception of *Capsicum* fruit that contains the unusual cyclic ketocarotenoids capsanthin and capsorubin (Davies et al., 1970; Camara and Moneger, 1978). These are formed during fruit ripening by the action of the enzyme capsanthin–capsorubin synthase (CCS; Bouvier et al., 1994). This converts antheraxanthin into capsanthin and violaxanthin into capsorubin. ZEP converts capsanthin into capsorubin (Fig. 4.9). Using CCS as a model carotene cyclase, the Strasbourg group has shown that it contains one noncovalently bound FAD that is essential for enzyme activity only in the presence of NADPH, which functions as the FAD reductant. The reaction proceeds without transfer of hydrogen from the dinucleotide cofactors to β-carotene or capsanthin. They conclude that plant carotenoid cyclases are novel enzymes that combine the characteristics of nonmetal-assisted terpene cyclases with those of flavoenzymes that catalyze reactions with no net redox, such as type 2 IPP isomerase (Mialoundama et al., 2010).

The precise locations of these carotenogenic enzymes, all of which are encoded by nuclear genes (listed by Farré et al., 2010), within the plastid and whether they aggregate as a functional complex, have yet to be unequivocally established and are probably different in chloroplasts and chromoplasts. For example, GGPPS is located in the plastoglobuli of *Capsicum* chromoplasts, but in the stroma of chloroplasts (Cheniclet et al., 1992). A GGPPS/PSY complex that converts IPP into phytoene has been reported in chromoplasts (Maudinas et al., 1977; Islam et al., 1977; Camara, 1993) and chloroplasts (Fraser et al., 2000). It is thought that PSY is in the stroma, as shown in *Capsicum* chloroplasts and chromoplasts (Dogbo et al., 1987), but as a peripheral membrane protein on the inner membrane of *Narcissus* chromoplasts (Kreutz et al., 1982). PSY1 and 2 have been partially purified from the stroma of ripe and green tomato fruits, respectively, and shown to have different substrate specificities and cofactor requirements (Fraser et al., 2000). The enzymes beyond PSY are hydrophobic (as are their substrates) and are therefore likely to be integral membrane proteins. The hypothesis that a membrane bound complex catalyzing the reaction from phytoene to cyclic carotenes, proposed by Cunningham and Gannt (1998), has received experimental evidence from studies on PDS in cauliflower, carrot, and tomato (Lopez et al., 2008a). These workers found two protein complexes, by Western blotting, of different molecular masses, both containing PDS; one in the stroma and the other in the membranes of chromoplasts. Whether either of these complexes contains other carotenogenic enzymes, or is functional, awaits further studies. Earlier work by Beyer and coworkers with daffodil chromoplasts and pea chloroplasts showed complexes of PDS with chaperonins, thought to have a role in transporting PDS to plastid membranes and subsequent activation (Al-Babili et al., 1996; Bonk et al., 1997).

Apocarotenoids

Apocarotenoids are formed from the oxidative cleavage of carbon–carbon double bonds in carotenoids, catalyzed by CCDs. Definitive proof that both oxygen atoms are incorporated into apocarotenoids has yet to be shown and so it may be more accurate to refer to these enzymes as carotenoid cleavage oxygenases (CCOs; Kloer and Schulz, 2006; Ohmiya, 2009). The resultant apocarotenoid aldehydes or ketones have rigid shapes and amphipathic properties, consistent with their role as signaling molecules, and many are colored due to the remaining conjugated double bonds and hence function as photoprotectants and accessory light-harvesting pigments. These reactions have been referred to as oxidative remodeling (Camara and Bouvier, 2004). All CCDs are nonheme iron (Fe^{2+}) oxygenases that contain four conserved histidines that are thought to coordinate iron binding. They are present in bacteria, plants, and animals (Bouvier et al., 2005; Kloer and Schulz, 2006; Auldridge et al., 2006b), with over 200 putative CCDs being identified from DNA databases (Kloer and Schulz, 2006; Vallabhaneni et al., 2010). Catalytically, they are specific with respect to the carbon–carbon double bonds they cleave, but many have low substrate specificities. Some only accept apocarotenoids as substrates, for example, CCD8 from *Arabidopsis* (Schwartz et al., 2004). Based on their substrate specificities, they are divided into the CCDs that result in the formation of flavor volatiles and a miscellany of other apocarotenoids such as bixin and the NCEDs, which utilize only 9-*cis* epoxycarotenoids (Auldridge et al., 2006a, 2006b). To date, only the ACO CCD from *Synechocystis* has been structurally characterized (Kloer et al., 2005; reviewed by Kloer and Schulz (2006)).

The first of these enzymes to be discovered was called viviparous14 (VP14), following identification from a viviparous ABA-deficient mutant of *Zea mays* (Schwartz et al., 1997; Tan et al., 2001). It cleaves the 11, 12 (11′, 12′) double bonds of the 9-*cis* neoxanthin and violaxanthin to yield xanthoxin (see Section Abscisic acid and Table 4.1). Subsequently, related enzymes in many plant species and other organisms have been discovered. For example, analysis of the genome sequence of *A. thaliana* has identified nine clades of dioxygenases (Tan et al., 2003). Five of these are NCEDs (NCED2, NCED3, NCED5, NCED6, and NCED9), and are closely related to and have the same activity as VP14. The remaining four (CCD1, CCD4, CCD7, and

Table 4.1 Examples of CCD enzymes in higher plants.

Name	Substrate	Product	Functional end product	Double bond cleaved	Origin	Reference
VP14	9-*cis* viola, anthera, neo, zea, luteo	Xanthoxin + C_{25}	ABA	9′10′	*Zea mays*	Schwartz et al., 1997
NCED1	9-*cis* viola, 9-*cis*-neo	Xanthoxin + C_{25}	ABA	11,12	*Phaseolus vulgaris*	Qin and Zeevart, 1999
NCED2	9-*cis* viola, neo	Xanthoxin + C_{25}	ABA	11,12	*Arabidopsis*	Tan et al., 2003
NCED3	9-cis viola, neo	Xanthoxin + C_{25}	ABA	11,12	*Arabidopsis*	Tan et al., 2003
NCED5	9-cis viola, neo	Xanthoxin + C_{25}	ABA	11,12	*Arabidopsis*	Tan et al., 2003
NCED6	9-cis viola, neo	Xanthoxin + C_{25}	ABA	11,12	*Arabidopsis*	Tan et al., 2003
NCED9	9-cis viola, neo	Xanthoxin + C_{25}	ABA	11,12	*Arabidopsis*	Tan et al., 2003
CCD1	β-car, lutein, zea, viola	C_{13} + C_{14} apocarotenoids	ABA	9,10	*Arabidopsis*	Auldridge et al., 2006a, 2006b
CCD1	Lycopene β-car	6 methyl-5-hepten-2-one C_{14} dialdehyde and 2 × C_{13}	β-ionone	5,6 (5′6′) 9,10 (9′10′)	*Rosa damascena*	Huang et al., 2009a
CCD1	β-car	C_{14} dialdehyde + 2 × C_{13} cyclohexones	β-ionone, geranylacetone, pseudoionone	9,10 (9′10′)	*Solanum esculentum*	Simkin et al., 2004
CCD1	lycopene	6-methyl-5-hepten-2-one		5,6	*Arabidopsis, Zea mays, Solanum esculentum*	Vogel et al., 2008
CCD4	β-car		β-ionone	9,10 (9′10′)	*Malus*	Huang et al., 2009b
CCD7	β-car	10-apocaroten-10′-al + β-ionone		9,10 (9′10′)	*Arabidopsis*	Schwartz et al., 2004
CCD8	10-apocaroten-10′-al	13-apo-β-caroten-13-one		13,14	*Arabidopsis*	Schwartz et al., 2004; Auldridge et al., 2006a

ABA, abscisic acid; β-car, β-carotene; luteo, luteoxanthin; neo, neoxanthin; viola, violaxanthin; zea, zeaxanthin.

CCD8) have low sequence substrate specificity also differ from those of the NCEDs. Examples of CCDs, with their substrate specificities and origins, are shown in Table 4.1.

The genes are nuclear encoded, but the enzymes are located in the plastids, with the exception of CCD1, which is cytosolic (Floss and Walter, 2009). VP14 from maize exists in two forms, one which is soluble in stroma and the other bound to thylakoid membranes (Tan et al., 2001). In *Medicago* mycorrhizal hairy roots, it appears that the substrate is a C_{27} apocarotenoid, produced in the plastid that is converted into C_{13} and C_{14} products following export to the cytoplasm (Floss and Walter, 2009).

There are several reviews on CCDs, encompassing biosynthesis, structure, catalytic activities, and function, including those of Giuliano et al. (2003), Auldridge et al. (2006b), Kloer and Schulz (2006), Ohmiya (2009), Floss and Walter, (2009), and Maraco and Schmidt-Dannert (2010). The structure of VP14 from maize has been determined to 3.2 Å resolution and

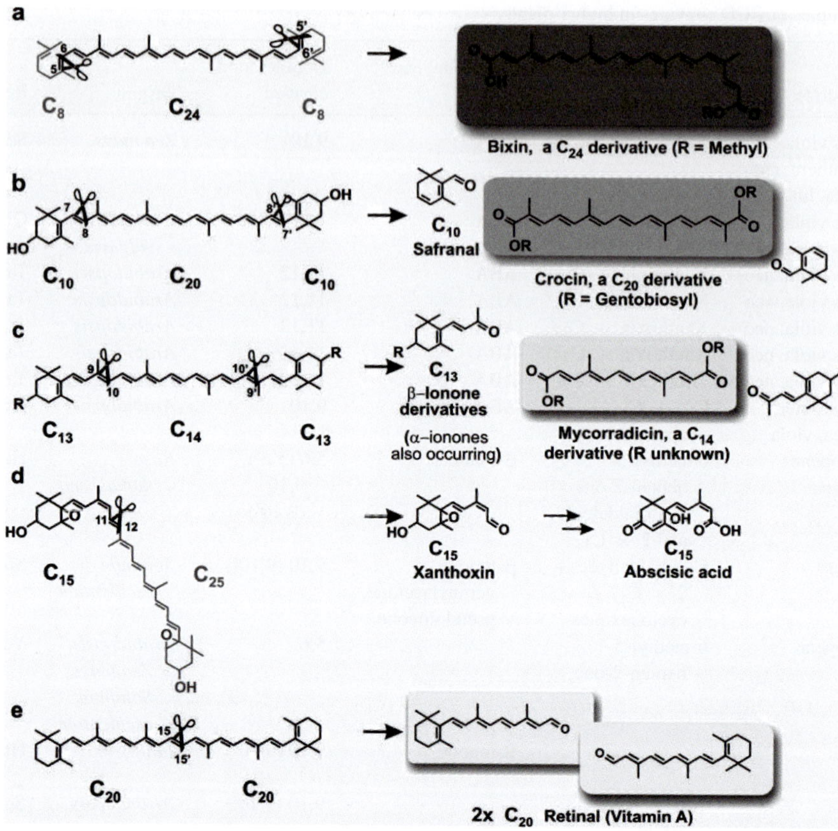

Figure 4.10 The biosynthesis of abscisic acid and related apocarotenoids. Cleavage of carotenoids to apocarotenoids. The double bond cleaved for specific products is indicated in the C_{40} carotenoid precursors. Moieties of unknown metabolic fate are indicated in gray. Taken from Walter et al., 2010 with permission from Springer.

is thought to represent a template for all plant carotenoid cleavage dioxygenases (Messing et al., 2010).

Abscisic Acid CCD7 and 8 act sequentially to cleave the 11, 12 (11′, 12′) double bond of 9-*cis*-violaxanthin or 9-*cis*-neoxanthin to form a C_{27} intermediate which is then cleaved to C_{18} and C_9. The former is exported, catalyzing the first step in ABA biosynthesis from C_{40} carotenoids to form *cis*-xanthoxin and a C_{25} metabolite (Schwartz et al., 2003). Xanthoxin is converted by a short-chain alcohol dehydrogenase (ABA2) into abscisic aldehyde, which is then oxidized into ABA by an abscisic aldehyde oxidase (AAO3), as outlined in Figure 4.10.

Flavor Volatiles The biosynthesis of volatile apocarotenoids, catalyzed by CCD1, and their influence of fruit flavor is described in Chapter 5 of this book.

Strigolactones The pathway of strigolactone biosynthesis has yet to be fully elucidated. It has been assumed that β-carotene is cleaved consecutively within the plastid by two CCOs (CCD7 and CCD8), as demonstrated in *Arabidopsis* (Auldridge et al., 2006b). The C_{18} intermediate (β-apo-13-carotenone) is then thought to be exported into the cytosol for further metabolism to the strigolactones (reviewed by Floss and Walter, 2009). However, this must involve the addition of a butenolide group. The assumption has been that these later steps require additional, unknown, enzymes. Although these reactions have yet to be fully established, it has recently been reported that the combined activities of CCD7 and 8, *in vitro*, lead to a product similar to strigolactones in structure and biological activity. This suggests that CCD8 catalyzes several reactions, including isomerization, a Baeyer–Villiger-like rearrangement, and repeated dioxygenation (Alder et al., 2012).

Regulation of Carotenoid Biosynthesis During Fruit Ripening

The regulation of carotenoid biosynthesis and metabolism to apocarotenoids is highly complex and yet to be fully elucidated in any higher plant species. That this is so is due to a number of compounding factors, which include the following:

- The pathway is lengthy, branched and occurs in different cell compartments, with multiple regulatory steps.
- The enzymes are both soluble and membrane bound, in both the cytosol and plastids, some in complexes, making *in vitro* assays and flux measurements extremely difficult.
- Some of the encoding genes are in families, with different roles and expression patterns in tissues.
- Regulatory mechanisms differ in green and nongreen (ripe fruit) tissues and between types of plastids that are inert-related from a regulatory perspective.
- The dynamics of fruit-ripening link change in carotenoids with other ripening-related events.

It is well established that gene expression at the transcriptional level is the prime regulatory mechanism for carotenoid biosynthesis, with many external factors causing changes to transcription, including plastid development and differentiation, hormone signaling, and environmental stimuli such as light, temperature, and stress. However, post-transcriptional controls are also present in many tissues, including carotenoid sequestration, redox balance, cross talk between isoprenoid pathways for common metabolites and post-translational effects on enzyme complexes. In addition, substrate promiscuity and silent metabolism (Lewinsohn and Gijzen, 2009) probably play unpredictable roles in the steady-state levels of carotenoids, especially in transgenic plants. To some extent these control mechanisms are inter-related, either sequentially or simultaneously. For example, plastid differentiation leads to changes in gene expression. Recent reviews on the topic include those of Cazzonelli et al. (2006), Lu and Li (2008), Fraser et al. (2009), Cordoba et al. (2009), Cazzonelli and Pogson (2010), Rodríguez-Concepción (2010), and Cazzonelli (2011).

Transcriptional Regulation

It is well established that differential carotenoid gene expression causes carotenoid changes in ripe fruit compared to green (Hirschberg, 2001). Changes in the abundance of transcripts are caused by a variety of stimuli. The formation of phytoene is believed to be the key regulatory step of the whole pathway, with *PSY* gene expression being controlled by a wide range of factors such as ABA, light, drought, temperature, and photoperiods. Other carotenogenic genes do not appear to have this degree of control over their transcription. Several of them, including *PSY*, belong to gene families that are expressed in a tissue-specific manner. For example, tomato contains two *Psy* genes. *PSY1* is expressed in ripening fruit, while *PSY2* encodes the enzyme active in photosynthetic tissue and flowers (Ray et al., 1987; Bartley and Scolnik, 1993; Fraser et al., 1999; Giorio et al., 2008). In contrast, a second *PSY* has yet to be found in *Capsicum* (Rodriguez-Uribe et al., 2012).

Little is known about the role or identity of transcription factors in controlling carotenogenic gene expression. The promoter of PSY contains two main regulatory regions: a G-box-containing region located near the TATA box, and a TATA box distal region containing the *cis*-acting element ATCTA, which mediates strong basal promoter activity. This second element is also present in the promoter of PDS. In *Arabidopsis*, the APETALA2 (AP2)/ethylene-responsive element-binding protein binds to the *PSY* promoter (Welsch et al., 2007). The chromatin-modifying histone methyltransferase, SDG8, is required to maintain expression of the *CRTISO* gene in developing seeds, leaves, and the shoot apex of *Arabidopsis* (Cazzonelli et al., 2010). Regulation of the *CRTISO* promoter was dependent on the presence or absence of SDG8. It will be interesting to discover if this also applies to crop plants.

The following factors have been shown to affect transcription of carotenogenic genes, although in some examples multiple factors come into play. In many cases, our understanding has come from studies on green tissues rather than ripening fruits and therefore may not be applicable to the latter. Only genes from PSY onward are discussed in this section; those of the MEP pathway are described below.

Light Studies on a wide range of photosynthetic and nonphotosynthetic organs have shown that light induces carotenoid gene expression, with photoreceptors being involved in the process (Simkin et al., 2003; Woitsch and Romer, 2003; Toledo-Ortiz et al., 2010). It has been demonstrated that the phytochromes (PHYs) and cryptochromes (CRYs) are the photoreceptors involved in light signal transduction leading to induction of carotenoid gene transcription, both requiring transcription factors (TFs). PHYA, but not PHYB, induces *Psy* in *Arabidopsis* (von Lintig et al., 1997; Welsch, et al., 2000, 2008) by promoting the binding of the TF HY5 to far red light-responsive elements (LREs) on the promoter. The involvement of HY5 in tomato carotenoid biosynthesis was established in RNAi *hy5* transgenic lines, which contain significantly lower levels of carotenoids than the wild type (Liu et al., 2004). Evidence for the involvement of other regulators in the control of light-induced carotenoid gene expression includes that of tomato COP1 (a ring finger ubiquitin ligase), a negative regulator (Liu et al., 2004), and *DET1* (de-etiolated-1), which, on silencing, causes overaccumulation of carotenoids in tomato fruit (Davuluri et al., 2004). The tomato mutants *hp-1* and *hp-2* have mutations in the orthologs of *DDB1* and *DET1* of *Arabidopsis*, respectively, and show a characteristic fruit phenotype of elevated carotenoid levels and an excessive response to light (Mustilli et al., 1999). Diurnal variations of *PSY* mRNA and MEP genes have been documented, consistent with the involvement of phytochromes and cryptochromes (Giliberto et al., 2005; Facella et al., 2008;

Cordoba et al., 2009). The involvement of phytochrome-interacting factor 1 (PIF1) and other PIFs in downregulating the accumulation of carotenoids through repressing *PSY* expression in *Arabidopsis* by binding to the promoter has been demonstrated. Light-triggered degradation of PIFs during de-etiolation thus causes rapid de-repression and carotenoid synthesis (Toledo-Ortiz, et al., 2010). Phytochrome is also thought to be involved in the regulation of the expression of various xanthophyll biosynthetic genes in tobacco (Woitsch and Romer, 2003). The transcript level of BCH was affected by different light qualities. The intricacies of the interactions between photoreceptors and transcription factors during light-mediated activation of carotenoid gene transcription are detailed in the review of (Pizarro and Stange (2009). Light intensity changes carotenoids levels. Hirschberg (2001) reports that shifting Arabidopsis or tomato plants from low light to strong light causes a five-fold increase in the ratio of Lcy-b mRNA and Lcy-e mRNA.

Hormones It is known that *PSY* is strongly induced by ethylene during ripening, indicating a major control point for total fruit carotenoid accumulation (Lois et al., 2000). Specifically, ethylene induces *PSY1* gene expression and downregulation of ethylene synthesis results in decreased *PSY1* expression and reduced carotenoids (Picton et al., 1993). Brassinosteroids induce fruit ripening by stimulating ethylene production and hence trigger lycopene formation in tomato pericarp (Vardhini and Rao, 2002).

In rare cases, ripe fruits re-green, for example, citrus. This is accompanied by decreased carotenoid gene expression, mediated by gibberellins (Alos et al., 2006; Rodrigo and Zacarias, 2007).

Chromoplast-specific Genes and Plastid Differentiation As expected, genes involved in carotenoid biosynthesis of fruit-related carotenoids are upregulated during chromoplast formation in many plants, including *PSY1* and *CYC-b* of tomato (Ray et al., 1987; Ronen et al., 2000), and *CYC-b* of citrus fruit (Alquezar et al., 2009), the wild species of tomato, *Solanum habrochaites* (Dalal et al., 2010), kiwifruit (Ampomah-Dwamena et al., 2009), and of papaya fruit (Skelton et al., 2006; Blas et al., 2010). Differential gene expression has been implicated in the accumulation of δ-carotene in fruits of the Delta tomato mutant, which results from increased transcription of LCY-e (Ronen et al., 1999) and in the formation of β-carotene rather than lycopene in the high-β mutant due to the upregulation of the CYC-b gene (Ronen et al., 2000). There appears to be a central role for gene duplication in the development of a chromoplast-specific carotenoid biosynthesis pathway (Galpaz et al., 2006; Giorio et al., 2008).

Reactive Oxygen Species and Nutrient Supply Alterations to the nutrient supply of certain plants result in changes in gene expression and carotenoids. For example, sucrose deficiency in tomato fruit discs caused delayed lycopene accumulation, but had no effect on other carotenoids, and appeared to modulate *Psy1* gene expression (Telef et al., 2006). The precise mechanism of action that causes these effects has yet to be elucidated, but it may be a result of insufficient cellular levels of the MEP precursors, pyruvate and glyceraldehyde 3-phosphate, which are derived from hexoses. The metabolism of starch to sugars in nectary tissue is associated with increases in β-carotene during chromoplast development (Horner et al., 2007).

Plastid-generated reactive oxygen species (ROS) are known to upregulate several carotenogenic genes in ripening *Capsicum* fruit (Bouvier et al., 1998b). ROS can be generated by light and stress (Elstner et al., 1998) and this may account for increases in carotenoids in their role as antioxidants. The mechanism of action is unknown.

Feedback Regulation Metabolite feedback regulation, at the transcriptional level, within the carotenoid pathway and also in the supply of precursors via the MEP pathway, has been shown by several groups. Both the positive and the negative feedback has been reported (Römer et al., 2000; Beyer et al., 2002; Qin et al., 2007). It has been suggested that transcription of *CRTISO* and ε-*LCY* are rate limiting with respect to lutein formation in *Arabidopsis* (Cuttriss et al., 2007), while the β-hydroxylase and ZEP of maize are probably subject to negative transcriptional regulation by a carotenoid or carotenoid metabolite (Bai et al., 2009). In transgenic tomatoes containing the bacterial *crtI* gene, an increase in β-carotene was found, but total carotenoids were reduced (Römer et al., 2000), perhaps by negative feedback regulation. However, other transgenic tomatoes with elevated β-carotene showed no such reduction (Ronen et al., 2000; Rosati et al., 2000). In green tissues, *Pds* expression is thought to be affected by end-product regulation (Corona et al., 1996).

Nontranscriptional Regulation

While regulation of transcription is recognized as the prime mechanism for changes in carotenoids in ripening fruits, there are several post-transcriptional controls that are physiologically important in the overall regulation of steady-state levels in plants.

Substrate Availability, Pathway Flux, and Turnover Since phytoene synthesis is the first committed step in carotenoid biosynthesis, it is reasonable to consider those compounds preceding it as precursors. Therefore, their concentrations and availability in the cell and in subcellular compartments are important when considering the control of the pathway and its flux. The flux through a metabolic pathway is determined by the activities of its enzymes and by their interactions with other enzymes. Although single rate-limiting enzymes exist, control of flux is quite often distributed over several enzymes. In the latter case, modulation of a single enzyme is likely to be an ineffective mechanism for changing a pathway's flux. This appears to be the case in the carotenogenic pathway and includes steps in the MEP pathway that supply isoprenoid precursors. Flux control coefficients in transgenic tomatoes, that define the ratio between the fractional change in flux and the fractional change in enzyme activity have been measured in transgenic and wild-type tomato fruits (Fraser et al., 2002). Determination of flux control coefficients in the wild-type fruit indicated that PSY has a value that is three-fold greater than other enzymes in the pathway, supporting previous evidence that PSY is the most influential step in the pathway. However, in the *crtB* transgenic fruit, the flux coefficients changed, indicating a shift in the regulatory step of carotenogenesis between phytoene synthesis and lycopene formation. A review on how to identify components of regulatory metabolic networks has been published by Stitt et al. (2010).

Studies on tomato fruit downregulated with respect to the *DET1* gene have revealed new insights into cross talk between the primary and the secondary metabolism that affects antioxidant levels (Enfissi et al., 2010). Correlation analyses of transcripts and metabolites have revealed coordinated activation of core metabolic pathways in developing fruit that lead to an increase in the precursor pool that is used in secondary metabolism within ripe fruit.

In addition, analysis of quantitative trait loci associated with tomato fruit carotenoid metabolism indicates that multiple loci, in addition to known structural components, contribute to carotenoid flux (Liu et al., 2003).

Within the MEP pathway (Fig. 4.7), DXS and to a lesser extent DXR and HDR have roles in controlling the flux through the pathway (Estévez et al., 2001; Botella-Pavía et al., 2004; Carretero-Paulet et al., 2006; Flores-Pérez et al., 2008). Most organisms have multiple copies of *DXS* (except *Arabidopsis*) and the expression of "class 1" genes correlates with carotenoid accumulation in ripening fruit (Lois et al., 2000). There may be multiple copies of *DXR* and *HDR* in gymnosperms, but their roles have yet to be elucidated (Kim et al., 2008). Similarly, there are multiple copies of *IPP isomerase* and *GGPP synthase* in plants. For example, there are 10 of the latter genes reported in *Arabidopsis*, of which *GGPPS1* is associated with carotenoid biosynthesis in endosperm (Vallabhaneni and Wurtzel, 2009).

It is well established that the two pathways are coordinately regulated, at least at the transcriptional level (reviewed by Giuliano et al., 2008; Rodríguez-Concepción, 2006, 2010). This is presumably is to ensure a sufficient, but not over-abundant, supply of prenyl phosphates for the carotenoid pathway, so that deleterious phenotypes are prevented, for example, dwarfing and abnormal development (Fray et al., 1995; Fraser et al., 2007). The transcription of the MEP pathway genes either precedes or parallels carotenoid formation in pepper and tomato fruit (Bouvier et al., 1998b; Lois et al., 2000; Botella-Pavía et al., 2004). The *cis* motif ATCTA has been found on the promoters of *DXS*, *HDR*, and *PSY* (Botella-Pavía et al., 2004), but *trans*-acting factors are to be elucidated.

Post-transcriptional Modifications There is a report of phytochrome-regulating PSY enzyme activity during tomato fruit ripening. A comparison of transcript levels with PSY *in vitro* activity in ripening fruit slices under different light regimes showed the reversibility of PSY activity under red and far red light, which was not found for transcript levels, suggesting that phytochrome-mediated effects are translational, or post-translational, modifications of the PSY protein (Schofield and Paliyath, 2005). Post-transcriptional regulation of carotenogenic enzymes has been found in chromoplasts of Narcissus. Both PSY and PDS were detected in inactive forms in the soluble fraction, but in active forms when membrane bound (Al-Babili et al., 1996).

CCS from the fruit of the canary (orange) cultivar of *Capsicum* is transcribed in ripe fruit, but not translated, suggesting some form of post-transcriptional control (Rodriguez-Uribe, et al., 2012).

Intracellular Location and Storage The accumulation of carotenoids within the plastid can be viewed as a metabolic sink for these pigments and so increasing this subcellular compartment could lead to increased levels of carotenoids. Studies on the *hp-1* and *hp-2* mutants of tomato have shown that they contain larger chromoplast compartments, which has been correlated to their increased pigmentation (Cookson et al., 2003; Kolotilin et al., 2007). Transgenic tomato lines with downregulated *DET1* have elevated carotenoid (and other isoprenoid) levels, with a larger plastid volume per cell (Enfissi et al., 2010). The *hp-3* mutant, with a mutation in *ZEP*, and consequently low ABA levels, has a two-fold increase in the plastid compartment compared to the wild type and also increased carotenoid content. Two tomato ABA-deficient mutants, *flacca* and *sitens*, show a similar phenotype (Galpaz et al., 2008). This concept has proved attractive to groups attempting to increase carotenoid levels in plants by genetic engineering of nuclear genes (reviewed by Farré et al., 2010, 2011) and transplastomic transformations (Apel and Bock, 2009). In addition, such an inert pool would reduce the potential for feedback inhibition on the pathway.

Carotenoids are known to accumulate in plastoglobules, in crystalline and microfibrillar structures and internal plastid membranes (Marano et al., 1993), with ripening tomato fruit being one of the best-studied tissues of chloroplast to chromoplast transitions. The type of carotenoid-containing bodies depends on the lipid:protein ratio and the proteins that facilitate assembly of such structures, for example, fibrillin. Early microscopy studies found red fruits with crystals of lycopene in the chromoplasts, leading to the idea of a metabolically inert end product (Harris and Spurr, 1969). It is now known that new membranes are formed in chromoplasts from vesicles of the inner plastid membrane and these are the site for formation of the crystals (Simkin et al., 2007). The chloroplast to chromoplast transition has been widely studied, including the use of confocal fluorescence microscopy to show that chromoplasts are exclusively derived from pre-existing chloroplasts, with lycopene levels increasing three-fold faster than chlorophyll degradation (Egea et al., 2011). Austin et al. (2006) have reported that plastoglobules contain biosynthetic enzymes. Our understanding of how carotenoids are sequestered in chromoplasts has been advanced with the cloning and characterization of the *Or* gene from *Brassica oleracea* (cauliflower). Following pioneering work by Vishnevetsky et al. (1999) it has been shown to induce the formation of chromoplasts and facilitates the accumulation of β-carotene in the cauliflower florets and in transgenic plants (Li et al., 2001; Lu et al., 2006; Li and Van Eck, 2007; Giuliano and Diretto, 2007; Lopez et al., 2008b). In *Capsicum* fruit, the 32 kDa protein fibrillin comprises some 80% of the chromoplast protein and is responsible for the sequestration of carotenoids in the center of fibrils (Deruère et al., 1994). Further details of chromoplast differentiation can be found in the review of Egea et al. (2010).

A schematic showing the various regulatory steps in the carotenoid pathway is shown in Figure 4.11.

Future Perspectives

Advances in our understanding of the chlorophyll degradation and carotenoid biosynthetic pathways have advanced significantly over the past 5 years. However, much of this has been derived from photosynthetic tissues and model plants. It cannot be assumed that these pathways and indeed regulation of the pathways are the same in fruits and so efforts now need to be directed to fruits of crop plants. The increasing maturity of "omic" technologies and systems biology makes such strategies experimentally feasible, as does the utilization of transgenic plants and crop plant QTLs to discover novel genes and regulatory cascades. However, these

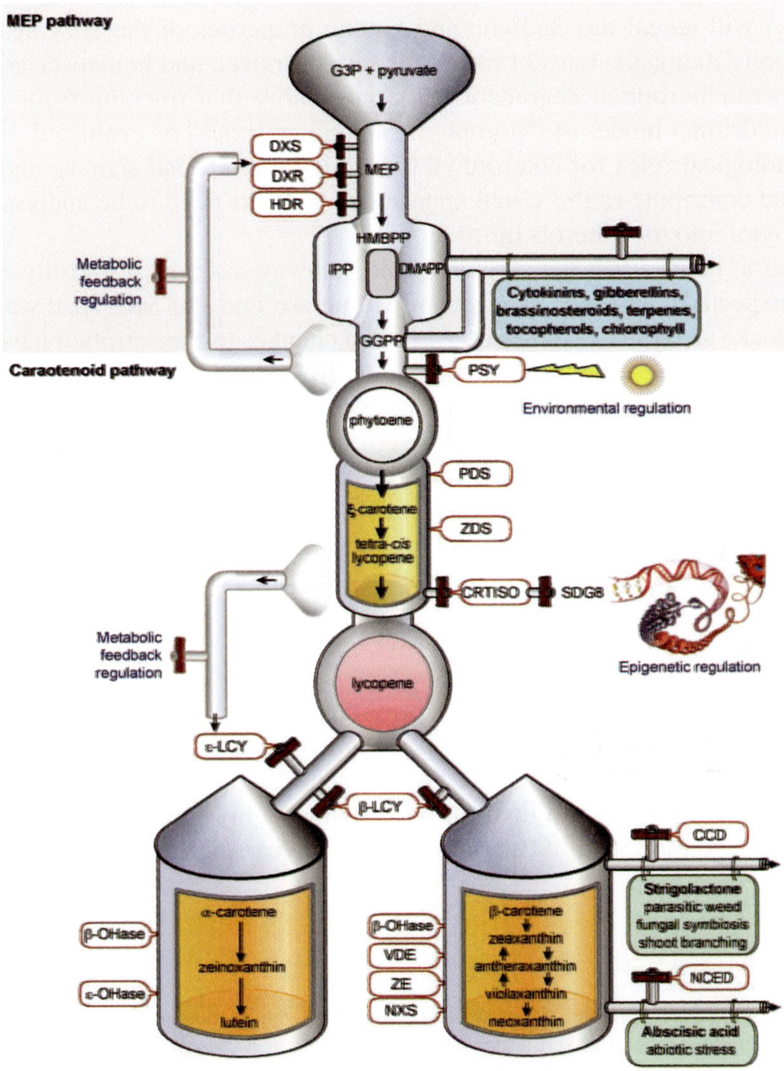

Figure 4.11 Regulation of carotenoid biosynthesis in plants. Major reactions in the higher plant carotenoid biosynthetic pathway showing enzymes, carotenoids and their precursors (pipes), carotenoid sinks (barrels), carotenoid-derived signaling hormones (green signs), and other MEP-isoprenoid-derived metabolites (blue sign). The windows displayed within the chrome pipes indicate abundant carotenoid pigments found in photosynthetic tissues and also represent key nodes for regulation in the pathway. Carotenoid biosynthesis is modulated by environmental factors (light), chromatin modification, and metabolic feedback regulation. The side funnels represent examples of metabolic feedback control mechanisms acting upon biosynthetic gene expression as a result of altered PSY and CRTISO enzymatic activity, respectively. First, the bottleneck in phytoene biosynthesis is regulated by PSY and its overexpression increased *DXS* and *DXR* mRNA levels post-transcriptionally in etiolated tissues. Second, loss-of-function CRTISO mutants show reduced *eLCY* transcript levels in etiolated tissues. βLCY, β-cyclase; βOHase, β-hydroxylase; CCD, carotenoid cleavage dioxygenase; CRTISO, carotenoid isomerase; DXR, 1-Deoxy-D-xylulose 5-phosphate reductoisomerase; DXS, 1-deoxyxylulose-5-phosphate synthase; εLCY, ε-cyclase; εOHase, ε-hydroxylase; GGPP, geranylgeranyl diphosphate; HDR, 1-hydroxy-2-methyl-2-(E)-butenyl 4-diphosphate reductase; NCED, 9-*cis*-epoxycarotenoid dioxygenase; NXS, neoxanthin synthase; PDS, phytoene desaturase; PSY, phytoene synthase; SDG8, histone methyltransferase; VDE, violaxanthin de-epoxidase; ZDS, ζ-carotene desaturase; ZE, zeaxanthin epoxidase. Reprinted from Cazzonelli et al., 2010. Copyright 2010, with permission from Elsevier.

newer protocols should be used in combination with biochemical and analytical chemistry approaches that will reveal the catalytic and kinetic properties of the enzymes and protein–ligand interactions, alongside rates of metabolic flux, turnover, and homeostasis.

With respect to chlorophyll degradation, the possibility that ripening fruits and senescing leaves represent distinct modes of chlorophyll catabolism should be evaluated. In addition, the potential physiological roles for chlorophyll catabolites, as internal signals, antioxidants, and as pigments that contribute to the visual appearance of fruits need to be analysed, as does the recycling of phytol into tocopherols in fruit.

It is tempting to believe that the carotenogenic pathway has now been fully elucidated, but the recent, unexpected discovery of a second isomerase and the MEP pathway a few years earlier should make us more circumspect. Although changes to transcription have been studied thoroughly and show these to be important in regulating carotenoid formation and accumulation in green and ripening fruit tissues, other mechanisms of regulation demand thorough study. In particular, we lack knowledge of which transcription factors and their binding proteins are involved in the regulation of carotenoid gene expression, the mechanism of protein targeting within the plastid, the nature of metabolic feedback, which signaling molecules are utilized during fruit ripening and how plants control homeostasis and turnover of these molecules.

Acknowledgments

I would like to thank Stefan Hortensteiner for providing Figures 4.3 and 4.4, Manuel Rodríguez-Concepción for Figure 4.7, Michael Walter for Figure 4.10, and Barry Pogson and Christopher Cazzonelli for Figure 4.11. I am grateful to Paul Fraser, Royal Holloway, for his critical reading of the manuscript. I would like to dedicate this review to my PhD supervisor, Brian Davies, who introduced me to the world of carotenoids over 40 years ago.

Bibliography

Akhtar, M.S., Goldschmidt, E.E., John, I., Rodoni, S., Matile, P., and Grierson, D. (1999) Altered patterns of senescence and ripening in *gf*, a stay-green mutant of tomato (*Lycopersicon esculentum* Mill.). *Journal of Experimental Botany*, 50, 1115–1122.

Akiyama, K., Matsuzaki, K., and Hayashi, H. (2005) Plant sesquiterpenes induce hyphal branching in the development of arbuscular mycorrhizal symbiosis. *Bioscience Biotechnology Biochemistry*, 71, 1405–1414.

Al-Babili, S., Hugueney, P., Schledz, M., Welsch, R., Frohnmeyer, H., Laule, O., and Beyer, P. (2000) Identification of a novel gene coding for neoxanthin synthase from *Solanum tuberosum*. *FEBS Letters*, 485, 168–172.

Al-Babili, S., Von Lintig, J., Haubruck, H., and Beyer, P. (1996) A novel, soluble form of phytoene desaturase from *Narcissus pseudonarcissus* chromoplasts is Hsp70-complexed and competent for flavinylation, membrane association and enzymatic activation. *Plant Journal*, 9, 601–612.

Alder, A., Jamil, M., Marzorati, M., Bruno, M., Vermathen, M., Bigler, P., Ghisla, S., Bouwmeester, H., Beyer, P., and Al-Babili, S. (2012) The path from β–carotene to carlactone, a strigolactone-like plant hormone. *Science*, 335, 1348–1351.

Alos, E., Cercos, M., Rodrigo, M.J., Zacarias, L., and Talon, M. (2006) Regulation of color break in citrus fruit. Changes in pigment profiling and gene expression induced by gibberellins and nitrate, two ripening retardants. *Journal of Agricultural and Food Chemistry*, 54, 4888–4895.

Alquezar, B., Zacarias, L., and Rodrigo, M.J. (2009) Molecular and functional characterization of a novel chromoplast-specific lycopene β-cyclase from Citrus and its relation to lycopene accumulation. *Journal of Experimental Botany*, 60, 1783–1797.

Ampomah-Dwamena, C., McGhie, T., Wibisono, R., Montefiori, M., Hellens, R.P., and Allan, A.C. (2009) The kiwifruit lycopene beta-cyclase plays a significant role in carotenoid accumulation in fruit. *Journal of Experimental Botany*, 60, 3765–3779.

Apel, W., and Bock, R. (2009) Enhancement of carotenoid biosynthesis in transplastomic tomatoes by induced lycopene to pro-vitamin A conversion. *Plant Physiology*, 151, 59–66.

Auldridge M.E., Block, A., Vogel, J.T., Dabney-Smith, C., Mila, I., Bouzayen, M., Magallanes-Lundback, M., DellaPenna, D., McCarty, D.R., and Klee, H.J. (2006a) Characterization of three members of the Arabidopsis carotenoid cleavage dioxygenase family demonstrates the divergent roles of this multifunctional enzyme family. *Plant Journal*, 45, 982–993.

Auldridge M.E, McCarty, D.R., and Klee, H.J. (2006b) Plant carotenoid cleavage oxygenases and their apocarotenoid products. *Current Opinions in Plant Biology*, 9, 315–321.

Austin II, J.R., Frost, E., Vidi, P-E., Kessler, F., and Staehelin, L.A. (2006) Plastoglobules are lipoprotein subcompartments of the chloroplast that are permanently coupled to thylakoid membranes and contain biosynthetic enzymes. *The Plant Cell*, 18, 1693–1703.

Azoulay Shemer, T., Harpaz-Saad, S., Belausov, E., Lovat, N., Krokhin, O., Spicer, V., Standing, K.G., Goldschmidt, E.E., and Eyal, Y. (2008) *Citrus* chlorophyllase dynamics at ethylene-induced fruit color-break; a study of chlorophyllase expression, post-translational processing kinetics and *in-situ* intracellular localization. *Plant Physiology*, 148, 108–118.

Bai, L., Kim, E.H., DellaPenna, D., and Brutnell, T.P. (2009) Novel lycopene epsilon cyclase activities in maize revealed through perturbation of carotenoid biosynthesis. *The Plant Journal*, 59, 588–599.

Barry, C.S., and Pandy, P. (2009) A survey of cultivated heirloom tomato varieties identifies four new mutant alleles at the *green flesh* locus. *Molecular Breeding*, 24, 269–276.

Bartley, G.E., and Scolnik, P.A. (1993) cDNA cloning, expression during development, and genome mapping of *PSY2*, a second tomato gene encoding phytoene synthase. *Journal of Biological Chemistry*, 268, 25718–25721.

Bartley, G.E., and Scolnik, P.A. (1995) Plant carotenoids: pigments for photoprotection, visual attraction and human health. *The Plant Cell*, 7, 1027–1038.

Bertrand, E., Sakai, R., Rozhkova-Novosad, E., Moe, L., Fox, B.G., Groves, J.T., and Austin, R.N. (2005) Reaction mechanisms of non-heme diiron hydroxylases characterized in whole cells. *Journal of Inorganic Biochemistry*, 99, 1998–2006.

Beyer, P., Al-Babili, S., Ye, X., Lucca, P., Schaub, P., Welsch, R., and Potrykus, I. (2002) Golden rice: introducing the beta-carotene biosynthesis pathway into rice endosperm by genetic engineering to defeat vitaminA deficiency. *Journal of Nutrition*, 132, 506S–510S.

Biswell, B. (1995) Carotenoid catabolism during leaf senescence and its control by light. *Journal of Photochemistry and Photobiology*, 30, 3–14.

Blankenship, R.E. (2002) *Molecular Mechanism of Photosynthesis*, Blackwell Science, Oxford.

Blas, A.L., Ming, R., Liu, Z, Veatch, O.J., Paull, R.R., Moore, P.H., and Yu, Q. (2010) Cloning of the papaya chromoplast-specific lycopene β–cyclase, *CpCYC-b*, controlling fruit flesh color reveals conserved microsynteny and recombination hot spots. *Plant Physiology*, 152, 2013–2022.

Bonk, M., Hoffmann B., Von Lintig, L.J., Schledz, M., Al-Babili, S., Hobeika, E., Kleinig, H., and Beyer, P. (1997) Chloroplast import of four carotenoid biosynthetic enzymes in vitro reveals differential fates prior to membrane binding and oligomeric assembly. *European Journal of Biochemistry*, 247, 942–950.

Borovsky, Y., and Paran, I. (2008) Chlorophyll breakdown during pepper fruit ripening in the chlorophyll retainer mutation is impaired at the homolog of the senescence-inducible stay-green gene. *Theoretical Applied Genetics*, 117, 235–240.

Botella-Pavía, P., Besumbes, O., Phillips, M.A., Carretero-Paulet, L., Boronat, A., and Rodríguez-Concepción, M. (2004) Regulation of carotenoid biosynthesis in plants: evidence for a key role of hydroxymethylbutenyl diphosphate reductase in controlling the supply of plastidial isoprenoid precursors. *Plant Journal*, 40, 188–199.

Botella-Pavia, P., and Rodriguez-Concepcion, M. (2006) Carotenoid biotechnology in plants for nutritionally improved foods. *Physiologia Plantarum*, 126, 369–381.

Bouvier, F., Backhaus, R.A., and Camara, B. (1998a) Induction and control of chromoplast-specific carotenoid genes by oxidative stress. *Journal of Biological Chemistry*, 273, 30651–30659.

Bouvier, F., D'Harlingue, A., Suire, C., Backhaus, R.A., and Camara, B. (1998b) Dedicated roles of plastid transketolases during the early onset of isoprenoid biogenesis in pepper fruits. *Plant Physiology*, 117, 1423–1431.

Bouvier, F., D'Harlingue, A, Backhaus, RA, Kumaga, M.H., and Camara, B. (2000) Identification of neoxanthin synthase as a carotenoid cyclase paralog. *European Journal of Biochemistry*, 267, 6346–6352.

Bouvier, F., Hugueney, P., d'Harlingue, A., Kuntz, M., and Camara, B. (1994) Xanthophyll biosynthesis in chromoplasts: isolation and molecular cloning of an enzyme catalyzing the conversion of 5,6-epoxycarotenoid into ketocarotenoid. *The Plant Journal*, 6, 45–54.

Bouvier, F., Mialoundama, A.S., and Camara, B. (2009) A sentinel role for plastids. In: *The Chloroplast. Interactions with the Environment* (eds. A.S. Sandelius, and H. Aronson), pp. 267–292. Springer-Verlag, Berlin.

Bouvier, F., Rahier, A., and Camara, B. (2005) Biogenesis, molecular regulation and function of plant isoprenoids. *Progress in Lipid Research*, 44, 357–429.

Bramley, P.M. (1985) The *in vitro* biosynthesis of carotenoids. *Advances in Lipid Research*, 21, 243–279.

Bramley, P.M., Bird, C.R., and Schuch, W. (1993) Carotenoid biosynthesis and manipulation. In: *Biosynthesis and Manipulation of Plant Products* (ed. D. Grierson), Chapter 4, pp. 139–177. Blackie, Glasgow.

Breitenbach, J., and Sandmann, G. (2005) ζ-Carotene cis isomers as products and substrates in the plant poly-cis carotenoid biosynthetic pathway to lycopene.*Planta*, 220, 785–793.

Britton, G. (1985) General carotenoid methods. *Methods in Enzymology*, 111, 113–149.

Britton, G. (1990) Carotenoid biosynthesis- an overview. In: *Carotenoids: Chemistry and Biology* (eds N.R. Krinsky, M.M. Mathews-Roth, and R.F. Taylor), pp. 167–184. Plenum Press, New York.

Britton, G. (1991) Carotenoids. In: *Methods in Plant Biochemistry*, Vol. 7 (eds. B.V. Charlwood, and D.V. Banthorpe), pp. 473–518. Academic Press, London.

Britton, G., Liaaen-Jensen, S., and Pfander, H. (2004) *Carotenoids Handbook*. Birkhauser, Basel.

Britton, G., Liaanen-Jensen, S., and Pfander, H. (2008) *Carotenoids*, Vol. 4: *Natural Functio*ns. Birkhauser, Basel.

Brown, S.B., Houghton, J.D., and Hendry, G.A.F. (1991) *Chlorophylls*. (H. Scheer), pp. 465–489. CRC Press, Boca Raton, USA.

Camara, B. (1993) Plant phytoene synthase complex – component enzymes, immunology and biogenesis. *Methods in Enzymology*, 214, 352–365.

Camara, B., and Bouvier, F. (2004) Oxidative remodeling of plastid carotenoids. *Archives of Biochemistry and Biophysics*, 430, 16–21.

Camara, B., and Moneger, R. (1978) Free and esterified carotenoids in green and red fruits of *Capsicum annuum*. *Phytochemistry*, 17, 91–93.

Carretero-Paulet, L., Cairo, A., Botella-Pavia, P., Besumbes, O., Campos, N., Boronat, A., and Rodríguez-Concepción, M. (2006) Enhanced flux through the methylerythritol 4-phosphate pathway in Arabidopsis plants overexpressing deoxyxylulose 5-phosphate reductoisomerase. *Plant Molecular Biology*, 62, 683–695.

Cazzonelli, C. I. (2011) Carotenoids in nature: insights from plants and beyond. *Functional Plant Biology*, 38, 833–847.

Cazzonelli, C.I., Nisar, N., Hussain, D., Carmody, M.E., and Pogson, B.J. (2006) Biosynthesis and regulation of carotenoids in plants—micronutrients, vitamins and health benefits. In: *Plant Developmental Biology-Biotechnology Perspectives* (eds. E.C. Pua, and M.R. Davey), Vol. 2, pp. 117–137. Springer-Verlag, Berlin.

Cazzonelli, C.I., and Pogson, B.J. (2010) Source to sink: regulation of carotenoid biosynthesis in plants. *Trends in Plant Science*, 15, 266–274.

Cazzonelli, C.I., Roberts, A.C., Carmody, M.E., and Pogson, B.J. (2010) Transcriptional control of *SET DOMAIN GROUP 8* and *CAROTENOID ISOMERASE* during *Arabidopsis* development. *Molecular Plant*, 3, 174–191.

Chappell, J. (1995) Biochemistry and molecular biology of the isoprenoid pathway in plants. *Annual Review of Plant Molecular Biology*, 46, 521–547.

Chasse, G.A., Mak, M.L., Deretey, E., Farkas, I., Torday, L.L., Papp, J.G., Sarma, D.S.R., Agarwal, A., Chakravarthi, S., Agarwal, S., and Rao, A.V. (2001) An *ab initio* computational study on selected lycopene isomers. *Journal of Molecular Structure*, 571, 27–37.

Chen, C.Y., Zou, J.H., Zhang, S.Y., Zaitlin, D., and Zhu, L.H. (2009) Strigolactones are a new-defined class of plant hormones which inhibit shoot branching and mediate the interaction of plant-AM fungi and plant-parasitic weeds. *Science in China Series C*, 52, 693–700.

Chen, M., Schliep, M., Willows, R.D., Cai, Z-L., Neilan, B.A., and Scheer, H. (2010a) A red-shifted chlorophyll. *Science*, 329, 1318–1319.

Chen, Y., Li, F. and Wurtzel, E.T. (2010b) Isolation and characterization of the *Z-ISO* gene encoding a missing component of carotenoid biosynthesis in plants. *Plant Physiology*, 153, 66–79.

Cheniclet, C.F., Rafia, F., Saint-Guily, A., Verna, A., and Carde, J.P. (1992) localization of the enzyme geranylgeranyl pyrophosphate synthase in Capsicum fruits by immunogold cytochemistry after conventional chemical fixation or quick freezing followed by freeze-substitution. Labelling evolution during fruit ripening. *Biology of the Cell*, 75, 145–154.

Cheung, A.Y., McNellis, T., and Piekos, B. (1993) Maintenance of chloroplast components during chromoplast differentiation in the tomato mutant *green flesh*. *Plant Physiology*, 101, 1223–1229.

Clough, J.M., and Pattenden, G. (1979) Naturally occurring poly-*cis* carotenoids. Stereochemistry of poly-*cis* lycopene and in congeners in 'tangerine' tomato fruits. *Journal of the Chemical Society Chemical Communications*, 14, 616–619.

Clough, J.M., and Pattenden, G. (1983) Stereochemical assignment of prolycopene and other poly-*Z*-isomeric carotenoids in fruits of the tangerine tomato *Lycopersicon esculentum* var. 'Tangella'. *Journal of the Chemical Society Perkin Transactions*, 1, 3011–3018.

Cook, C.E., Whichard, L.P., Turner, B., Wall, M.E., and Egley, G.H. (1966) Germination of witchweed (*Striga lutea* Lour.): isolation and properties of a potent stimulant. *Science*, 154, 1189–1190.

Cookson, P.J., Kiano, J.W., Shipton, C.A., Fraser, P.D., Romer, S., Schuch, W., Bramley, P.M., and Pyke, K.A. (2003) Increases in cell elongation, plastid compartment size and phytoene synthase activity underlie the phenotype of the high pigment-1 mutant of tomato. *Planta*, 217, 896–903.

Cordoba, E., Salmi, M, and Leon, P. (2009) Unravelling the regulatory mechanisms that modulate the MEP pathway in higher plants. *Journal of Experimental Botany*, 60, 2933–2943.

Corona, V., Aracri, B., Kosturkova, G., Bartley, G.E., Pitto, L., Giorgetti, L., Scolnik, P.A., and Giuliano, G. (1996) Regulation of a carotenoid biosynthesis gene promoter during plant development. *The Plant Journal*, 9, 505–512.

Csukasi, F., Merchante, C., and Valpuesta, V. (2009) Modification of plant hormone levels and signaling as a tool in plant biotechnology. *Biotechnology Journal*, 4, 1293–1304.

Cunningham Jr., F.X. (2002) Regulation of carotenoid synthesis and accumulation in plants. *Pure and Applied Chemistry*, 74, 1409–1417.

Cunningham Jr., F.X., and Gantt, E. (1998) Genes and enzymes of carotenoid biosynthesis in plants. *Annual Review of Plant Physiology and Plant Molecular Biology*, 49, 557–583.

Cutler, S.R., Rodriguez, P.L., Finkelstein, R.R., and Abrams, S.R. (2010) Abscisic acid: emergence of a core signaling network. *Annual Reviews of Plant Biology*, 61, 651–679.

Cuttriss, A.J., Chubb, A.C., Alawady, A., Grimm, B., and Pogson, B.J. (2007) Regulation of lutein biosynthesis and prolamellar body formation in *Arabidopsis*. *Functional Plant Biology*, 34, 663–672.

Dalal, M., Chinnusamy, V., and Bansal, K.C. (2010) Isolation and functional characterization of *lycopene β-cyclase* (*CYC-B*) promoter from *Solanum habrochaites*. *BMC Plant Biology*, 10, 61–76.

Davies, B.H. (1976) Carotenoids. In: *Chemistry and Biochemistry of Plant Pigments*, Vol. 2, 2nd ed. (ed. T.W. Goodwin), pp. 38–165. Academic Press, London.

Davies, B.H., Matthews, S., and Kirk, J.T.O. (1970) The nature and biosynthesis of the carotenoids of different colour varieties of *Capsicum annuum*. *Phytochemistry*, 9, 797–805.

Davuluri, G.R., Van Tuinen, R., Mustilli, A.C., Manfredonia, A., Newman, R., Burgess, D., Brummell, D.A., King, S.K., Plays, J., Uhlig, J., Pennings, H.M., and Bowler, C. (2004) Manipulation of DET1 expression in tomato results in photomorphogenic phenotypes caused by post-transcriptional gene silencing. *Plant Journal*, 40, 344–354.

Demmig-Adams, B., and Adams III, W.W. (1996) The role of xanthophyll cycle carotenoids in the protection of photosynthesis. *Trends in Plant Science*, 1, 21–26.

Demmig-Adams, B., Gillmore, A.M., and Adams III, W.W. (1996) In vivo functions of carotenoids in higher plants. *FASEB Journal*, 10, 403–412.

Deruère, J., Römer, S., D'Harlingue, A., Backhaus, R.A., Kuntz, M., and Camara, B. (1994) Fibril assembly and carotenoid overaccumulation in chromoplasts: a model for supramolecular lipoprotein structures. *Plant Cell*, 6, 119–133.

Devini, M., and Steinmuller, D. (1985) Composition and function of plastoglobule. *Planta*, 163, 91–96.

Dogbo, O., Bardat, F., Laferriere, A., Quennemet, J., Brangeon, J., and Camara, B. (1987) Metabolism of plastid terpenoids. I. Biosynthesis in plastid stroma isolated from higher plants. *Plant Science*, 49, 89–101.

Dun, E.A., Brewer, P.B., and Beveridge, C.A. (2009) Strigolactones: discovery of the elusive shoot branching hormone. *Trends in Plant Science*, 14, 364–372.

Eisenreich, W., Rohdich, F., and Bacher, A. (2001) Deoxyxylulose phosphate pathway to terpenoids. *Trends in Plant Science*, 6, 78–84.

Eisenreich, W., Schwarz, M., Cartayrade, A., Arigoni, D., Zenk, M.H., and Bacher, A. (1998) The deoxyxylulose phosphate pathway of terpenoid biosynthesis in plants and microorganisms. *Chemical Biology*, 5, R221–R233.

Egea, I., Barsan, C., Bian, W., Purgatto, E., Latche, A., Chervin, C., Bouzayen, M., and Pech, J.-C. (2010) Chromoplast differentiation: current status and perspectives. *Plant Cell Physiology*, 51, 1601–1611.

Egea, I., Bian, W., Barsan, C., Jauneau, A., Pech, J.-C., Latche, A., Li, Z., and Chervin, C. (2011) Chloroplast to chromoplast transition in tomato fruit: spectral confocal microscopy analyses of carotenoids and chlorophylls in isolated plastids and time-lapse recording on intact live tissue. *Annals of Botany*, 108, 291–297.

Elstner, E.F., Wagner, G.A., and Schultz, W. (1998) Activated oxygen in green plants in relation to stress situation. *Current Topics in Plant Biochemistry and Physiology*, 7, 159–187.

Enfissi, E.M.A., Barneche, F., Ahmed, I., Lichtle, C., Gerrish, C., McQuinn, R.P., Giovannoni, J.J., Lopez-Juez, E., Bowler, C., Bramley, P.M., and Fraser, P.D. (2010) Integrative transcript and metabolite analysis of nutritionally enhanced DE-ETIOLATED1 downregulated tomato fruit. *Plant Cell*, 22, 1190–1215.

Estévez, J.M, Cantero, A., Reindl, A., Reichler, S., and León, P. (2001) 1-Deoxy-D-xylulose-5-phopsahte synthase, a limiting enzyme for plastidic isoprenoid biosynthesis in plants. *Journal of Biological Chemistry*, 276, 22901–22909.

Facella, P., Lopez, L., Carbone, F., Galbraith, D.W., Giuliano, G., and Perrotta, G. (2008) Diurnal and circadian rhythms in the tomato transcriptome and their modulation by cryptochrome photoreceptors. *PLoS ONE*, 3, e2798–2813.

Farré, G., Bai, C., Twyman, R.M., Capell, T., Christou, P., and Zhu, C. (2011) Nutritious crops producing multiple carotenoids - a metabolic balancing act. *Plant Science*, 179, 28–48.

Farré, G., Sanahuja, G., Naqvi, S., Bai, C., Capell, T., Zhu, C., and Christou, P. (2010) Travel advice on the road to carotenoids in plants. *Plant Science*, 179, 28–48.

Ferruzzi, M.G., and Blakeslee, J. (2007) Digestion, absorption, and cancer preventative activity of dietary chlorophyll derivatives. *Nutrition Research*, 27, 1–12.

Flores-Pérez, U., Pérez-Gil, J., Rodríguez-Villalón, A., Gil, M.J., Vera, P., and Rodríguez-Concepción, M. (2008) Contribution of hydroxymethylbutenyl diphosphate synthase to carotenoid biosynthesis in bacteria and plants. *Biochemical and Biophysical Research Communications*, 371, 510–514.

Floss, D.S., and Walter, M.H. (2009) Role of carotenoid cleavage dioxygenase 1 (CCD1) in apocarotenoid biogenesis revisited. *Plant Signaling and Behavior*, 4, 172–175.

Fraser, P.D., and Bramley, P.M. (2004) The biosynthesis and nutritional uses of carotenoids. *Progress in Lipid Research*, 43, 228–265.

Fraser, P.D., and Bramley, P.M. (2009) Genetic manipulation of carotenoid content and composition in crop plants. In: *Carotenoids* (eds. G. Britton, S. Liaanen-Jensen, and H. Pfander), Vol. 5: *Nutrition and Health*, pp. 99–114. Springer-Verlag, Berlin.

Fraser, P.D., Enfissi, E.M.A., and Bramley, P.M. (2009) Genetic engineering of carotenoid formation in tomato fruit and the potential application of systems and synthetic biology approaches. *Archives of Biochemistry and Biophysics*, 483, 196–204.

Fraser, P.D., Enfissi, E.M.A., Halket, J.M., Truesdale, M.R., Yu, D., Gerrish, C., and Bramley, P.M. (2007) Manipulation of phytoene levels in tomato fruit: effects on isoprenoids, plastids, and intermediary metabolism. *Plant Cell*, 19, 3194–3211.

Fraser, P.D., Kiano, J.W., Truesdale, M.R., Schuch W., and Bramley P.M. (1999). Phytoene synthase-2 enzyme activity in tomato does not contribute to carotenoid synthesis in ripening fruit. *Plant Molecular Biology*, 40, 687–698.

Fraser, P.D., Romer, S, Shipton, C.A., Mills, P.B., Kiano, J.W., Misawa, N., Drake, R.G., Schuch, W., and Bramley, P.M. (2002) Evaluation of transgenic tomato plants expressing an additional phytoene synthase in a fruit-specific manner. *Proceedings of the National Academy of Sciences*, 99, 1092–1097.

Fraser, P.D., Schuch, W., and Bramley, P.M. (2000) Phytoene synthase from tomato (*Lycopersicon esculentum*) chloroplasts - partial purification and biochemical properties. *Planta*, 211, 361–369.

Fraser, P.D., Truesdale, M.R., Bird, C.R., Schuch, W., and Bramley, P.M. (1994) Carotenoid biosynthesis during tomato fruit development. *Plant Physiology*, 105, 405–413.

Fray, R.G., Wallace, A., Fraser, P.D., Valero, D., Hedden, P., Bramley, P.M., and Grierson, D. (1995) Constitutive expression of a fruit phytoene synthase gene in transgenic tomatoes causes dwarfism by redirecting metabolites from the gibberellin pathway. *The Plant Journal*, 8, 693–701.

Galpaz, N., Ronen, G., Khalfa, Z., Zamir, D., and Hirschberg, J. (2006) A chromoplast-specific carotenoid biosynthesis pathway is revealed by cloning of the tomato *white-flower* locus. *Plant Cell*, 18, 1947–1960.

Galpaz, N., Wang, Q., Menda, N., Zamir, D., and Hirschberg, J. (2008) Abscisic acid deficiency in the tomato mutant high-pigment 3 leading to increased plastid number and higher fruit lycopene content. *Plant Journal*, 53, 717–730.

Ganjewala, D., Kumar, S., and Luthra, R. (2009) An account of cloned genes of methyl erythritol -4- phosphate pathway of isoprenoid biosyntheses in plants. In: *Plant Genomics* (ed. H.A. El-Shemy), pp. 35–45. Horizon Scientific Press, Norwich, U.K.

Giliberto, L., Perrotta, G., Pallara, P., Weller, J.L., Fraser, P.D., Bramley, P.M., Fiore, A, Tavazza, M., and Giuliano, G. (2005) Manipulation of the blue light photoreceptor cryptochrome 2 in tomato affects vegetative development, flowering time and fruit.

Giorio, G., Stigliani, A.L., and D'Ambrosio, C. (2008) Phytoene synthase genes in tomato (*Solanum lycopersicum* L.) – new data on the structures, the deduced amino acid sequences and the expression patterns. *FEBS Journal*, 275, 527–535.

Giovannoni, J. (2001) Molecular biology of fruit maturation and ripening. *Annual Review of Plant Physiology and Plant Molecular Biology*, 52, 725–749.

Giovannoni, J. (2004) Genetic regulation of fruit development and ripening. *Plant Cell*, 16, 170–180.

Giuliano, G., and Diretto, G. (2007) Of chromoplasts and chaperones. *Trends in Plant Science*, 12, 529–531.

Giuliano, G., Al-Babili, S., and von Lintig, J. (2003) Carotenoid oxygenases: cleave it or leave it. *Trends in Plant Science*, 8, 145–149.

Giuliano, G., Pollock, D., and Scolnik, P.A. (1986) The gene *crtI* mediates the conversion of phytoene into colored carotenoids in *Rhodopseudomonas capsulata*. *Journal of Biological Chemistry*, 261, 12925–12929.

Giuliano, G., Tavazza, R., Diretto, G., Beyer, P., and Taylor, M.A. (2008) Metabolic engineering of carotenoid biosynthesis in plants. *Trends in Plant Science*, 26, 139–145.

Goffman, F.D., Velasco, L., and Becker, H.C. (1999) Tocopherols accumulation in developing seeds and pods of rapeseed (*Brassica napus* L.). *Lipid/Fett*, 101, 400–403.

Gomez-Roldan, V., Fermas, S., Brewer, P.B., Puech-Pages, V., Dun, E.A., Pillot, J.-P., Letisse, F., Matusova, R., Danoun, S., Portais, J.-C., Bouwmeester, H., Becard, G., Beveridge, C.A., Rameau C., and Rochange, S.F. (2008) Strigolactone inhibition of shoot branching. *Nature*, 455, 189–194.

Gong, Y.H., Zhang, J., Gao, J.-F., Lu, J.-Y., and Wang, J.-R. (2005) Slow export of photoassimilate from stay-green leaves during late grain-filling stage in hybrid winter wheat (*Triticum aestivum* L.). *Journal of Agronomy and Crop Science*, 191, 292–299.

Goodwin, T.W. (1980) *The Biochemistry of Carotenoids*, Vol. 1: *Plants,* 2nd ed. pp. 305–368. Chapman and Hall, London.

Goodwin, T.W., and Goad, L.J. (1970) Carotenoids and triterpenoids. In: *The Biochemistry of Fruits and Their Products*, Vol. 1 (ed. A.C. Hulme). Academic Press, London and New York.

Goss, J. (1991) *Pigments in Vegetables. Chlorophylls and Carotenoids*. Van Nostrand Reinhold, New York.

Harborne, J.B., and Williams, C.A. (2000) Advances in flavonoid research since 1992. *Phytochemistry*, 55, 481–504.

Harjes, C.E., Torbert, R., Rocheford, T.R., Bai, L., Thomas P. Brutnell, T.P., Bermudez Kandianis, C., Sowinski, S.G., Stapleton, A.E., Vallabhaneni, R., Williams, M., Wurtzel, E.T., Yan, J., and Buckler, E.S. (2008) Natural genetic variation in lycopene epsilon cyclase tapped for maize biofortification. *Science*, 319, 330–333.

Harris, W.M., and Spurr, A.R. (1969) Chromoplasts of tomato fruits. II. The red tomato. *American Journal of Botany*, 56, 380–389.

Hauser, F., Waadt, R., and Schroeder, J.I. (2011) Evolution of abscisic acid synthesis and signaling mechanisms. *Current Biology*, 21, R346–R355.

Havaux, M., and Niyogi, K.K. (1999) The violaxanthin cycle protects plants from photooxidative damage by more than one mechanism. *Proceedings of the National Academy of Science*, 96, 8762–8767.

Hendry, G.A.F., Houghton, J.D., and Brown, S.B. (1987) The degradation of chlorophyll – a biological enigma. *New Phytologist*, 107, 255–302.

Hirschberg, J. (2001) Carotenoid biosynthesis in flowering plants. *Current Opinion in Plant Biology*, 4, 210–218.

Horner, H.T., Healy, R.A., Ren, G., Fritz, D., Klyne, A., Seames, C., and Thornburg, R.W. (2007) Amyloplast to chromoplast conversion in developing ornamental tobacco floral nectaries provides sugar for nectar and antioxidants for protection. *American Journal of Botany*, 94, 12–24.

Hörtensteiner, S. (2004) The loss of green colour during chlorophyll degradation – a prerequisite to prevent cell death? *Planta*, 219, 191–194.

Hörtensteiner, S. (2006) Chlorophyll degradation during senescence. *Annual Reviews Plant Biology*, 57, 55–77.

Hörtensteiner, S. (2009) Stay-green regulates chlorophyll and chlorophyll-binding protein degradation during senescence. *Trends in Plant Science*, 14, 155–162.

Hörtensteiner, S., and Krautler, B. (2011) Chlorophyll breakdown in higher plants. *Biochimica et Biophysica Acta*, 1807, 977–988.

Hu, Z.-L., Den, L., Yan, B., Pan, Y., Luo, M., Chen, X.-Q, Hu, T.-Z., and Chen, G.-P. (2011) Silencing of the *LeSGR1* gene in tomato inhibits chlorophyll degradationand exhibits a stay-green phenotype. *Biologia Plantarum*, 55, 27–34.

Huang, F.-C., Horváth, G., Molnár, P., Turcsi, E., Deli, J., Schrader, J., Sandmann, G., Schmidt, H., and Schwab, W. (2009a) Substrate promiscuity of RdCCD1, a carotenoid cleavage oxygenase from *Rosa damascena*. *Phytochemistry*, 70, 457–464.

Huang, F.-C., Molnár, P., and Schwab, W. (2009b) Cloning and functional characterization of carotenoid cleavage dioxygenase 4 genes. *Journal of Experimental Botany*, 60, 3011–3022.

Hunter, W.N. (2007) The non-mevalonate pathway of non-isoprenoid precursor biosynthesis. *Journal of Biological Chemistry*, 282, 21573–21577.

Isaacson, T., Ohad, I., Beyer, P., and Hirschberg, J. (2004) Analysis in vitro of the enzyme CRTISO establishes a poly-*cis* carotenoid biosynthesis pathway in plants. *Plant Physiology*, 136, 4246–4255.

Isaacson, T., Ronen, G., Zamir, D., and Hirschberg, J. (2002) Cloning of *tangerine* from tomato reveals a carotenoid isomerase essential for production of β-carotene and xanthophylls in plants. *Plant Cell*, 14, 333–342.

Ischebeck, T., Zbierzak, A.M., Kanwischer, M., and Dormann, P. (2006) A salvage pathway for phytol metabolism in *Arabidopsis*. *Journal of Biological Chemistry*, 281, 2470–2477.

Islam, M., Lyrene, S.S., Miller, E.M., and Porter, J.W. (1977) Dissociation of prelycopersene pyrophosphate synthetase from phytoene synthetase complex of tomato fruit plastids. *Journal of Biological Chemistry*, 252, 1523–1525.

Jia, H.-F., Chai, Y.-M., Li, C.-L., Lu, D., Luo, J.-J., Qin, L., and Shen, Y.-Y. (2011) Abscisic acid plays an important role in the regulation of strawberry fruit ripening. *Plant Physiology*, 157, 188–199.

Jeffrey, S.W., Douce, R., and Benson, A.A. (1974) Carotenoid transformations in the chloroplast envelope. *Proceedings of the National Academy of Sciences*, 71, 807–810.

de Jesus Ornelas-Paz, J., Yahia, E.M., and Gardea-Bejar, A. (2007) Identification and quantification of xanthophyll esters, carotenes and tocopherols in the fruit of seven Mexican mango cultivars by liquid chromatography–atmospheric pressure chemical ionization–time-of-flight mass spectrometry [LC-(APcI +)-MS]. *Journal of Agricultural and Food Chemistry*, 55, 6628–6635.

Johnson, E.J., and Krinsky, N.I. (2009) Carotenoids and coronary heart disease. In: *Carotenoids*, Vol. 5 (eds. G. Britton, S. Liaanen-Jensen, and H. Pfander), pp 287–300. Birkhauser, Basel.

Jones, B.L., and Porter, J.W. (1986) Biosynthesis of carotenoids in higher plants. *CRC Critical Reviews in Plant Science*, 3, 295–324.

Joyard, J., Ferro, M., Masselon, C., Seigneurin-Berry, D., Salvi, D., Garin, J., and Rolland, N. (2009) Chloroplast proteomics and the compartmentation of plastidial isoprenoid biosynthetic pathways. *Molecular Plant*, 2, 1154–1180.

Kapulnik, Y., Resnick, N., Mayzlish-Gati, E., Kaplan, Y., Wininger, S., Hershenhorn, J., and Koltai, H. (2011) Strigolactones interact with ethylene and auxin in regulating root-hair elongation in *Arabidopsis*. *Journal of Experimental Botany*, 62, 2915–2924.

Kerr, E.A. (1956) Linkage relations of *gf*. *Tomato Genetics Cooperative Report*, 8, 21.

Khachik, F., Beecher, G.R., Goli, M.B., and Lusby, W.R. (1991) Separation, identification, and quantification of carotenoids in fruits, vegetables and human plasma by high performance liquid chromatography. *Pure and Applied Chemistry*, 63, 71–80.

Kim, J., and DellaPenna, D. (2006) Defining the primary route for lutein synthesis in plants: The role of Arabidopsis carotenoid beta-ring hydroxylase CYP97A3. *Proceedings of the National Academy of Sciences*, 103, 3474–3479.

Kim, S.-M., Kuzuyama, T., Kobayashi, A., Sando, T., Chang, Y.-J., and Kim, S.-U. (2008) 1-Hydroxy-2-methyl-2-(*E*)-butenyl 4-diphosphate reductase (IDS) is encoded by multicopy genes in gymnosperms *Ginkgo biloba* and *Pinus taeda*. *Planta*, 227, 287–298.

Kloer, D.P., Ruch, S., Al-Babili, S., Beyer, P., and Schulz, G.E. (2005) The structure of a retinal-forming carotenoid oxygenase. *Science*, 308, 267–269.

Kloer, D.P., and Schulz, G.E. (2006) Structural and biological aspects of carotenoid cleavage. *Cellular and Molecular Life Sciences*, 63, 2291–2303.

Kolotilin, I., Koltai, H., Tadmor, Y., Bar-Or, C., Reuveni, M., Meir, A., Nahon, S., Shlomo, H., Chen, L., and Levin, I. (2007) Transcriptional profiling of high pigment-2^{dg}tomato mutant links early fruit plastid biogenesis with its overproduction of phytonutrients. *Plant Physiology*, 145, 389–401.

Koltai, H. (2011) Strigolactones are regulators of root development. *New Phytologist*, 190, 545–549.

Krautler, B. (2008) Chlorophyll breakdown and chlorophyll catabolites in leaves and fruit. *Photochemistry Photobiology Science*, 7, 1114–1120.

Kreutz, K., Beyer, P., and Kleinig, H. (1982) The site of carotenogenic enzymes in chromoplasts from *Narcissus pseudonarcissus L. Planta*, 154, 66–69.

Kumar, S., Hahna, F.M., Baidooc, E., Kahlona, T.S., Wood, D.F., McMahana, C.M., Cornish, K., Keasling, J.D., Daniell, H., and Whalen, M.C. (2011) Remodeling the isoprenoid pathway in tobacco by expressing 1 the cytoplasmic mevalonate pathway in chloroplasts. *Metabolic Engineering*, 14(1), 19–28.

Kuntz, M. (2004) Plastid terminal oxidase and its biological significance. *Planta*, 218, 896–899.

Kushwaha, S.C., Suzue, G., Subbarayan, C., and Porter, J.W. (1970) The conversion of phytoene-^{14}C to acyclic, monocyclic and dicyclic carotenes and the conversion of lycopenene-15, 15′-^{3}H to mono- and dicyclic carotenes by soluble enzyme systems obtained from plastids of tomato fruits. *Journal of Biological Chemistry*, 245, 4708–4717.

Landrum, J.T. (2010) *Carotenoids. Physical, Chemical and Biological Functions and Properties*. CRC Press, Boca Raton.

Lange, B.M., Rujan, T., Martin, W., and Croteau, R. (2000) Isoprenoid biosynthesis: the evolution of two ancient and distinct pathways across genomes. *Proceedings of the National Academy of Sciences*, 97, 13172–13177.

Leon, P., Avendano-Vazquez, O., Cordoba, E., Nisar, N., San Roman, C., Ramos-Vega, M., and Pogson, B. J. (2011) A novel carotenoid-derived signal modulates chloroplast biogenesis and leaf development. *Plant Biology Congress*, Minneapolis, Minnesota.

LeRosen, A.L., and Zechmeister, L. (1942) Prolycopene. *Journal of the American Chemical Society*, 64, 1075–1079.

Lewinsohn, E., and Gijzen, M. (2009) Phytochemical diversity: the sounds of silent metabolism. *Plant Science*, 176, 161–169.

Li, C., Jia, H., Chai, Y., and Shen, Y. (2011) Abscisic acid perception and signaling transduction in strawberry: a model for non-climacteric fruit ripening. *Plant Signaling and Behavior*, 6, 1950–1953.

Li, L., Paolillo, D.J., Parthasarathy, M.V., DiMuzio, E.M., and Garvin, D.F. (2001) A novel gene mutation that confers abnormal patterns of β–carotene accumulation in cauliflower (*Brassica oleracea* var. *botrytis*). *Plant Journal*, 26, 59–67.

Li, L., and Van Eck, J. (2007) Metabolic engineering of carotenoid accumulation by creating a metabolic sink. *Transgenic Research*, 16, 581–585.

Liaaen-Jensen, S., and Lutnoes, B.F. (2008) E/Z isomers and isomerization. In: *Carotenoids*, Vol. 4 (eds. G. Britton, S. Liaaen-Jensen, and H. Pfander), pp. 15–36. Birkhauser Verlag, Basel.

Lichtenthaler, H.K. (1968) Plastoglobuli and fine structure of plastids. *Endeavour*, 27, 144–148.

Lichtenthaler, H.K. (1999) The 1-deoxy-d-xylulose 5-phosphate pathway of isoprenoid biosynthesis in plants. *Annual Reviews of Plant Physiology and Plant Molecular Biology*, 50, 47–65.

Liu, R.H. (2003) Health benefits of fruit and vegetables are from additive and synergistic combinations of phytochemicals. *American Journal of Clinical Nutrition*, 78, 517S–520S.

Liu, Y., Gur, A., Ronen, G., Causse, M., Damidaux, R., Buret, M., Hirschberg, J., and Zamir, D. (2003) There is more to tomato fruit colour than candidate carotenoid genes. *Plant Biotechnology Journal*, 1, 195–208.

Liu, Y.S., Roof, S., Ye, Z., Barry, C., van Tuinen, A, Vrebalov, J., Bowler, C., and Giovannoni, J. (2004) Manipulation of light signal transduction as a means of modifying fruit nutritional quality in tomato. *Proceedings of the National Academy of Sciences*, 101, 9897–9902.

Lois, L., Rodriguez, C., Gallego, F., Campos, N., and Boronat, A. (2000) Carotenoid biosynthesis during tomato fruit development: regulatory role of 1-deoxy-D-xylulose 5-phosphate synthase. *Plant Journal*, 22, 503–513.

Lopez, A.B., Van Eck, J., Conlin, B.J., Paolillo, D.J., O'Neill, J., and Li, L. (2008a) Effect of the cauliflower *Or* transgene on carotenoid accumulation and chromoplast formation in transgenic potato tubers. *Journal of Experimental Botany*, 59, 213–223.

Lopez, A.B., Yang, Y., Thannhauser, T.W., and Li, L. (2008b) Phytoene desaturase is present in a large protein complex in the plastid membrane. *Physiologia Plantarum*, 133, 190–198.

Lopez-Raez, J.A., Charnikhova, T., Mulder, P., Kohlen, W., Bino, R., Levin I., and Bouwmeester, H. (2008) Susceptibility of the tomato mutant *High Pigment-2*^{dg} (*hp-2*^{dg}) to *Orobanche* spp. infection. *Journal of Agricultural and Food Chemistry*, 56, 6326–6332.

Lu, S., and Li, L. (2008) Carotenoid metabolism: biosynthesis, regulation, and beyond. *Journal of Integrative Plant Biology*, 50, 778–785.

Lu, S., Van Eck, J., Zhou, X., Lopez, A.B., O'Halloran, D.M., Cosman, K.M., Conlin, B.J., Paolillo, D.J., Garvin, D.F. Vrebalov, J., Kochian, L.V., Küpper, H., Earle, E.D., Cao, J., and Li, L. (2006) The cauliflower *Or* gene encodes a DnaJ cysteine-rich domain-containing protein that mediates high levels of beta-carotene accumulation. *Plant Cell*, 18, 3594–3605.

Lu, Y.M., Zhang, D.P., and Yan, H.Y. (1999) Sugar unloading mechanism in the developing apple fruit. *Acta Horticulturae Sinica*, 26, 147–151.

MacDougall, D. (2002) *Colour in food: Improving quality*. Woodhead Publishing Series in Food Science, Technology and Nutrition No. 75. Sawston, Cambridge, UK.

Maoka, T. (2009) Recent progress in structural studies of carotenoids in animals and plants. *Archives of Biochemistry and Biophysics*, 483, 191–195.

Maraco, E.K., and Schmidt-Dannert, C. (2010) Diverse activities of carotenoid cleavage enzymes. In: *Carotenoids. Physical, Chemical and Biological Functions and Properties* (ed. J.J. Landrum), Chapter 19, pp. 389–416. CRC Press, Boca Raton, Florida.

Marano, M.R., Serra, E.C., Orellano, G., and Carrillo, N. (1993) The path of chromoplast development in fruits and flowers. *Plant Science*, 94, 1–17.

Markwell, J., Bruce, B.D., and Keegstra, K. (1992) Isolation of a carotenoid-containing sub-membrane particle from the chloroplastic envelope outer membrane of pea (*Pisum sativum*). *Journal of Biological Chemistry*, 267, 13933–13937.

Matsumato, H., Ikoma, Y., Kato, M., Kuniga, T., Nakajima, N., and Yoshida, T. (2007) Quantification of carotenoids in citrus fruit by LC-MS and comparison of patterns of seasonal changes for carotenoids among citrus varieties. *Journal of Agricultural and Food Chemistry*, 55, 2356–2368.

Maudinas, B., Bucholtz, M.L., Papastephanou, C., Katiyar, S.S., Briedis, A.V., and Porter, J.W. (1977) The partial purification and properties of a phytoene synthesizing enzyme system. *Archives of Biochemistry and Biophysics*, 180, 354–362.

Mayer, M.P., Beyer, P., and Kleinig, H. (1990) Quinone compounds are able to replace molecular oxygen as terminal electron acceptor in phytoene desaturation in chromoplasts of *Narcissus pseudonarcissus* L. *European Journal of Biochemistry*, 191, 359–363.

Merzlyak, M.N., and Gitelson, A.A. (1995) Why and what for the leaves are yellow in autumn? On the interpretation of optical spectra of senescing leaves (*Acer platanoides* L.). *Journal of Plant Physiology*, 145, 315–320.

Merzlyak, M.N., Gitelson, A.A., Chivkunova, O.B., and Rakitin, V.Y. (1999) Non-destructive optical detection of pigment changes during leaf senescence and fruit ripening. *Physiologia Plantarum*, 106, 135–141.

Merzlyak, M.N., Melo, T.B., and Naqvi, K.R. (2008) Effect of anthocyanins, carotenoids and flavonols on chlorophyll fluorescence excitation spectra in apple fruit: signature analysis, assessment, modelling and relevance to photoprotection. *Journal of Experimental Botany*, 59, 349–359.

Merzlyak, M.N., and Solovchenko, A.E. (2002) Photostability of pigments in ripening apple: a possible photoprotective role of carotenoids during plant senescence. *Plant Science*, 163, 881–888.

Messing, S.A.J., Gabelli, S.B., Echeverria, I., Vogel, J.T., Guan, J.C., Tan, B.C., Klee, H.J., McCarty, D.R., and Amzel, L.M. (2010) Structural insights into maize *viviparous14*, a key enzyme in the biosynthesis of the phytohormone abscisic acid. *The Plant Cell*, 22, 2970–2980.

Mialoundama, A.S., Heintz, D., Jadid, N., Nkeng, P., Rahier, A., Deli, J., Camara, B., and Bouvier, F. (2010) Characterization of plant carotenoid cyclases as members of the flavoprotein family functioning with no net redox change. *Plant Physiology*, 153, 970–979.

Milborrow, B.V. (1974) The chemistry and physiology of abscisic acid. *Annual Reviews of Plant Physiology*, 25, 259–307.

Misawa, N., Nakagawa, M., Kobayashi, K., Yamano, S., Izawa, Y., Nakamura, K., and Harashima K. (1990) Elucidation of the *Erwinia uredovora* carotenoid biosynthetic pathway by functional analysis of gene products expressed in *Escherichia coli*. *Journal of Bacteriology*, 172, 6704–6712.

Mochizuki, N., Tanaka, R., Grimm, B., Masuda, T., Moulin, M., Smith, A.G., Tanaka, A., and Terry, M.J. (2010) The cell biology of tetrapyrroles: a life and death struggle. *Trends in Plant Science*, 15, 488–498.

Morstadt, L., Graber, P., De Pascalis, L., Kleinig, H., Speth, V., and Beyer, P. (2002) Chemiosmotic ATP synthesis in photosynthetically inactive chromoplasts from *Narcissus pseudonarcissus* L. linked to a redox pathway potentially also involved in carotene desaturation. *Planta*, 215, 134–140.

Mortain-Bertrand, A., Stammitti, L., Telef, N., Colardelle, P., Brouquisse, R., Rolin, D., and Gallusci, P. (2008) Effects of glucose on carotenoid accumulation in tomato leaves. *Physiologia Plantarum*, 134, 246–256.

Moser, D., and Matile, P. (1997) Chlorophyll breakdown in ripening fruits of *Capsicum annuum*. *Journal of Plant Physiology*, 150, 759–761.

Moser, S., Müller, T., Ebert, M.O., Jockusch, S., Turro, N.J., and Kräutler, B. (2008) Blue luminescence of ripening bananas. *Angewante Chemie International Edition*, 47, 8954–8957.

Moser, S., Muller, T., Oberhuber, M., and Kräutler. B. (2009) Chlorophyll catabolites – chemical and structural footprints of a fascinating biological phenomenon. *European Journal of Organic Chemistry*, 1, 21–31.

Mur, L.A.J., Aubry, S., Mondhe, M., Kingston-Smith, A., Gallagher, J., Timms-Taravella, E., James, C., Papp, I., Hörtensteiner, H., Thomas, H., and Ougham, H. (2010) Accumulation of chlorophyll catabolites photosensitizes the hypersensitive response elicited by *Pseudomonas syringae* in *Arabidopsis*. *New Phytologist*, 188, 161–174.

Mustilli, A., Fenzi, F., Ciliento, R., Alfano, F., and Bowler, C. (1999) Phenotype of tomato high pigment-2 is caused by a mutation in the tomato homolog of De-etiolated 1. *Plant Cell*, 11, 15–157.

Nambara, E., and Marion-Poll, A. (2005) Abscisic acid biosynthesis and catabolism. *Annual Reviews of Plant Biology*, 56, 165–185.

Nitsch, L.M.C., Oplaat, C., Feron, R., Ma, Q., Wolters-Arts, M., Hedden, P., Mariani, C., and Vriezen, W.H. (2009) Abscisic acid levels in tomato ovaries are regulated by LeNCED1and SlCYP707A1. *Planta*, 229, 1335–1346.

Niyogi, K.K. (2000) Safety valves for photosynthesis. *Current Opinions in Plant Biology*, 3, 455–460.

Norris, S.R., Barrette, T.R., and DellaPenna, D. (1995) Genetic dissection of carotenoid synthesis in Arabidopsis defines plastoquinone as an essential component of phytoene desaturation. *The Plant Cell*, 7, 2139–2149.

North, H.M., Almeida, A.D., Boutin, J.-P., Frey, A., To, A., Botran, L., Sotta, B., and Marion-Poll, A. (2007) The *Arabidopsis* ABA-deficient mutant *aba4* demonstrates that the major route for stress-induced ABA accumulation is via neoxanthin isomers. *The Plant Journal*, 50, 810–824.

Ohmiya, A. (2009) Carotenoid cleavage dioxygenases and their apocarotenoid products in plants. *Plant Biotechnology*, 26, 351–358.

Patterson, G.W., Hugly, S., and Harrison, D. (1993) Sterols and phytyl esters of *Arabidopsis thaliana* under normal and chilling temperatures. *Phytochemistry*, 33, 1381–1383.

Peisker, C., Duggelin, T., Rentsch, D., and Matile, P. (1989) Phytol and the breakdown of chlorophyll in senescent leaves. *Journal of Plant Physiology*, 135, 428–432.

Peralta, I.E., Knapp, S., and Spooner, D.M. (2006) Nomenclature for wild and cultivated tomatoes. *Report of the Tomato Genetics Cooperative*, 56, 6–12.

Peralta, I.E., and Spooner, D.M. (2007) History, origin and early cultivation of tomato (*Solanaceae*). In: *Genetic Improvement of Solanaceous Crops*, Vol. 2: Tomato (eds. M.K. Razdan, and A.K. Mattoo), pp. 1–27. Science Publishers, Enfield, CT.

Phillips, M.A., Leon, P., Boronat, A., and Rodríguez-Concepción, M. (2008) The plastidial MEP pathway: unified nomenclature and resources. *Trends in Plant Science*, 13, 619–623.

Picton, S., Gray, J., Barton, S., AbuBakar, U., Lowe, A., and Grierson, D. (1993) cDNA cloning and characterisation of novel ripening-related mRNAs with altered patterns of accumulation in the ripening inhibitor (rin) tomato ripening mutant. *Plant Molecular Biology*, 23, 193–207.

Pizarro, L., and Stange, C. (2009) Light – dependent regulation of carotenoid biosynthesis in plants. *Ciencia e Investigación Agraria*, 36, 143–162.

Pogson, B., McDonald, K.A., Truong, M., Britton, G., and DellaPenna, D. (1996) *Arabidopsis* carotenoid mutants demonstrate that lutein is not essential for photosynthesis in higher plants. *The Plant Cell*, 8, 1627–1639.

Pogson, B.J., Rissler, H.M., and Frank, H.A. (2005) The roles of carotenoids in photosystem II of higher plants. In *Photosystem II: the light-driven water: plastoquinone oxidoreductase* (eds. T. Wydrzynski, and K. Satoh), pp. 515–537. Springer-Verlag, Dordrecht, The Netherlands.

Polivka, T., and Frank, H.A. (2010) Molecular factors controlling photosynthetic light harvesting by carotenoids. *Accounts of Chemical Research*, 43, 1125–1134.

Porter, J.W., and Anderson, D.G. (1962). The biosynthesis of carotenes. *Archives of Biochemistry and Biophysics*, 97, 520–528.

Porter, J.W., and Lincoln, R.E. (1950) Lycopersicon selections containing a high content of carotenes and colourless polyenes. II. The mechanism of carotene biosynthesis. *Archives of Biochemistry and Biophysics*, 27, 390–403.

Qin, G., Gu, H., Ma, L., Peng, Y., Deng, X.W., Chen, Z., and Qu, L.J. (2007) Disruption of phytoene desaturase gene results in albino and dwarf phenotypes in *Arabidopsis* by impairing chlorophyll, carotenoid and gibberellin biosynthesis. *Cell Research*, 17, 471–482.

Qin, X., and Zeevaart, J.A. (1999) The 9-*cis*-epoxycarotenoidcleavage reaction is the key regulatory step of abscisic acid biosynthesis in water-stressed bean. *Proceedings of the National Academy of Sciences*, 96, 15354–15361.

Quinlan, R.R., Jaradat, T.T., and Wurtzel, E.T. (2007) *Escherichia coli* as a platform for the functional expression of plant P450 carotene hydroxylases. *Archives of Biochemistry and Biophysics*, 458, 146–157.

Qureshi, A., Andrewes, A.G., Qureshi, N., and Porter, J.W. (1974) The enzymatic conversion of cis-^{14}C phytofluene, trans- ^{14}C phytofluene and trans- ^{14}C zeta-carotene to more unsaturated acyclic, monocyclic and dicyclic carotenes by a cell – free preparation of red tomato fruit. *Archives of Biochemistry and Biophysics*, 162, 93–107.

Rameau, C., Pillot, J.-P., Becard, G., Gomez-Roldan, V., Puech-Pages, V., Rochange, F., Beveridge, C., Dun, E., and Brewer, P. (2011) Treatment process for a superior plant in order to control its growth and architecture. US Patent 20110230352.

Ray, J., Bird, C., Maunders, M., Grierson, D., and Schuch, W. (1987) Sequence of pTOM5, a ripening related cDNA from tomato. *Nucleic Acids Research*, 15, 10587.

Reinbothe, C., El Bakkouri, M., Buhr, F., Muraki, N., Nomata, J., Kurisu, G., Fujita Y., and Reinbothe S. (2010) Chlorophyll biosynthesis: spotlight on protochlorophyllide reduction. *Trends in Plant Science*, 15, 1360–1385.

Rise, M., Cojocaru, M., Gottlieb, H.E., and Goldschmidt, E.E. (1989) Accumulation of α-tocopherol in senescing organs as related to chlorophyll degradation. *Plant Physiology*, 89, 1028–1030.

Robles-Sardin, A.E., Bolanos-Villar, A.V., Gonzalez-Aguilar G.A., and de la Rosa L.A. (2009) Flavonoids and their relation to human health. *Fruits and Vegetable Phytonutrients: Chemistry, Nutritional Value and Stability* (L.A. De la Rosa, E. Alvarez-Parrilli, and G.A. Gonzalez-Aguilar), Chapter 6. Wiley-Blackwell, Oxford.

Roca, M., and Minguez-Mosquera, M.I. (2006) Chlorophyll catabolism pathway in fruits of *Capsicum annuum* (L.): stay-green versus red fruits. *Journal of Agricultural Food Chemistry*, 54, 4035–4040.

Rodrigo, M.J., and Zacarias, L. (2007) Effect of post harvest ethylene treatment on carotenoid accumulation and the expression of carotenoid biosynthetic genes in the flavedo of orange (*Citrus sinesis* L. Osbeck) fruit. *Postharvest Biology and Technology*, 43, 14–22.

Rodríguez-Concepción, M. (2006) Early steps in isoprenoid biosynthesis: multilevel regulation of the supply of common precursors in plant cells. *Phytochemistry Reviews*, 5, 1–15.

Rodríguez-Concepción, M. (2010) Supply of precursors for carotenoid biosynthesis in plants. *Archives of Biochemistry and Biophysics*, 504, 118–122.

Rodríguez-Concepción, M., and Boronat, A. (2002) Elucidation of the methylerythritol phosphate pathway for isoprenoid biosynthesis in bacteria and plastids. A metabolic milestone achieved through genomics. *Plant Physiology*, 130, 1079–1089.

Rodríguez-Concepción, M., Forés, O., Martínez-García, J.F., González, V., Phillips, M.A., Ferrer, A., and Boronat, A. (2004) Distinct light-mediated pathways regulate the biosynthesis and exchange of isoprenoid precursors during Arabidopsis seedling development. *Plant Cell*, 16, 144–156.

Rodriguez-Uribe, L., Guzman, I., Rajapakse, W., Richins, R.D., and O'Connell, M.A. (2012) Carotenoid accumulation in orange-pigmented *Capsicum annuum* fruit, regulated at multiple levels. *Journal of Experimental Botany*, Jan;63(1), 517–526.

Rohmer, M., Knani, M., Simonin, P., Sutter, B., and Sahm, H. (1993) Isoprenoid biosynthesis in bacteria: a novel pathway for the early steps leading to isopentenyl diphosphate. *Biochemical Journal*, 295, 517–524.

Römer, S., and Fraser, P.D. (2005) Recent advances in carotenoid biosynthesis, regulation and manipulation. *Planta*, 221, 305–308.

Römer, S., Fraser, P.D., Kiano, J.W., Shipton, C.A., Misawa, N., Schuch, W., and Bramley, P.M. (2000) Elevation of the pro-vitamin A content of transgenic tomato plants. *Nature Biotechnology*, 18, 666–669.

Ronen, G., Carmel-Goren, L., Zamir, D., and Hirschberg, J. (2000) An alternative pathway to β-carotene formation in plant chromoplast discovered by map-based cloning of beta and old-gold color mutations in tomato. *Proceedings of the National Academy of Sciences*, 97, 11102–11107.

Ronen, G., Cohen, M., Zamir D., and Hirschberg, J. (1999) Regulation of carotenoid biosynthesis during tomato fruit development: expression of the gene for lycopene epsilon cyclase is down regulated during ripening and is elevated in the mutant delta. *The Plant Journal*, 17, 341–351.

Rontani, J.-F., Cuny, P., and Grossi, V. (1996) Photodegradation of chlorophyll phytol chain in senescent leaves of higher plants. *Phytochemistry*, 42, 347–351.

Rosati, C., Aquilani R., Dharmapuri, S., Pallara, P., Marusic, C., Tavazza, R., Bouvier, F., Camara, B., and Giuliano, G. (2000) Metabolic engineering of beta-carotene and lycopene content in tomato fruit. *The Plant Journal*, 24, 413–419.

Rosati, C., Direto, G., and Giuliano, G. (2009) Biosynthesis and genetic engineering of carotenoids and apocarotenoids in plants: state of the art and future prospects. *Biotechnology and Genetic Engineering Reviews*, 26, 139–161.

Sandmann, G. (1991) Biosynthesis of cyclic carotenoids: biochemistry and molecular genetics of the reaction sequence. *Physiologia Plantarum*, 83, 186–193.

Sandmann, G., Romer, S., and Fraser, P.D. (2006) Understanding carotenoid metabolism as a necessity for genetic engineering of crop plants. *Metabolic Engineering*, 8, 291–302.

Schelbert, S., Aubry, S., Burla, B., Agne, B., Kessler, F., Krupinska, K., and Hörtensteiner, S. (2009) Pheophytin pheophorbide hydrolase (Pheophytinase) is involved in chlorophyll breakdown during leaf senescence in *Arabidopsis*. *The Plant Cell*, 21, 767–785.

Schmidt, H.O. (1988) The structure and function of grana free thylakoids membranes in gerontoplasts of senescing leaves of *Vicia faba* L. *Zeitschrift fur Naturforschung Teil C*, 43, 149–154.

Schofield, A., and Paliyath, G. (2005) Modulation of carotenoid biosynthesis during tomato fruit ripening through phytochrome regulation of phytoene synthase activity. *Plant Physiology and Biochemistry*, 43, 1052–1060.

Schwartz, S.H., Tan, B.C., Gage, D.A., Zeevaart, J.A.D., and McCarty, D.R. (1997) Specific oxidative cleavage of carotenoids by VP14 of maize. *Science*, 276, 1872–1874.

Schwartz, S.H., Qin, X., and Loewen, M.C. (2004) The biochemical characterization of two carotenoid cleavage enzymes from Arabidopsis indicates that a carotenoid-derived compound inhibits lateral branching. *Journal of Biological Chemistry*, 279, 46940–46945.

Schwartz, S.H., Qin, X., and Zeevaart, J.A.D. (2003) Elucidation of the indirect pathway of abscisic acid biosynthesis by mutants, genes, and enzymes. *Plant Physiology*, 131, 1591–1601.

Schwartz, S.H., and Zeevart, J.A.D. (2010) ABA biosynthesis and metabolism. In: *Plant Hormones. Biosynthesis, Signal Transduction, Action!* (ed. P.J. Davies), Chapter B7, pp. 137–155. Springer, The Netherlands.

Shahbazi, M., Gilbert, M., Laboure, A.-M., and Kuntz, M. (2007) Dual role of the plastid terminal oxidase in tomato. *Plant Physiology*, 145, 691–702.

Shirley, B.W. (1996) 'New' functions for an 'old' pathway. *Trends in Plant Science*, 1, 377–382.

Siefermann-Harms, D., Joyard, J., and Douce, R. (1978) Light-induced changes of the carotenoid levels in chloroplast envelopes. *Plant Physiology*, 61, 530–533.

Simkin, A.J., Gaffe, J., Alcaraz, J.P., Carde, J.P., Bramley, P.M., Fraser, P.D., and Kuntz, M. (2007) Fibrillin influence on plastid ultrastructure and pigment content in tomato fruit. *Phytochemistry*, 68, 1545–1556.

Simkin, A.J., Schwartz, S.H., Auldridge, M.E., Taylor, M.G., and Klee, H.J. (2004) The tomato carotenoid cleavage dioxygenase 1 genes contribute to the formation of the flavor volatiles β-ionone, pseudoionone and geranylacetone. *The Plant Journal*, 40, 882–892.

Simkin, A.J., Zhu, C., Kuntz, M., and Sandmann, G. (2003) Light-dark regulation of carotenoid biosynthesis in pepper (*Capsicum annuum*) leaves. *Journal of Plant Physiology*, 160, 439–443.

Skelton, R.L., Yu, Q., Srinivasan, R., Manshardt, R., Moore, P.H., and Ming, R. (2006) Tissue differential expression of lycopene β-cyclase gene in papaya. *Cell Research*, 16, 731–739.

Smith, P.G. (1950) Inheritance of brown and green mature fruit color in peppers. *Journal of Heredity*, 41, 138–140.

Solovchenko, A.E., Merzlyak, M.N., and Pogosyan, S.I. (2010) Light-induced decrease of reflectance provides an insight in the photoprotective mechanism of ripening apple fruit. *Plant Science*, 178, 281–288.

Spurgeon, S.L., and Porter, J.W. (1983) Biosynthesis of carotenoids. In: *Biosynthesis of Isoprenoids*, Vol. 2 (eds. J.W. Porter, and L. Spurgeon), pp. 1–122. Wiley, New York.

Stigliani, A.L., Giorio, G., and D'Ambrosio, C. (2011) Characterization of P450 carotenoid β–and ε–hydroxylases of tomato and transcriptional regulation of xanthophyll biosynthesis in root, leaf, petal and fruit. *Plant Cell Physiology*, 52, 851–865.

Stitt, M., Sulpice, R., and Keurentjes, J. (2010) Metabolic networks: how to identify key components in the regulation of metabolism and growth. *Plant Physiology*, 152, 428–444.

Sun, Z., Gannt, E., and Cunningham Jr., F.X. (1996) Cloning and functional analysis of the β-carotene hydroxylase of *Arabidopsis thaliana*. *The Journal of Biological Chemistry*, 271, 24349–24352.

Tan, B.C., Cline, K., and McCarthy, D.R. (2001) Localization and targeting of the VP14 epoxy-carotenoid dioxygenase to chloroplast membranes. *Plant Journal*, 27, 373–382.

Tan, B.C., Joseph, L.M., Deng, W.T., Liu, L., Li, Q.B., Cline, K., and McCarthy, D.R. (2003) Molecular characterization ofthe *Arabidopsis* 9-*cis* epoxycarotenoid dioxygenase gene family. *Plant Journal*, 35, 44–56.

Tanaka, R., and Tanaka, A. (2007) Tetrapyrrole biosynthesis in higher plants. *Annual Review Plant Biology*, 58, 321–346.

Tanaka, R., and Tanaka, A. (2011) Chlorophyll cycle regulates the construction and destruction of the light-harvesting complexes. *Biochimica et Biophysica Acta*, 1087, 968–976.

Telef, N., Stammitti-Bert, L., Mortain-Bertrand, A., Maucourt, M., Carde, J.P., Rolin, D., and Gallusci, P. (2006) Sucrose deficiency delays lycopene accumulation in tomato fruit pericarp discs. *Plant Molecular Biology*, 62, 453–469.

Thomas, H., and Howarth, C.J. (2000) Five ways to stay green. *Journal Experimental Botany*, 51, 329–337.

Thompson, A.J., Andrews, J., Mulholland, B.J., McKee, J.M.YT., Hilton, H.W., Horridge, J.S., Farquhar, G.D., Smeeton, R.C., Smillie, I.R.A., Black, C.R., and Taylor, I.B. (2007) Overproduction of abscisic acid in tomato increases transpiration efficiency and root hydraulic conductivity and influences leaf expansion. *Plant Physiology*, 143, 1905–1917.

Tian, L., and DellaPenna, D. (2004) Progress in understanding the origin and functions of carotenoid hydroxylases in plants. *Archives of Biochemistry and Biophysics*, 430, 22–29.

Tian, L., Musetti, V., Kim J., Magallanes-Lundback, M., and DellaPenna, D. (2004) The *Arabidopsis* LUT1 locus encodes a member of the cytochrome P450 family that is required for carotenoid epsilon-ring hydroxylation activity. *Proceedings of the National Academy of Sciences*, 101, 402–407.

Toledo-Ortiz, G., Huq, E., and Rodríguez-Concepción, M. (2010) Direct regulation of phytoene synthase gene expression and carotenoid biosynthesis by phytochrome -interacting factors. *Proceedings of the National Academy of Sciences*, 107, 11626–11631.

Tonetto de Freitas, S., Shackel, K.A., and Mitcham, E.J. (2011) Abscisic acid triggers whole-plant and fruit-specific mechanisms to increase fruit calcium uptake and prevent blossom end rot development in tomato fruit. *Journal of Experimental Botany*, 62, 2645–2656.

Tung, S.A., Smeeton, R., White, C.A., Black, C.R., Taylor, I.B., Hilton, H.W., and Thompson, A.J. (2008) Overexpression of LeNCED1 in tomato(*Solanum lycopersicum* L.) with the rbcS3C promoter allows recovery of lines that accumulate very high levels of abscisic acid and exhibit severe phenotypes. *Plant, Cell and Environment*, 31, 968–981.

Valentin, H.E., Lincoln, K., Moshiri, F., Jensen, P.K., Qi, Q., Venkatesh, T.V., Karunanandaa, B., Baszis, S.R., Norris, S.R., Savidge, B., Gruys, K.J., and Last, R.L. (2006) The *Arabidopsis* vitamin E pathway gene5-1 mutant reveals a critical role for phytol kinase in seed tocopherol biosynthesis. *Plant Cell*, 18, 212–224.

Vallabhaneni, R., Bradbury, L.M.T., and Wurtzel, E.T. (2010) The carotenoid dioxygenase gene family in maize, sorghum and rice. *Archives of Biochemistry and Biophysics*, 504, 104–111.

Vallabhaneni, R., and Wurtzel, E.T. (2009) Timing and biosynthetic potential for carotenoid accumulation in genetically diverse germplasm of maize. *Plant Physiology*, 150, 562–572.

Vardhini, B.V., and Rao, S.S.R. (2002) Acceleration of ripening of tomato pericarp discs by brassinosteroids. *Phytochemistry*, 61, 843–847.

Vavilin, D., and Vermaas, W. (2007) Continuous chlorophyll degradation accompanied by chlorophyllide and phytol reutilization for chlorophyll synthesis in *Synechocystis* sp. PCC 6803. *Biochimica et Biophysica Acta*, 1767, 920–929.

Vishnevetsky, M., Ovadis, M., Zuker, A., and Vainstein, A. (1999) Molecular mechanisms underlying carotenogenesis in the chromoplast: multilevel regulation of carotenoid-associated genes. *Plant Journal*, 20, 423–431.

Vogel, J.T., Tan, B.-C., McCarty, D.R., and Klee, H.J. (2008) The carotenoid cleavage dioxygenase 1 enzyme has broad substrate specificity, cleaving multiple carotenoids at two different bond positions. *Journal of Biological Chemistry*, 283, 11364–11373.

Vogel, J.T., Tieman, D.M., Sims, C.A., Odabasi, A.Z., Clark, D.G., and Klee, H.J. (2010) Carotenoid content impacts flavour and acceptability in tomato (*Solanum lycopersicum*). *Journal of Science of Food and Agriculture*, 90, 2233–2240.

Volkman, J.K., and Maxwell, J.R. (1986) Acyclic isoprenoids as biological markers. In: *Biological Markers in the Sedimentary Record* (ed. R.B. Johns), pp 1–46, Elsevier, Amsterdam, The Netherlands.

Von Lintig, J., Welsch, R., Bonk, M., Giuliano, G., Batschauer, A., and Kleinig, H. (1997) Light-dependent regulation of carotenoid biosynthesis occurs at the level of phytoene synthase expression and is mediated by phytochrome in *Sinapis alba* and *Arabidopsis thaliana* seedlings. *The Plant Journal*, 12, 625–634.

Wackenroder, H.W.F. (1831) Ueber das Oleum radicis Dauci aetherum, das Carotin, den Carotenzucker und den officinellensuccus Dauci; so wie auch über das Mannit, welches in dem Möhrensafte durch eine besondere Art der Gährung gebildet wird. *Gieger's Magazin Pharm*, 33, 144.

Walter, M.H., Floss, D.S., and Strack, D. (2010) Apocarotenoids: hormones, mycorrhizal metabolites and aroma volatiles. *Planta*, 232, 1–17.

Walter, M.H., and Strack, D. (2011) Carotenoids and their cleavage products: biosynthesis and functions. *Natural Product Reports*, 28, 663–692.

Welsch, R., Beyer, P., Hugueney, P., Kleinig, H., and von Lintig, J. (2000) Regulation and activation of phytoene synthase, a key enzyme in carotenoid biosynthesis, during photomorphogenesis. *Planta*, 211, 846–854.

Welsch, R., Mass, D., Voegel, T., DellaPenna, D., and Beyer, P. (2007) Transcription factor RAP2.2 and its interacting partner SINAT2: stable elements in the carotenogenesis of Arabidopsis leaves. *Plant Physiology*, 145, 1073–1085.

Welsch, R., Wust, F., Bar, C., Al-Babili, S., and Beyer, P. (2008) A third phytoene synthase is devoted to abiotic-stress induced abscisic acid formation in rice and defines functional diversification of phytoene synthase genes. *Plant Physiology*, 147, 367–380.

von Wettstein, D., Gough, S., and Kannangara, C.G. (1995) Chlorophyll biosynthesis. *Plant Cell*, 7, 1039–1057.

Willson, M.F., and Whelan, C.J. (1990) The evolution of fruit colour in fleshy fruited plants. *American Naturalist*, 136, 790–809.

Woitsch, S., and Romer, S. (2003) Expression of xanthophyll biosynthetic genes during light-dependent chloroplast differentiation. *Plant Physiology*, 132, 1508–1517.

Xie, X., Yoneyama, K., and Yoneyama, K. (2010) The strigolactone story. *Annual Review of Phytopathology*, 48, 93–117.

Yamamoto, H.Y. (2004) Carotenoids: localization and function. *Advances in Photosynthesis and Respiration*, 4, 539–563.

Yamamoto, H.Y. (2006) Functional roles of the major chloroplast lipids in the violaxanthin cycle. *Planta*, 224, 719–724.

Yamamoto, H.Y., Bugos, R.C., and Hieber, A.D. (1999) Biochemistry and molecular biology of the xanthophyll cycle. In: *Advances in photosynthesis. The photochemistry of carotenoids*, Vol. 8 (eds. H.A. Frank, A.J. Young, G. Britton and R.J. Cogdell), pp. 293–303. Kluwer, Dordrecht.

Yang, X.-T., Zhang, Z.-Q., Joyce, D, Huang, X.-M., Xu, L.-Y., and Pang, X.-Q. (2009) Characterisation of chlorophyll degradation in banana and plantain during ripening at high temperature. *Food Chemistry*, 114, 383–390.

Ye, H., Pilon, M., and Pilon-Smits, E.A.H. (2006) CpNifS-dependent iron-sulfur cluster biogenesis in chloroplasts. *New Phytologist*, 171, 285–292.

Yoneyama, K., Xie, X., Yoneyama, K., and Takeuchi, Y. (2009) Strigolactones: structures and biological activities. *Pest Management Science*, 65, 467–470.

Zechmeister, L., LeRosen, A.L., Went, F.W., and Pauling, L. (1941) Prolycopene, a naturally occurring stereoisomer of lycopene. *Proceedings of the National Academy of Sciences*, 27, 468–474.

Zhang, M., Yuan, B., and Leng, P. (2009a) The role of ABA in triggering ethylene biosynthesis and ripening of tomato fruit. *Journal of Experimental Botany*, 60, 1579–1588.

Zhang, M., Yuan, B., and Leng, P. (2009b) Cloning of 9-*cis*-epoxycarotenoid dioxygenase (NCED) gene and the role of ABA on fruit ripening. *Plant Signalling Behaviour*, 4, 460–463.

Zhu, C., Naqvi, S., Gomez-Galera, S., Pelacho, A.M., Capell, T., and Christou, P. (2007) Transgenic strategies for the nutritional enhancement of plants. *Trends in Plant Science*, 12, 548–555.

Zimmermann, P., Hirsch-Hoffmann, M., Hennig, L., and Gruissem, W. (2004) GENEVESTIGATOR. *Arabidopsis* microarray database and analysis toolbox. *Plant Physiology*, 136, 2621–2632.

5 Phenylpropanoid Metabolism and Biosynthesis of Anthocyanins

Laura Jaakola

Introduction

Polyphenols are a large group of organic compounds found in plants, characterized by the presence of phenol units in the structure. Polyphenols are generally divided into two subgroups, hydrolyzable tannins and phenylpropanoids. All phenylpropanoids are biosynthesized from the amino acid phenylalanine. They have a variety of functions including protection from UV light, defense against pathogens or herbivores, as structural components of cell wall, or as pigments and signaling molecules (Weisshaar and Jenkins, 1998; Vogt, 2009). In fruits, phenylpropanoids constitute one of the most important groups of secondary metabolites. Phenylpropanoid compounds offer protection during the early stages of fruit development and act as pigments serving visual signals for seed dispersers in ripe fruits. Phenylpropanoids have a major role in quality characteristics of ripe fruits and fruit products. They contribute not only to the characteristic flavor and color but also to unfavorable traits such as browning of fruit tissues via enzymatic oxidation of phenolic compounds by polyphenol oxidases (Macheix et al., 1990).

Fruit phenylpropanoids form an important part of the recommended human diet. Phenylpropanoid compounds have long been assumed to be antioxidants that scavenge excessive damaging free radicals, which arise from normal metabolic processes (Stevenson and Hurst, 2007). There is increasing evidence for many potential benefits of phenylpropanoids in regulation of cellular processes such as inflammation from human trials (Badimon et al., 2010). Phenylpropanoid compounds can also have indirect antioxidant effects through the induction of endogenous protective enzymes (Stevenson and Hurst, 2007). The beneficial effect of diverse phenylpropanoids for human health is at present generally accepted. However, the metabolism and mechanisms of action of these compounds in the human system are not well understood as there is high variation in absorption of different phenylpropanoids through the human gut. Moreover, thus far the evidence for the health benefits of phenolic compounds comes primarily from *in vitro* tests or animal trials. Due to the complexity of human metabolism, additional

The Molecular Biology and Biochemistry of Fruit Ripening, First Edition.
Edited by Graham B. Seymour, Mervin Poole, James J. Giovannoni and Gregory A. Tucker.

evidence on bioavailability and metabolism of diverse phenolic compounds in the human system is still needed (D'Archivio et al., 2010).

Phenylpropanoid metabolism in plants comprises of a complex series of branching biochemical reactions (Fig. 5.1). The plant shikimate pathway is the entry to the biosynthesis of phenylpropanoids (Vogt, 2009). The general phenylpropanoid pathway begins from phenylalanine, which is converted to cinnamate (*trans*-Cinnamic acid) by phenylalanine ammonia lyase

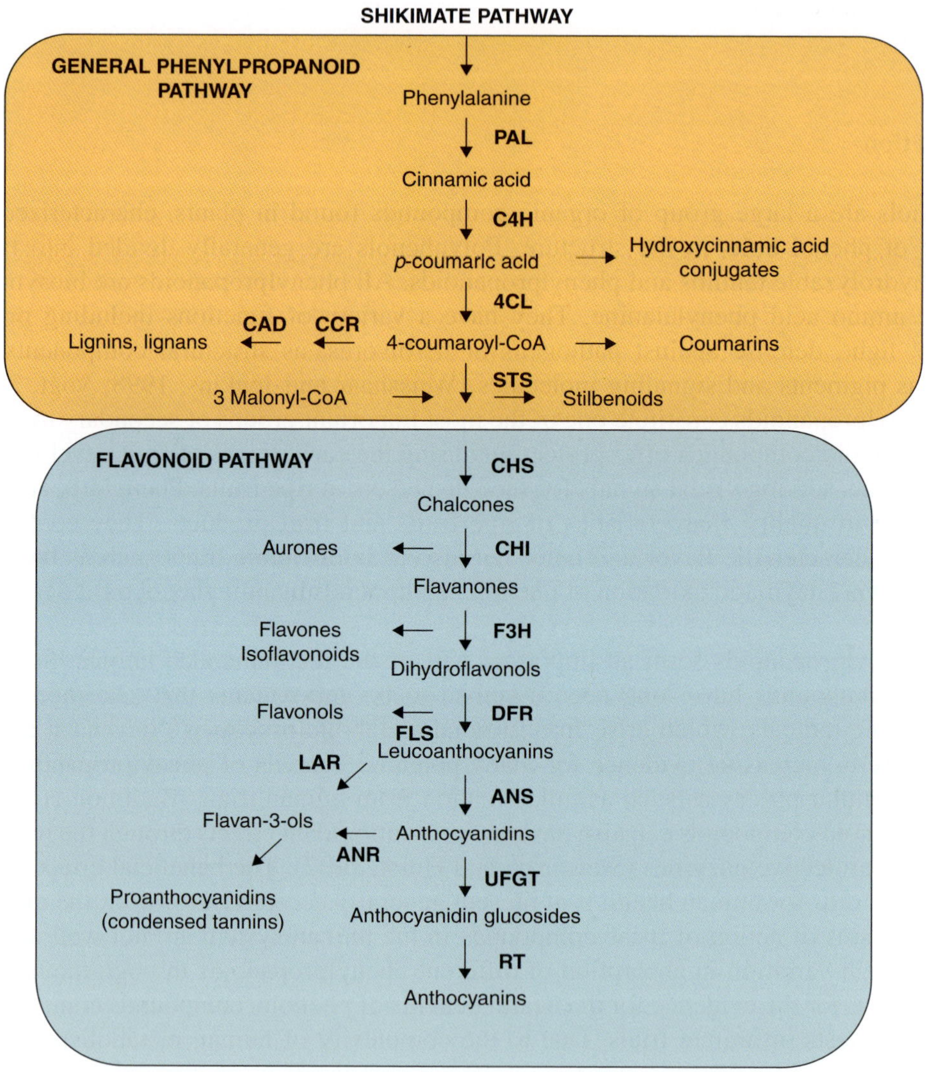

Figure 5.1 A schematic presentation of the phenylpropanoid and flavonoid biosynthetic pathways. PAL, phenylalanine ammonia-lyase; C4H, cinnamate 4-hydroxylase; 4CL, 4-coumaroyl:CoA ligase; CCR, cinnamoyl-CoA reductase; CAD, cinnamoyl alcohol dehydrogenase; STS, stilbene synthase; CHS, chalcone synthase; CHI, chalcone isomerase; F3H, flavanone 3-hydroxylase; FLS, flavonol synthase; DFR, dihydroflavonol 4-reductase; LAR, leucoanthocyanidin reductase; ANR, anthocyanidin reductase; ANS, anthocyanidin synthase; UFGT, UDP glucose-flavonoid 3-o-glucosyl transferase; RT, rhamnosyl transferase.

(PAL). PAL also converts tyrosine to *p*-coumarate, although with lower efficiency (Weisshaar and Jenkins, 1998).

Several copies of PAL genes are found in all plant species, comprising four genes in Arabidopsis and 20 putative PAL genes in tomato (*Solanum Lycoperscium*) (Chang et al., 2008; Vogt, 2009). The expression of the individual PAL genes may be developmentally and spatially different, and they also may respond differentially to biotic and abiotic stressors. In fruits, the expression of PAL genes may be activated, for instance, at the period of ripening related to the biosynthesis of anthocyanin pigments, as a response to pathogens or other stress-causing factors and in the context of enzymatic browning (He and Luo, 2007).

Cinnamic Acids

Cinnamic acid is a key intermediate in phenylpropanoid and shikimate pathways and the precursor of its hydroxyl derivatives such as *p*-coumaric acid, caffeic acid, ferulic acid, sinapic acid, or chlorogenic acid, compounds that are common in many fruits. Hydroxycinnamic acids (HCAs) are present in fruits in combined forms involving either an ester bond between the carboxylic group and one of the alcohol groups of an organic compound or a bond with one of the phenolic groups (Macheix et al., 1990; Lafay and Gil-Izquierdo, 2008). Only exceptional situations, such as attack of fungal pathogens or anaerobiosis, can cause the accumulation of the free forms. HCAs can also combine, for example, with flavonoids, lignin, suberin, and cutin.

Caffeic acid is the most common hydroxyl derivative and accounts for up to 70% of total hydroxycinnamic acids in fruits (Macheix et al., 1990). *P*-coumaric acid is present in most fruits but is often found to be less abundant than caffeic acid. However, in some *Citrus* fruits, such as pineapple and certain raspberry varieties it can be predominant. *Citrus* fruits have also been described as one of the best dietary sources of sinapic and ferulic acids (Clifford, 2000). High ferulic acid levels have also been detected, for example, in peppers and raspberries (Macheix et al., 1990).

Biosynthesis of cinnamic acids occur via cinnamate 4-hydroxylase (C4H), which catalyzes the synthesis of *p*-hydroxycinnamate from cinnamate. 4-coumarate:CoA ligase (4CL) converts *p*-coumarates to corresponding coenzyme A esters (Bohm, 1998).

C4H, which is a cytochrome P450 enzyme, has been found to exist in two distinct families, class I and class II. These classes are functionally identical but show only 60% sequence similarity. Millard et al. (2007) introduced sense constructs of C4H to create transgenic tomatoes. Some of the transgenic plants showed an increase in the levels of flavonols, naringenin and rutin in the fruit and all lines tested increased in leaf chlorogenic acid. Interestingly, almost all transgenic lines showed considerable reductions in stem lignin correlated with partial sense suppression of C4H.

In many plants, *4CL* occurs as gene families consisting of two to three members. In some species 4CL appears to possess identical catalytic properties, whereas in *Arabidopsis* and soybean *4CL* genes are associated with the unique phenylpropanoid end products. In raspberry (*Rubus idaeus*), three *4CL* genes have been characterized, each of which has differential expression pattern (Kumar and Ellis, 2003).

Monolignols, Lignans, and Lignin

Monolignols (coniferyl alcohol, sinapyl alcohol, paracoumaryl alcohol) are source materials for biosynthesis of lignans and lignin. Monolignols are synthesized in the cytosol and polymerized into lignin (Umezawa, 2010).

Lignans are dimers $(C_6–C_3)_2$, formed by tail-to-tail linkage of two coniferyl or sinapyl alcohol units. These compounds belong to the group of phytoestrogens, which are defined as a group of nonsteroidal plant secondary metabolites with structural and functional similarity to 17β-estradiol. Besides the ability to bind to estrogen receptors, phytoestrogens can act as antioxidants and inhibitors of enzymes such as tyrosine kinase and DNA topoisomerase. Based on this bioactivity, they have potentially beneficial effects for human health, for instance, in preventing breast cancer (Adolphe et al., 2010). In addition to lignans, isoflavonoids that are found in leguminous plants belong to phytoestrogens. The best sources of lignans are flax and sesame seeds in addition to rye and other cereals. In fruits, lignans are not abundant, except for some lignans like secoisolariciresinol that are found in substantial concentrations in apricots and berries, such as bearberries (*Rubus fruticosus*), strawberries (*Fragaria*), and many *Vaccinium* berries. The lignan, matairesinol is also found in many fruits and berries, albeit at lower levels (Mazur et al., 2000; Kuhnle et al., 2009).

Lignin is the most prominent polymer on earth, forming an integral part of the secondary cell walls of plants (Vogt, 2009). Lignin is not generally found in the pulp of fleshy fruits, with the exception of special tissues like sclereids or "stone cells" in some fruits, for example, pear. However, lignin is abundant in the integuments of dry fruits and in certain seeds. It also forms the major part of the stone of drupes. Mechanical wounding or pathogen attack can cause the formation of lignin-like compounds in fruit tissues (Macheix et al., 1990).

Lignin is synthesized via oxidative coupling of *p*-hydroxycinnamyl alcohols (monolignols) and related compounds that are formed in the cinnamate/monolignol pathway (Umezawa, 2010). Cinnamoyl-CoA (CCoA) reductase (CCR) and cinnamyl-alcohol dehydrogenase (CAD) are the key enzymes in the formation of monolignols. However, information on enzymatic reactions from coniferyl alcohol leading to various types of lignans is limited (Vogt, 2009). In lignin biosynthesis, hydroxylation and methylation steps catalyzed by cytochrome P450 enzymes (e.g., ferulate-5 hydroxylase) and S-adenosyl-L-methionine (AdoMet)-dependent *O*-methyltransferases (OMT) determine the contribution of guaiacyl and syringyl monomers (units resulting from monolignols, when incorporated into the lignin polymer) to gymnosperm and angiosperm pattern. In *Arabidopsis*, cation-dependent CCoAOMT1 and cation-independent caffeic acid OMT (COMT) are involved in the formation of hydroxylated monolignol precursors (Vogt, 2009). According to the current model, the lignification process involves peroxidases and/or laccases to provide the oxidative capacity in the cell wall (Vanholme et al., 2010).

Coumarins

Coumarins (1,2-benzopyrones) are lactones derived from *o*-hydroxycinnamic acids by cyclization and ring closure between the o-hydroxy and carboxy groups (Macheix et al., 1990).

Figure 5.2 Rutaceae family including *Citrus* species contains specific phenylpropanoids of coumarins and flavanones that are not present in other fruits.

Classification of coumarins can be made according to the chemical structure. The most important classes are the simple coumarins (benzo-α-pyrones syn 1,2-benzopyrone), including coumarin itself, the 7-oxygenated coumarins (furanocoumarins), pyranocoumarins (benzodipyran-2-ones), and phenylcoumarins (benzo-benzopyrones) (Bourgaud et al., 2006). Most of the coumarins are present in the free form in cells, although glycosylated forms also exist.

Coumarins can be found in many fruits. For instance, the Rutaceae family including *Citrus* species contains over 350 different coumarins, many of which are not detected in other fruits (Fig. 5.2). Also, plants belonging to the Apiaceae (Umbelliferae) family, for example, angelica (*Angelica archangelica*) and parsnip (*Pastinaca sativa*), are rich in coumarins, especially furanocoumarins (Zobel, 1997).

Quantitative and qualitative changes in coumarins during fruit development have been detected. For instance, in Japanese bitter orange (*Poncirus trifoliate*), unripened fruit contained predominantly simple coumarins, but ripened fruit furanocoumarins (Guiotto et al., 1977).

It has been suggested that large concentrations of furanocoumarins within dry fruit and seed tissues would have a role in autoinhibition of germination (Zobel, 1997). For human health, coumarin and several of its derivatives have been found to possess anticancer properties (Singh et al., 2011; Harada et al., 2010). The name coumarin is derived from tonka bean tree (*Coumarouna odorata* Aubl.), from which coumarin was first isolated (Zobel, 1997).

Coumarins are synthesized via hydroxylation, glycolysis, and cyclization of cinnamic acid. The 2*H*-1benzopyran-2-one core structure of coumarins is formed via the *ortho*-hydroxylation of cinnamates, *trans/cis* isomerization of the side chain, and lactonization. However, the major details of the biosynthesis of coumarins are still unknown, especially many P450-dependent enzymatic steps (Bourgaud et al., 2006). Recently, knockout of CCoAOMT1, which has a major role in lignin biosynthesis, led to a reduction of the coumarin scopoletin in *Arabidopsis* roots (Kai et al., 2008). The results suggest that feruloyl-CoA is a key precursor in scopoletin biosynthesis.

Stilbenoids

Stilbenoids are hydroxylated derivates of stilbene. The best-known stilbene is resveratrol that is found, for example, in wine grapes (*Vitis* sp.) and in *Vaccinium* berries, especially lingonberries (*Vaccinium vitis-idaea*). In general, higher resveratrol concentrations are found in red grape cultivars compared to white or pink varieties, but the variation is high also between red varieties (Gatto et al., 2008).

Resveratrol has been postulated to be involved in the health benefits associated with a moderate consumption of red wine (Chong et al., 2009). Several studies have demonstrated that resveratrol together with proanthocyanidins are the major compounds present in grapes and wines responsible for cardioprotection (Bertelli and Das, 2009). There is also an increasing body of evidence that resveratrol could have a role in preventing obesity and obesity-related syndromes, such as type 2 diabetes, by boosting energy metabolism (Dal-Pan et al., 2010).

In the phenylpropanoid pathway, stilbene synthase (STS) catalyzes the biosynthesis of the stilbene backbone from three malonyl-CoA and one CoA ester of a cinnamic acid derivate. STS shares extensive homology with chalcone synthase (CHS)—the first key enzyme of the flavonoid pathway. Both CHS and STS use *p*-coumarouyl-CoA and malonyl-CoA as substrates and synthesize the same linear tetraketide intermediate. However, STS uses a different cyclization mechanism involving an aldol condensation with an additional decarboxylation leading to the loss of one carbon as CO_2 (Chong et al., 2009).

Stress-related regulation of stilbene biosynthesis has been well documented in grapevine and also, for example, in peanuts (Sobolev, 2008; Chong et al., 2009). The expression of STS genes and accumulation of stilbenes are induced upon infection with different fungal pathogens, including powdery mildew (*Erysiphe necator*), downy mildew (*Plasmopara viticola*), gray mold (*Botrytis cinerea*), and various *Aspergillus* species (Sobolev, 2008; Chong et al., 2009). Also UV light, ozone, methyl jasmonate, and ethylene-induced regulation of stilbene biosynthesis have been detected (Chong et al., 2009). Gatto et al. (2008) demonstrated ripening and genotype specific control of stilbene biosynthesis in grapes. However, no specific transcription factor responsible for the regulation of the stilbene biosynthetic pathway has been reported thus far.

Flavonoids

Flavonoids are one of the best-studied secondary metabolites in plants. Over 10 000 naturally occurring flavonoids have been described and in addition to hydroxycinnamic acids, flavonoids possess the most important and common phenolic compounds in fruits (Harborne and Williams, 2000; Crozier et al., 2009; Martens et al., 2010).

All flavonoids are built upon the same C_{15} (C_6-C_3-C_6) basic skeleton, in which the three carbon bridge is commonly cyclized with oxygen. The flavonoid classes differ in the oxidation

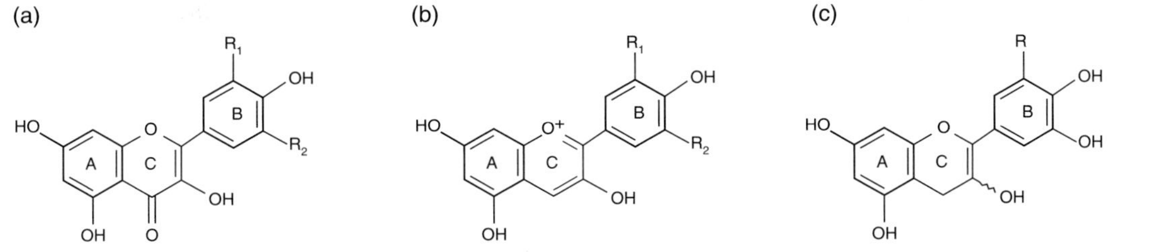

Flavonol aglycons:
kaempferol $R_1 = R_2 = H$
Quercetin $R_1 = OH, R_2 = H$
Myrisetin $R_1 = R_2 = OH$
Isorhamnetin R_1 $OCH_3 = R_2 = H$

Anthocyanidins:
Pelargonidin $R_1 = R_2 = H$
Cyanidin $R_1 = OH, R_2 = H$
Delphinidin $R_1 = R_2 = OH$
Peonidin R_1 $OCH_3 = R_2 = H$
Petudinin R_1 $OCH_3 = R_2 = OH$
Malvidin $R_1 = R_2 = OCH_3$

Flavan-3-ols:
(+)-Catechin (2R,3S) R = H
(-)-Epicatechin (2R,3R) R = H
(+)-Gallocatechin (2R,3S) R = OH
(-)-Epigallocatechin (2R,3R) R = OH

Figure 5.3 Chemical structures of the most common flavonoids in fruits: flavonol aglycons (a), anthocyanidins (b), and flavan-3-ols (c).

level of the central pyran ring except chalcones, in which it is open. Most flavonoids reported in the literature are glycosides of a relatively small number of flavonoid aglycons (Bohm, 1998). Three of the numerous flavonoid classes, including anthocyanins, flavonols, and flavan-3-ols, are highly widespread in fruits. The other classes are less important in fruits, except in special cases like flavanone hesperidin in *Citrus* fruits. Certain classes of flavonoids are only found in particular groups of plants, for example, isoflavonoids in leguminous plants (Macheix et al., 1990). Fruits, in general, are among the best dietary sources of flavonoids (Fig. 5.3).

Flavonols (3-hydroxyflavones) are characterized by a double bond between C-2 and C-3 and the presence of a hydroxyl group at C-3 (Bohm, 1998). The most common flavonols are quercetin, kaempferol, myricetin, and isorhamnetin (Fig. 5.2). In fruits, flavonols are almost exclusively present as glycosides, mainly as monoglycosides but also in di- or rarely triglycosides. Glycosylation occurs preferentially on the hydroxyl group in 3-position. The possible sugar residues are D-glucose, D-galactose, L-rhamnose, L-arabinose, D-xylose, and D-glucuronic acid.

The health benefits of flavonols are related to their high antioxidant activity similar to other flavonoid and phenylpropanoid compounds (Stevenson and Hurst, 2007). Knekt et al. (2002) analysed the relationship of the flavonoid intake with the risk of chronic diseases. They found that the persons with higher quercetin intake had lower mortality from ischemic heart disease, lung cancer, asthma, and type 2 diabetes. Also other flavonols (kaempferol, myricetin, naringenin, and hesperitin) were connected with the reduced risk on the chronic diseases.

Flavonol quercetin is present in all fruits and kaempferol, in approximately 80% of fruits, whereas myricetin and isorhamnetin are less frequent. Myricetin is often found in dark-colored fruits, which also contain delphinidin, anthocyanins, and its derivatives. The flavonol glycosides are usually most abundant in the skin and peel fo the fruit, and decrease in concentration toward the central core (Aherne and O'Brien, 2002). Onions are recognized as one of the best sources of flavonols, but also fruits such as apples and apricots in addition to tomato and

Figure 5.4 Anthocyanins are the main pigments coloring fruits and berries with reddish and bluish hues.

many berries are good sources of flavonol quercetin (Aherne and O'Brien, 2002; Naczk and Shahidi, 2006).

Anthocyanins are the most conspicuous group of flavonoids. They are the main pigments in fruits and berries, generating the characteristic blue, purple, or reddish hues (Fig. 5.4). Besides anthocyanins, carotenoids and betacyanins are also common pigments in plants. Anthocyanins are often markers of ripening as many fruits accumulate the anthocyanins only during the ripening phase (Macheix et al., 1990).

Anthocyanins consist of an anthocyanidin aglycon bound to one or more sugar moieties. Anthocyanidins are hydroxylated and methoxylated derivatives of phenyl-2-benzopyrylium and they are present in cation form in acid media. Thus they are charged molecules and the structure of which displays stabilization by resonance with several mesomeric forms (Macheix et al., 1990; Bohm, 1998). The commonly occurring anthocyanidins are cyanidin (Cy), delphinidin (Dp), pelargonidin (Pg), peonidin (Pn), petunidin (Pt), and malvidin (Mv) (Fig. 5.2). In fruits, cyanidin is the most common anthocyanidin as it is found in over 82% of fruits (Macheix et al., 1990). The next most common anthocyanidins are delphinidin and pelargonidin. Delphinidins and their derivatives, petunidins and malvidins, are responsible for dark bluish colors whereas pelargonidins together with cyanidins are found in bright-red-colored berries such as strawberries, raspberries, and cherries (Macheix et al., 1990; Jaakola et al., 2002).

Anthocyanins always have a sugar at the 3-position, but additional sugar residues at 5-position and 7-position are also possible. The possible sugar moieties are D-glucose, D-galactose, L-rhamnose, L-arabinose, D-xylose, and D-glucuronic acid. The anthocyanin glycosides can be modified through acylation and through complexation with other flavonoids or metal ions. Also differences in cell pH, copigmentation, and ionic complexation affect the final hue of the tissue (Bohm, 1998).

Acylation of anthocyanins by organic acids is common in fruits. In grapes, it has been shown that *p*-coumaric acid can acylate all the anthocyanins present, whereas caffeic acid only combines with malvidin 3-glucose (Macheix et al., 1990).

Health-related properties of anthocyanins have received a great deal of attention during the last decades. Anthocyanins are strong antioxidants and potent inhibitors of lipid peroxidation, which play a role in the prevention of cardiovascular and neuronal illnesses, cancer, and diabetes (Zafra-Stone et al., 2007). In recent years, many studies have shown evidence of the effect of anthocyanin-rich diet in prevention of age-related neural disorders such as Alzheimer's disease (Tarozzi et al., 2008; Darvesh et al., 2010).

Fruits and berries, especially dark-colored berries such as bilberry (*Vaccinium myrtillus*), chokeberry (*Aronia*), blueberries, and blackcurrant (*Ribes nigrum*) are the best dietary sources of anthocyanins (Määttä-Riihinen et al., 2004).

Flavan-3-ols are found as monomers as well as structural units in proanthocyanidin polymers. The most common flavan-3-ols are (+)-catechin (2,3-*trans*-), (−)-epicatechin (2,3-*cis*-), (+)-gallocatechin and (+)-epigallocatechin (Fig. 5.3c).

Proanthocyanidins (PAs) are the most complex members of the flavonoid family (Bohm, 1998). On acid hydrolysis, proanthocyanidins are converted to anthocyanidins, which form the basis of the name and the classical assays of these compounds (Xie and Dixon, 2005). The trivial name of PAs, condensed tannis, refers to the traditional use of PAs as tanning agents to complex with proteins.

Proanthocyanidin structures vary depending upon the stereochemistry and hydroxylation pattern of the flavan-3-ol starter and extension units (Xie and Dixon, 2005). The most common flavanol–flavanol linkages are carbon–carbon bonds between the 4-position of the upper unit and the 6- or 8-position of the lower unit (B-type PAs). In A-type proanthocyanidins, linkages occur between C4→C8 and an ether bond O7→C2 (Xie and Dixon, 2005). The length of the proanthocyanidin chains may vary from dimers to high-molecular-weight polymers.

Proanthocyanidins are known for contributing astringent flavor to food. In fruits, proanthocyanidins are often produced at the early phases of fruit development before the shift to anthocyanin biosynthesis that occurs at the onset of ripening. It has been suggested that as astringent and defensive compounds, they would provide protection against predation of unripe fruits and pathogens (Harborne, 1997). The PAs found in grape skins have been of considerable interest because of their importance for flavor and astringency of wine.

It has been hypothesized that the free radical scavenging properties of PAs may reduce the risk of cardiovascular diseases and cancer (de Pascual-Teresa et al., 2010; Song et al., 2010). Specific A-type trimeric PAs are found in high levels in cranberries (*Vaccinium macrocarpon, Vaccinium oxycoccos*) and lingonberries (*Vaccinium vitis-idaea*) (Määttä-Riihinen et al., 2005). Cranberries and lingonberries have also been shown to offer protection against urinary tract infections in clinical trials, and the A-type PAs have been suggested to be the active ingredients responsible for the effect (Howell et al., 1998; Kontiokari et al., 2001).

Best sources of proanthocyanidins are cocoa beans, certain berries like chokeberries (*Aronia*), rosehip (*Rosa* ssp.) and blueberries (*Vaccinium* ssp.).

Biosynthesis of flavonoids—Flavonoids are synthesized via the phenylpropanoid pathway. The specific flavonoid pathway begins with the condensation of one molecule of 4-coumaroyl-CoA and three molecules of malonyl-CoA, yielding naringenin chalcone. This reaction is carried out by a Type III polyketide synthase known as chalcone synthase CHS. Chalcone is isomerized to (2S)-flavanone by chalcone flavanone isomerase (CHI) and from these central intermediates, the pathway diverges into the side branches resulting in different classes of flavonoids. Flavanone 3-hydroxylase (F3H), a member of the 2-oxo-glutarate-dependent dioxygenase (2-ODD) class, catalyzes the stereospecific 3β-hydroxylation of (2S)-flavones to dihydrokaempferol (DHK), which can serve as precursor to flavonols, anthocyanins, flavan-3-ols, and proanthocyanidins.

Cytochrome P450 monooxygenases, flavonoid 3′hydroxylase (F3′H), and flavonoid 3′5′-hydroxylase (F3′5′H) are enzymes responsible for the hydroxylation of the B-ring of the DHK, leading to dihydroquercetin or dihydromyricetin, respectively (Han et al., 2010). Another member of the 2-ODD class, flavonol synthase (FLS) catalyzes the dehydrogenation of 3-hydroxyflavanones to corresponding flavonols. The B-ring hydroxylation pattern determines also the structures of corresponding anthocyanins and flavan-3-ols.

In competition with FLS, dihydroflavonol 4-reductase (DFR) can utilize dihydroflavonols in formation of leucoanthocyanins, which are further converted to anthocyanidins by anthocyanidin synthase (ANS). Anthocyanins are formed with glycosylation by UDP-glucose:flavonoid-O-glucosyltransferases (UFGTs) and methyltranferases (MTs) are responsible for the formation of O-methylated anthocyanins such as peonidin, petunidin, and malvidin.

Flavan-3-ols, the monomeric units of proanthocyanidins, are converted either from leucocyanidin by leucoanthocyanidin reductase (LAR) leading to (+)-catechins or from anthocyanidin by anthocyanidin reductase (ANR) leading to (−)-epicatechins. The polymerization of proanthocyanidins is believed to occur in vacuoles followed by conversion to brown oxidation products, quinones (Lepiniec et al., 2006; Zhao and Dixon, 2009). However, the nature of the intermediates transported to vacuole and the polymerization process is not fully understood (Xie and Dixon, 2005; Zhao and Dixon, 2009; Zhao et al., 2010). It is still not clear whether the polymerization proceeds enzymatically by laccase type of enzyme or nonenzymatically.

The enzymes involved in flavonoid biosynthesis are localized in the cytosol. After biosynthesis, flavonoids are transported to vacuoles or cell walls (Koes et al., 2005). In addition to this intracellular transport, there is also some evidence of the long-distance transport of flavonoids in plants (Buer et al., 2007). It has been suggested that the enzymes involved in flavonoid biosynthesis would be acting as membrane-associated multienzyme complex, which have an implication on overall efficiency, specificity, and regulation of the pathway (Winkel-Shirley, 1999; Jørgensen et al., 2005).

Regulation of the flavonoid biosynthesis—The biosynthesis of flavonoids in fruits is highly controlled at the developmental level. However, external factors like temperature, light conditions, and nutrition also play a role in the regulation of flavonoid accumulation (Koes et al., 2005). In general, the genetic background of the plant determines the content of phenolic compounds in fruits, whereas external factors can cause qualitative or quantitative changes in the composition of phenolic compounds in ripe fruits.

Several studies have shown that light exposure can increase the concentration of anthocyanins and flavonols in fruit skin (Spayd et al., 2002; Pereira et al., 2006; Takos et al., 2006) and that shading of fruits has an opposite effect (Bakhshi and Arakawa, 2006; Fujita et al., 2006). A high temperature (30°–35°C) has been shown to decrease the anthocyanin content in apple and grape skin tissues (Bakhshi and Arakawa, 2006; Mori et al., 2007). Moreover, an excess of nitrogen fertilizer is known to decrease the flavonoid content in many fruits (Stefanelli et al., 2010). Carbone et al. (2009) analysed the influence of genetic, developmental, and external factors on flavonoid biosynthesis in strawberry. They found out that the natural variation of flavan-3-ol and anthocyanin levels were affected more by genetic background than by environmental factors. However, flavonol levels, in addition to epicatechin/catechin ratio and degree of polymerization in proanthocyanidins, showed significant environmental effects.

Transcriptional regulation of the flavonoid biosynthesis—At the molecular level, anthocyanin biosynthesis is regulated via coordinated transcriptional control of the enzymes in the biosynthetic pathway by the interaction with DNA-binding R2R3 MYB transcription factors, MYC-like basic helix-loop-helix (bHLH), and WD40 repeat proteins (Broun, 2005; Ramsay and Glover, 2005; Allan et al., 2008).

Plant MYBs have been implicated in controlling pathways of diverse secondary metabolites, development, signal transduction, and disease resistance (Allan et al., 2008). *MYB* genes are characterized by a structurally conserved DNA-binding domain consisting of single or multiple imperfect repeats, 100–160 bp in length. Different MYB families have evolved after duplications of their DNA-binding domains (Matus et al., 2008). The MYBs associated with the flavonoid pathway are of two repeat (R2R3) classes, which is most abundant in plants. In *Arabidopsis thaliana*, 126 R2R3 subfamily members have been characterized, of which 13 are implicated in the regulation of flavonoids (Yanhui et al., 2006; Matus et al., 2008).

The R2R3 MYB transcription factors have been shown to interact closely with bHLH transcription factors, which are defined by a bHLH domain of 18 basic amino acids followed by two regions of hydrophobic α-helices separated by an intervening loop (Allan et al., 2008). The bHLH family has also been associated with range functions in plants. In Arabidopsis, 161 bHLH genes are found (Heim et al., 2003). The bHLH regulators have not been shown to bind DNA, but it has been suggested that they might activate transcription by recruiting an unknown factor that binds to a second *cis*-element promoter of the anthocyanin biosynthesis genes (Koes et al., 2005). The WD40 proteins are highly conserved, even between plants and animals. Their precise molecular function in regulation of anthocyanin biosynthesis is still unclear, but it has been suggested that WD40 proteins might activate the downstream MYB and bHLH protein complex posttranscriptionally (Koes et al., 2005).

Regulation of flavonoid biosynthesis in fruits has most intensively been studied in wine grapes. Matus et al. (2008) described 108 R2R3 MYB family members from grapes. Of those, 14 genes were associated with the regulation of anthocyanin or general flavonoid pathway. Biosynthesis of anthocyanins in grape berries is regulated by VvMYBA1, VvMYBA2, and VvMYBA3 transcription factors, homologs of *Arabidopsis* AtMYB75, AtMYB113, and AtMYB114 (Azuma et al., 2008). Another MYB transcription factor, VvMYBPA1, regulates specifically biosynthesis of proanthocyanidins during the development of grape berries (Bogs et al., 2007). Some grape MYB genes have been found to be involved in regulation of biosynthesis of both anthocyanins and proanthocyanidins such as VvMYB5a and VvMYB5b (Deluc et al., 2008). The ectopic expression of *VvMYB5* in tomato caused dwarfism, modified leaf structure, floral abnormalities, smaller fruit clusters, and pigmented and glossy fruits at the green mature stage (Mahjoub et al., 2009). Moreover, phenylpropanoid metabolism in transgenic tomatoes was downregulated and biosynthesis of β-carotene upregulated. Recently, Czemmel et al. (2009) reported the characterization of a light-inducible MYB (VvMYBF1) that controls specifically the biosynthesis of flavonols during grape berry development.

Also in apples (*Malus domestica*), light-induced expression of MYB genes involved in regulation of anthocyanin biosynthesis has been detected (Takos et al., 2006; Ban et al., 2007). These MYB genes involved in regulation of anthocyanin biosynthesis (*MdMYB1*, *MdMYB10*, and *MdMYBA*) have been characterized from different apple cultivars (Takos et al., 2006; Ban et al., 2007; Espley et al., 2007). The genes partially share similar functions, as all of them are expressed especially in red parts of the apple skin. *MdMYB10* is also involved in the determination of anthocyanin biosynthesis of fruit flesh and leaves. Recently, it has been proven that these genes are actually allelic (Lin-Wang et al., 2010). In the same study, the orthologs of these MYBs were isolated from several other commercially important rosaceous fruits—pears, plums, cherries, peaches, raspberries and strawberries (Lin-Wang et al., 2010).

Most of the MYB transcription factors that are in control of the biosynthesis of anthocyanins and other flavonoids are positive regulators that enhance the expression of the structural flavonoid pathway genes. However, in strawberry an R2R3 MYB family member, *FaMYB1*, acts to repress the transcription of the anthocyanin pathway genes (Aharoni et al., 2001). The ectopic expression of *FaMYB1* in tobacco flowers led to a reduction of anthocyanin pigments and flavonols as well as decrease in expression of the corresponding structural genes. It has been discovered that toward the end of ripening, the expression of both *FaMYB1* and *FaMYB10* (*MdMYB1*/*MdMYBA*/*MdMYB10* orthologs) increases along with the accumulation of anthocyanins (Aharoni et al., 2001; Lin-Wang et al., 2010). The upstream regulators that control the expression of these ripening-specific MYBs with opposite actions still remain elusive.

Some of the key genes that are involved in the regulation of fruit ripening have been identified (Giovannoni, 2007; Seymour et al., 2008). Many of those are *SEPALLATA*- and *SQUAMOSA*-class MADS box transcription factors. However, links between these regulators and the downstream effectors that control the changes in color have been poorly understood. Recently, a *SQUAMOSA*-class transcription factor (*VmTDR4*) involved in the regulation of anthocyanin accumulation has been identified (Jaakola et al., 2010). *VmTDR4* is a homolog of tomato *TDR4 and Arabidopsis FRUITFULL* genes and its expression was both spatially and temporally tightly associated with anthocyanin biosynthesis in bilberry fruits. The functional link between *VmTDR4* and anthocyanin biosynthesis was demonstrated using virus-induced gene silencing. The silencing of *VmTDR4* affected the expression of the MYB genes in bilberry fruits. However, the mechanism with which this MADS box gene is involved in regulation of anthocyanin biosynthesis is still not understood.

Engineering Elevated Levels of Flavonoids and Other Phenylpropanoids

With the diverse contribution to fruit quality and potential beneficial health properties of flavonoids and other phenylpropanoids, there is increasing interest in the engineering of their metabolism (Ververidis et al., 2007). In addition to metabolic engineering of fruit plants, there is also interest in transforming plant genes to microorganisms, with focus on developing bioreactor-based production systems for specific flavonoids by heterologous protein expression (Fowler and Koffas, 2009).

In fruits, several attempts have been made to increase the content of phenylpropanoid compounds in tomato. Increase in tomato flavonoid levels has been achieved with genetic engineering using the structural genes from the flavonoid pathway either from tomato or from other species (Verhoeyen et al., 2002; Schijlen et al., 2006). Accumulation of anthocyanins in tomato has been achieved with genetic engineering of flavonoid-biosynthesis-related transcription factors (Mathews et al., 2003). Butelli et al. (2008) transformed two transcription factors from snapdragon encoding bHLH (*Del*) and an R2R3 MYB family member (*Ros1*) to tomato. The transgenic fruits showed strong accumulation of anthocyanins. Moreover, feeding of these anthocyanin-rich tomatoes to cancer-susceptible mice resulted in prolonged life span.

Upregulation of proanthocyanidin biosynthesis has been achieved, for instance, with overexpression of vine grape VvMYBPA1 transcription factor in *Arabidopsis* (Bogs et al., 2007), or persimmon (*Diospyros kaki*) MYB genes (DkMyb2, DkMyb4) in persimmon or kiwifruit (*Aktinidia delicosa*) callus tissues (Akagi et al., 2009, 2010). When Maize bHLH transcription factor (*Lc*) was transformed to apple, increase in anthocyanins, proanthocyanidins and other phenylpropanoids were detected (Li et al., 2007).

The accumulation of new information on genome, transcriptome and metabolome levels will open new possibilities for biotechnological modification of fruit traits such as color, flavor, health properties, or pathogen resistance. The extent to which these approaches will be adopted for commercial production remains to be seen.

Conclusion

Phenylpropanoids constitute the majority of the secondary metabolites in fruits. Several thousand phenylpropanoid compounds have been described in plants, of which those belonging to classes of hydroxycinnamic acids, flavonols, anthocyanins and flavan-3-ols (proanthocyanidins) are highly abundant in fruits. Phenylpropanoids contribute in various ways to quality characteristics of ripe fruit and fruit products. The multiple health benefits these compounds possess are of major interest in the present-day fruit production and breeding programs. The recent advances in understanding the regulation of phenylpropanoid biosynthesis have opened new possibilities to modify the qualitative and the quantitative composition of these compounds in fruits, both with traditional and modern breeding and cultivation technologies. The simultaneous accumulation of new information on health benefits of specific phenylpropanoid compounds or groups for human health and the factors that modulate the biosynthesis of the same compounds in plants will give guidelines for the future fruit breeding work.

References

Adolphe, J.L., Whiting, S.J., Juurlink, B.H.J., Thorpe, L.U., and Alcorn, J. (2010) Health effects with consumption of the flax lignan secoisolariciresinol diglucoside. *The British Journal of Nutrition*, 103, 929–938.

Aharoni, A., De Vos, C.H.R., Wein, M., Sun, Z., Greco, R., Kroon, A., Mol, J.N.M., and O'Connell, A.P. (2001) The strawberry *FaMYB1* transcription factor suppresses anthocyanin and flavonol accumulation in transgenic tobacco. *The Plant Journal: for cell and molecular biology*, 28, 319–332.

Aherne, S.A., and O'Brien N.M. (2002) Dietary flavonols: chemistry, food content, and metabolism. *Nutrition*, 18, 75–81.

Akagi, T., Ikegami, A., Tsujimoto, T., Kobayashi, S., Sato, A., Kono, A., and Yonemori, K. (2009) DkMyb4 is a Myb transcription factor involved in proanthocyanidin biosynthesis in persimmon fruit. *Plant Physiology*, 151, 2028–2045.

Akagi, T., Ikegami, A., and Yonemori, K. (2010) DkMyb2 wound-induced transcription factor of persimmon (*Diospyros kaki* Thunb.) contributes to proanthocyanidin regulation. *Planta*, 232, 1045–1059.

Allan, A.C., Hellens, R.G., and Laing, W.A. (2008) MYB transcription factors that colour our fruit. *Trends in Plant Science*, 13, 99–102.

Azuma, A., Kobayashi, S., Mitani, N., Shiraishi, M., Yamada, M., Ueno, T., Kono, A., Yakushiji, H., and Koshita, Y. (2008) Genomic and genetic analysis of *Myb*-related genes that regulate anthocyanin biosynthesis in grape berry skin. *Theoretical and Applied Genetics*, 117, 1009–1019.

Badimon, L., Vilahur, G., and Pardo, T. (2010) Nutraceuticals and atherosclerosis: human trials. *Cardiovascular Therapeutics*, 28, 202–215.

Bakhshi, D., and Arakawa, O. (2006) Induction of phenolic compounds biosynthesis with light irradiation in the flesh of red and yellow apples. *Journal of Applied Horticulture*, 8, 101–104.

Ban, Y., Honda, C., Hatsuyama, Y., Igarashi, M., Bessho, H., and Moriguchi, T. (2007) Isolation and functional analysis of a MYB transcription factor gene that is a key regulator for the development of red coloration in apple skin. *Plant and Cell Physiology*, 48, 958–970.

Bertelli, A.A.A., and Das, D.K. (2009) Grapes, wines, resveratrol and heart health. *Journal of Cardiovascular Pharmacology*, 54, 468–476.

Bogs, J., Jaffe, F.W., Takos, A.M., Walker, A.R., and Robinson, S.P. (2007) The grapevine transcription factor VvMYBPA1 regulates proanthocyanidin synthesis during fruit development. *Plant Physiology*, 143, 1347–1361.

Bohm, B. (1998) *Introduction of Flavonoids*. Harwood Academic Publishers, Singapore.

Bourgaud, F., Hehn, A., Larbat, R., Doerper, S., Gontier, E., Kellner, S., and Matern U. (2006) Biosynthesis of coumarins in plants: a major pathway still to be unravelled for cytochrome P450 enzymes. *Phytochemistry Reviews*, 5, 293–308.

Broun, P. (2005) Transcriptional control of flavonoid biosynthesis: a complex network of conserved regulators involved in multiple aspects of differentiation in Arabidopsis. *Current Opinion in Plant Biology*, 8, 272–279.

Buer, C.S., Muday, G.K., and Djordjevic, M.A. (2007) Flavonoids are differentially taken up and transported long distance in Arabidopsis. *Plant Physiology*, 145, 478–490.

Butelli, E., Titta, L., Giorgio, M., Mock, H.P., Matros, A., Peterek, S., Schijlen, E.G.W.M., Hall, R.D., Bovy, A.G., Luo, J., and Martin, C. (2008) Enrichment of tomato fruit with health-promoting anthocyanins by expression of select transcription factors. *Nature Biotechnology*, 26, 1301–1308.

Carbone, F., Preuss, A., De Vos, R.C.H., D'Amico, E., Perotta, G., Bovy, A.G., Martens, S., and rosati, C. (2009) Developmental, genetic and environmental factors affect the expression of flavonoid genes, enzymes and metabolites in strawberry fruits. *Plant Cell & Environment*, 32, 1117–1131.

Chang, A., Lim, M.H., Lee, S.W., Robb, E.J., and Nazar, R.N. (2008) Tomato phenylalanine ammonia-lyase gene family, highly redundant but strongly underutilized. *Journal of Biological Chemistry*, 283, 33591–33601.

Chong, J., Poutaeaud, A., and Hugueney, P. (2009) Metabolism and roles of stilbenes in plants. *Plant Science*, 177, 143–155.

Clifford, M.N. (2000) Chlorogenic acids and other cinnamates—nature, occurrence, dietary burden absorption and metabolism. *Journal of the Science of Food and Agriculture*, 80, 1033–1043.

Crozier, A., Jaganath, I.B., and Clifford, M.N. (2009) Dietary phenolics: chemistry, bioavailabilty and effects on health. *Natural Product Reports*, 26, 1001–1043.

Czemmel, S., Stracke, R., Weisshaar, B., Cordon, N., Harris, N.N., Walker, A.R., Robinson, S.P., and Bogs, J. (2009) The grapevine R2R3-MYB transcription factor VvMYBF1 regulates flavonol synthesis in developing grape berries. *Plant Physiology*, 151, 1513–1530.

Dal-Pan, A., Blanc, S., and Aujard, F. (2010) Resveratrol suppresses body mass gain in a seasonal non-human primate model of obesity. *BMC Physiology*, 10, 11.

D'Archivio, M., Filesi, C., Vari, R., Scazzocchio, B., and Masella, R. (2010) Bioavailability of the polyphenols: status and controversies. *International Journal of Molecular Sciences*, 11, 1321–1342.

Darvesh, A.S., Carroll, R.T., Bishayee, A., Geldenhuys, W.J., and Van Der Schyf, C.J. (2010) Oxidative stress and Alzheimer's disease: dietary polyphenols as potential therapeutic agents. *Expert Review of Neurotherapeutics*, 10, 729–745.

Deluc L., Bogs, J., Walker, A.R., Ferrier, T., Decendit, A., Merillon, J.M., Robinson, S.P., and Barrieu, F. (2008) The transcription factor VvMYB5b contributes to the regulation of anthocyanin and proanthocyanidin biosynthesis in developing grape berries. *Plant Physiology*, 147, 2041–2053.

de Pascual-Teresa, S., Moreno, D.A., and Garsia-Viguera, C. (2010) Flavanols and anthocyanins in cardiovascular health: a review of current evidence. *International Journal of Molecular Sciences*, 11, 1679–1703.

Espley, R.V., Hellens, R.P., Putterill, J., Stevenson, D.E., Kutty-Amma, S., and Allan, A.C. (2007) Red colouration in apple fruits is due to the activity of the MYB transcription factor, MdMYB10. *The Plant Journal: for cell and molecular biology*, 49, 414–427.

Fowler, Z.L., and Koffas, M.A.G. (2009) Biosynthesis and biotechnological production of flavanones: current state and perspectives. *Applied Microbiology and Biotechnology*, 83, 799–808.

Fujita, A., Goto-Yamamoto, N., Aramaki, I., and Hashizume, K. (2006) Organ-specific transcription of putative flavonol synthase genes of grapevine and effects of plant hormones and shading on flavonol biosynthesis in grape berry skins. *Bioscience, Biotechnology, and Biochemistry*, 70, 632–638.

Gatto, P., Vrhovsek, U., Muth, J., Segala, C., Romualdi, C., Fontana, P., Pruefer, D., Stefanini, M., Moser, C., Mattivi, F., and Velasco, R. (2008) Ripening and genotype control stilbene accumulation in healthy grapes. *Journal of Agricultural and Food Chemistry*, 56, 11773–11785.

Giovannoni, J.J. (2007) Fruit ripening mutants yield insights into ripening control. *Current Opinion in Plant Biology*, 10, 283–289.

Guiotto, A., Rodighiero, P., Pastorni, G., and Celon, E. (1977) Coumarins from unripe fruits of Ponicus trifoliate. *Phytochemistry*, 16, 1257–1260.

Han, Y., Vimolmangkang, S., Soria-Guerra, R.E., Rosales-Mendoza, S., Zheng, D., Lygin, A.V., and Korban, S.S. (2010) Ectopic expression of apple F3′H genes contributes to anthocyanin accumulation in the Arabidopsis tt7 mutant grown under nitrogen stress. *Plant Physiology*, 153, 806–820.

Harada, K., Kubo, H., Tomighara, Y., Nishioka, K., Takahashi, J., Momose, M., Inoue, S., and Kojima, A. (2010) Coumarins as novel 17β-hydroxysteroid dehydrogenase type inhibitors for potential treatment of prostate cancer. *Bioorganic & Medicinal Chemistry Letters*, 20, 272–275.

Harborne, J. (1997) Phytochemistry of fruits and vegetables: an ecological overview. In: *Phytochemistry of Fruits and Vegetables* (eds. F.A. Tomas-Barberan, and R.J. Robins), pp. 335–367Oxford University Press Inc., New York, NY.

Harborne, J.B., and Williams, C. (2000) Advances in flavonoid research since 1992. *Phytochemistry*, 55, 481–504.

He, Q., and Luo, Y. (2007) Enzymatic browning and its control in fresh-cut produce. *Stewart Postharvest Review*, 3, 16.

Heim, M.A., Jakoby, M., Werber, M., Martin, C., Weisshaar, B., and Bailey P.C. (2003) The basic helix-loop-helix transcription factor family in plants: a genome-wide study of protein structure and functional diversity. *Molecular Biology and Evolution*, 20, 735–747.

Howell, A.B., Vorsa, N., Der Maderosian, A., and Foo, L.Y. (1998) Inhibition of the adherence of P-fimbriated Escheria coli to uroepithelial-cell surfaces by proanthocyanidin extract from cranberries. *The New England Journal of Medicine*, 339, 1085–1086.

Jaakola, L., Määttä, K., Pirttilä, A.M., Törrönen, R., Kärenlampi, S., and Hohtola, A. (2002) Expression of structural genes involved in anthocyanin biosynthesis in relation to anthocyanin and flavonoid levels during bilberry (*Vaccinium myrtillus* L.) fruit development. *Plant Physiology*, 130, 729–739.

Jaakola, L., Poole, M., Jones, M., Kämäräinen-Karppinen, T., Koskimäki, J.J., Hohtola, A., Häggman, H., Fraser, P., Manning, K., King, G.J. Thomson, H., and Seymour G.B. (2010) A SQUAMOSA MADS-box gene regulates anthocyanin accumulation in bilberry fruits. *Plant Physiology*, 153, 1619–1629.

Jørgensen, K., Rasmussen, A.V., Morant, M., Nielsen, A.H., Bjarnholt, N., Zagrobelny, M., Bak, S., and Møller, B.L. (2005) Metabolon formation and metabolic channeling in the biosynthesis of plant natural products. *Current Opinion in Plant Biology*, 8, 280–291.

Kai, K., Mizutani, M., Kawamura, N., Yamamoto, R., Tamai, M., Yamaguchi, H., Skata, K., and Shimizu, B. (2008) Scopoletin is biosynthesized via ortho-hydroxylation of feruloyl CoA by a 2-oxoglutarate-dependent dioxygenase in Arabidopsis thaliana. *The Plant Journal: for cell and molecular biology*, 55, 989–999.

Knekt, P., Kumpulainen, J., Järvinen, R., Rissanen, H., Heliövaara, M., Reunanen, A., Hakulinen, T., and Aromaa, A. (2002) Flavonoid intake and risk of chronic diseases. *American Journal of Clinical Nutrition*, 76, 560–568.

Koes, R., Verweij, W., and Quattrocchio, F. (2005) Flavonoids: a colorful model for the regulation and evolution of biochemical pathways. *Trends in Plant Science*, 10, 236–242.

Kontiokari, T., Sundqvist, K., Nuutinen, M., Pokka, T., Koskela, M., and Uhari, M. (2001) Randomised trial of cranberry-lingonberry juice and Lactobacillus GG drink for the prevention of urinary tract infections in women. *British Medical Journal*, 322, 1571–1573.

Kuhnle, G.G.C, Dell'Aguila, C., Aspinall, S.M., and Runswick, S.A. (2009) Phytoestrogen content of fruits and vegetables commonly consumed in the UK based on LC-MS and 13C-labelled standards. *Food Chemistry*, 116, 542–554.

Kumar, A., and Ellis, B.E. (2003) 4-Coumarate:CoA ligase gene family in Rubus idaeus: cDNA structures, evolution, and expression. *Plant Molecular Biology*, 31, 327–340.

Lafay, S., and Gil-Izquierdo, A. (2008) Bioavailability of phenolic acids. *Phytochemistry Reviews*, 7, 301–311.

Lepiniec, L., Debeaujon, I., Routaboul, J.M., Baudry, A., Pourcel, L., Nesi, N., and Caboche, M. (2006) Genetics and biochemistry of seed flavonoids. *Annual Review of Plant Biology*, 57, 405–430.

Li, H., Flachowsky, H., Fischer, T.C., Hanke, M.V., Forkmann, G., Treutter, D., Schwab, W., Hoffman, T., and Szankowski, I. (2007) Maize Lc transcription factor enhances biosynthesis of anthocyanins, distinct proanthocyanidins and phenylpropanoids in apple (*Malus domestica* Borkh.). *Planta*, 226, 1243–1254.

Lin-Wang, K., Bolitho, K., Grafton, K., Kortstee, A., Karunairetnam, S., McGhie, T.K., Espley, R.V., Hellens, R.P., and Allan, A.C. (2010) An R2R3 MYB transcription factor associated with regulation of the anthocyanin biosynthetic pathway in Rosaceae. *BMC Plant Biology*, 10, 50.

Määttä-Riihinen, K.R., Kamal-Eldin, A., Mattila, P.H., Gonzales-Paramas, A.M., Törrönen, R. (2004) Distribution and contents of phenolic compounds in eighteen Scandinavian berry species. *Journal of Agricultural and Food Chemistry*, 52, 4477–4486.

Määttä-Riihinen, K.R., Kähkönen, M.P., Törrönen, A.R., and Heinonen, I.M. (2005) Catechins and proanthocyanidins in berries of *Vaccinium* species and their antioxidant activity. *Journal of Agricultural and Food Chemistry*, 53, 8485–8491.

Macheix, J.J., Fleuriet, A., and Billot, J. (1990) *Fruit Phenolics*. CRC Press, Boca Raton, FL.

Mahjoub, A., Hernould, M., Joubes, J., Decendit, A., Mars, M., Barrieu, F., Hamdi, S., and Delrot, S. (2009) Overexpression of a grapevine R2R3-MYB factor in tomato affects vegetative development, flower morphology and flavonoid and terpenoid metabolism. *Plant Physiology and Biochemistry*, 47, 551–561.

Martens, S., Preuss, A., and Matern, U. (2010) Multifunctional flavonoid dioxygenases: flavonol and anthocyanin biosynthesis in Arabidopsis thaliana L. *Phytochemistry*, 71, 1040–1049.

Mathews, H., Clendennen, S.K., Caldwell, C.G., Liu, X.L., Connors, K., Matheis, N., Schuster, D.K., Menasco, D.J., Wagoner, W., Lightner, J., and Wagner, R.D. (2003) Activation tagging in tomato identifies a transcriptional regulator of anthocyanin biosynthesis, modification, and transport. *The Plant Cell*, 15, 1689–1703.

Matus, J.T., Aquea, F., and Acre-Johnson (2008) Analysis of the grape MYB R2R3 subfamily reveals expanded wine quality-related clades and conserved gene structure organization across *Vitis* and *Arabidopsis* genomes. *BMC Plant Biology*, 8, 83.

Mazur, W.M., Uehara, M., Wähälä, K., and Adlercreuz, H. (2000) Phyto-oestrogen content of berries, and plasma concentrations and urinary excretion of enterolactone after a single strawberry-meal in human subjects. *The British Journal of Nutrition*, 83, 381–387.

Millard, M.J., Long, M., Donovan, G., Fraser, P.D., Boudet, A.M., Bramley, P.M., and Bolwell, G.P. (2007) Introduction of sense constructs of cinnamate 4-hydroxylase (CYP73A24) in transgenic tomato plants shows opposite effects on flux into stem lignin and fruit flavonoids. *Phytochemistry*, 68, 1497–1509.

Mori, K., Goto-Yamamoto, N., Kitayama, M., and Hashizume, K. (2007) Loss of anthocyanins in red-wine grape under high temperature. *Journal of Experimental Botany*, 58, 1935–1945.

Naczk, M., and Shahidi, F. (2006) Phenolics in cereals, fruits and vegetables: occurrence, extraction and analysis. *Journal of Pharmaceutical and Biomedical Analysis*, 4, 1523–1542.

Pereira, G.E., Gaudillere, J.P., Pieri, P., Hilbert, G., Maucourt, M., Deborde, C., Moing, A., and Rolin, D. (2006) Microclimate influence on mineral and metabolic profiles of grape berries. *Journal of Agricultural and Food Chemistry*, 54, 6765–6775.

Ramsay, N.A., and Glover, B.J. (2005) MYB-bHLH-WD40 protein complex and the evolution of cellular diversity. *Trends in Plant Science*, 10, 63–70.

Schijlen, E., de Vos, C.H.R., Jonker, H., van den Broek, H., Molthoff, J., van Tunen, A., Martens, S., and Bovy, A. (2006) Pathway engineering for healthy phytochemicals leading to the production of novel flavonoids in tomato fruit. *Plant Biotechnology Journal*, 4, 433–444.

Seymour, G.B., Poole, M., Manning, K., and King, G. (2008) Genetics and epigenetics of fruit development and ripening. *Current Opinion in Plant Biology*, 11, 58–63.

Singh, R.K., Lange, T.S., Kim, K.K., Brard, L. (2011) A coumarin derivative (RKS262) inhibits cell-cycle progression, causes pro-apoptotic signaling and cytotoxicity in ovarian cancer cells. *Investigational New Drugs*, 29, 63–72.

Sobolev, V.S. (2008) Localized production of phytoalexins by peanut (*Arachis hypogaea*) kernels in response to invasion by *Aspergillus* species. *Journal of Agricultural and Food Chemistry*, 56, 1949–1954.

Song, X., Siriwardhana, N., Rathore, K., Lin, D., and Wang, H.C.R. (2010) Grape seed proanthocyanidin suppression of breast cell carcinogenesis induced by chronic exposure to combined 4-methylnitrosamino)-1-(3-pyridyl)-1-butanone and benzo[a]pyrene. *Molecular Carcinogenesis*, 49, 450–463.

Spayd, S.E., Tarara, J.M., Mee, D.L., and Ferguson, J.C. (2002) Separation of sunlight and temperature effects on the composition of *Vitis vinifera* cv. Merlot berries. *American Journal of Enology and Viticulture*, 53, 171–182.

Stefanelli, D., Goodwin, I., and Jones, R. (2010) Minimal nitrogen and water use in horticulture: effects on quality and content of selected nutrients. *Food Research International*, 43, 1833–1843.

Stevenson, D.E., and Hurst, R.D. (2007) Polyphenolic phytochemicals—just antioxidants or much more? *Cellular and Molecular Life Sciences*, 64, 2900–2916.

Takos, A.M., Jaffe, F.W., Jakob, S.R., Bogs, J., Robinson, S.P., and Walker A.R. (2006) Light-induced expression of a MYB gene regulates anthocyanin biosynthesis in red apples. *Plant Physiology*, 142, 1216–1232.

Tarozzi, A., Merlicco, A., Morroni, F., Franco, F., Cantelli-Forti, G., Teti, G., Falconi, M., Hrelia, P. (2008) Cyanidin 3-O-glucopyranoside protects and rescues SH-SY5Y cells against amyloid-beta peptide-induced toxicity. *NeuroReport*, 19, 1483–1486.

Umezawa, T. (2010) The cinnamate/monolignol pathway. *Phytochemistry Reviews*, 9, 1–17.

Vanholme, R., Demedts, B., Morreel, K., Ralph, J., and Boerjan, W. (2010) Lignin biosynthesis and structure. *Plant Physiology*, 153, 895–905.

Verhoeyen, M.E., Bovy, A., Collins, G., Muir, S., Robinson, S., de Vos, C.H.R., and Colliver, S. (2002) Increasing antioxidant levels in tomatoes through modification of the flavonoid biosynthetic pathway. *Journal of Experimental Botany*, 53, 2099–2106.

Ververidis, F., Trantas, E., Douglas, C., Vollmer, G., Kretzschmar, G., and Panopoulos, N. (2007) Biotechnology of flavonoids and other phenylpropanoid-derived natural products. Part II: reconstruction of multienzyme pathways in plants and microbes. *Biotechnology Journal*, 2, 1235–1249.

Vogt, T. (2009) Phenylpropanoid biosynthesis. *Molecular Plant*, 3, 2–20.

Weisshaar, B., and Jenkins, G.I. (1998) Phenylpropanoid biosynthesis and its regulation. *Current Opinion in Plant Biology*, 1, 251–257.

Winkel-Shirley, B. (1999) Evidence for enzyme complexes in the phenylpropanoid and flavonoid pathways. *Physiologia Plantarum*, 107, 142–149.

Xie, D.Y., and Dixon, R.A. (2005) Proanthocyanidin biosynthesis – still more questions than answers? *Phytochemistry*, 66, 212–2144.

Yanhui, C., Xiaoyuan, Y., Kun, H., Meihua, L., Zhaofeng, G., Zhiqiang, L., Yunfei, Z., Xiaoxiao, W., Xiaoming, Q., Yunping, S., Li, Z., Xiaohui, D., Jingchu, L., Xing-Wang, D., Zhangliang, C., Hongya, G., and Li-Jia, Q. (2006)

The MYB transcription factor superfamily of Arabidopsis: expression analysis and phylogenetic comparison with the rice MYB family. *Plant Molecular Biology*, 60, 107–124.

Zafra-Stone, S., Yasmin, T., Bagchi, M., Chatterjee, A., Vinson, J.A., and Bagchi, D. (2007) Berry anthocyanins as novel antioxidants in human health and disease prevention. *Molecular Nutrition and Food Research*, 51, 675–683.

Zhao, J., and Dixon, R.A. (2009) MATE transporters facilitate vacuolar uptake of epicatechin 3'-*O*-glucoside for proanthocyanidin biosynthesis in Medicago truncatula and Arabidopsis. *The Plant Cell*, 21, 2323–2340.

Zhao, J., Pang, Y., and Dixon, R.A. (2010) The mysteries of proanthocyanidin transport and polymerization. *Plant Physiology*, 153, 437–443.

Zobel, A.M. (1997) Coumarins in fruit and vegetables. In: *Phytochemistry of Fruits and Vegetables* (eds F.A. Tomas-Barberan and R.J. Robins), pp. 173–203. Oxford University Press Inc., New York, NY.

6 Biosynthesis of Volatile Compounds

Antonio Granell and José Luis Rambla

Introduction

Volatile compounds are typically low-polarity small molecules with a molecular weight below 250–300 Da and relatively high vapor pressures which can cross membranes and be released into the atmosphere in the absence of a diffusion barrier (Dudareva et al., 2006; Pickersky et al., 2006). Overall they constitute about 1% of the known plant secondary metabolites (Dudareva et al., 2004). It has been reported that volatiles emitted by plants are involved in the defense against pathogens and herbivores, abiotic stress, symbiosis in mycorrhiza, and also in the attraction of pollinators and seed dispersers (Pickersky and Gershenzon, 2002; Reinhard et al., 2004; Walter et al., 2010). The biosynthesis and emission of volatile compounds in fruit is a developmentally regulated process which is finely modulated during ripening. During this process, the fruit's volatile profile changes dramatically. Although the levels of some compounds remain almost constant, and some even decrease, most of them rise sharply during ripening, and this is thought to be a part of the mechanism for attracting seed dispersers. In this sense, it has been described that fruit bats were able to detect and locate fruits and assess their ripening stage exclusively by their olfaction (Luft et al., 2003). Furthermore, as many of the volatile compounds released from ripe fruits are synthesized from substrates with high nutritional value, such as amino acids, fatty acids, or carotenoids, it has been proposed that flavor volatiles may function as important cues to attract animals to nutritious fruits (Goff and Klee, 2006).

From the human perspective, volatile compounds are important contributors to fruit quality, as they play a critical role in flavor. Human perception of flavor involves the integration of many sensorial perceptions, some of physical nature such as texture and mouth feel (Causse et al., 2001) and many due to chemical properties, such as those mediated by the taste and olfactory detection of a large set of primary and secondary metabolites. While the foundation of good flavor relies on an appropriate balance of chemicals that influence taste (Klee, 2010),

The Molecular Biology and Biochemistry of Fruit Ripening, First Edition.
Edited by Graham B. Seymour, Mervin Poole, James J. Giovannoni and Gregory A. Tucker.

flavor complexity is built by the volatiles perceived by the olfactory system (Shepherd, 2006). Although characteristic odor descriptors have been associated with most of the aroma volatiles, defining their specific contribution to the overall aroma is very complex. Some volatiles can even alter the perception of a taste descriptor like sweetness and *vice versa*, revealing a complex interaction between both taste and olfactory systems (Baldwin et al., 2008; Vogel et al., 2010), and indicating that it is the integrated perception of volatile and nonvolatile compounds what we define as "flavor."

Fruits display a large chemical diversity in the nature of volatiles produced. These include compounds in many different chemical functional groups such as aldehydes, ketones, alcohols, esters, organic acids, furanones, or terpenoids. Differences in the volatile profile, that is, quantitative and qualitative chemical composition, between different species are also dramatic. Although fruits of even quite unrelated species can share many of the volatile compounds they produce, their relative abundance can vary widely, and often just a few (those above the odor threshold) out of hundreds of volatiles detected, really contribute to the characteristic aroma of each type of fruit. The diversity in the volatile fingerprint between some cultivated fruits is illustrated in Figure 6.1.

The volatiles emitted by a fruit are dependent on many factors. The production of volatiles is very dynamic, and processes like injuring or chewing a fruit produce a rapid and profound modification in the volatile compounds that are emitted. As a consequence, sample processing and the analytical method employed have a significant influence on both the volatiles detected and their abundance. In addition, results obtained from fruits analysed with different methodologies are often difficult to compare, as each technique tends to produce a particular bias on the volatile profile described. Therefore, when the aim is to compare between different samples, methodological coherence is an essential matter.

A common factor associated with many cultivated fruits is a general loss of flavor which occurred during domestication. This negative effect of human intervention has been particularly intense in the last decades, as breeding programs have focused on characteristics different to flavor, and is in part due to the difficulty to breed for complex, poorly known, multigenic traits such as aroma (Goff and Klee, 2006). Chromatograms in Figures 6.1c and 6.1d illustrate the impoverishment of the volatile profiles in the case of a commercial strawberry variety (*Fragaria x ananassa*) as compared to its close relative, the wild strawberry, *Fragaria vesca*.

Despite the overwhelming complexity of compounds and volatile profiles produced by fruits, most of them can be grouped according to specific metabolic pathways, type of compounds, etc. (Fig. 6.2). The aim of this chapter is to describe the state of the art of these important fruit compounds based on the information available for the most studied fruit models.

Metabolic Pathways

The biosynthesis of volatile compounds comprises of many interconnected metabolic pathways and a complex regulation network, much of which is still unknown. Transcriptional regulation during fruit ripening is a very important aspect but with a few exceptions such as the gene

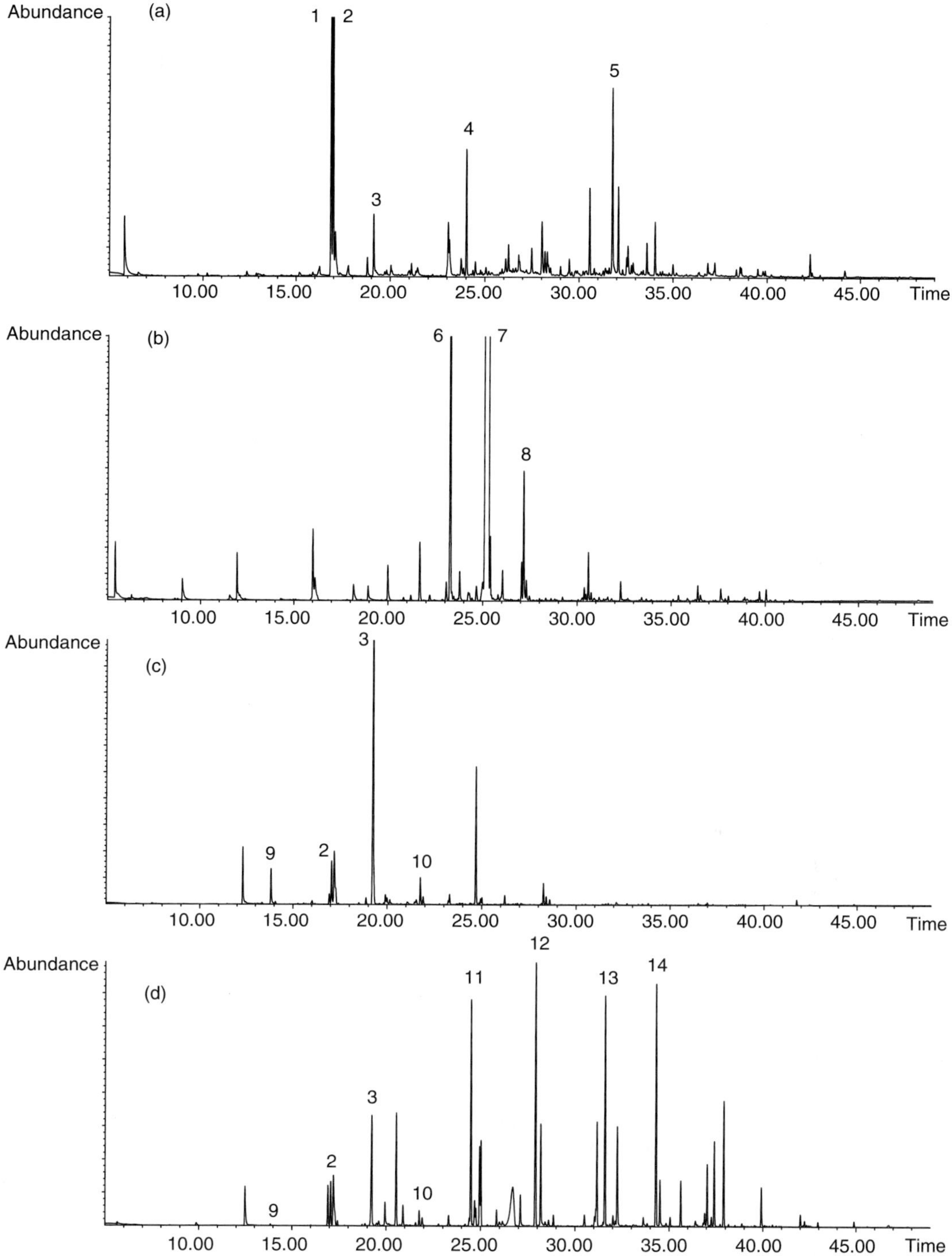

Figure 6.1 Fingerprint of the volatile compounds from homogenized ripe fruits of different species obtained by headspace solid phase microextraction coupled to gas chromatography–mass spectrometry. (a) Tomato (*Solanum lycopersicum*); (b) mandarine (*Citrus clementina*); (c) strawberry (*Fragaria* x *ananassa*); (d) wild strawberry (*Fragaria vesca*). Some of the differential peaks are highlighted: 1, (Z)-3-hexenal; 2, Hexanal; 3, (E)-2-hexenal; 4, 6-methyl-5-hepten-2-one; 5, Methyl salicylate; 6, Myrcene; 7, Limonene; 8, Linalool; 9, Methyl butanoate; 10, Methyl hexanoate; 11, Ethyl hexanoate; 12, 2-nonanone; 13, Octyl acetate; 14, 2-undecanone.

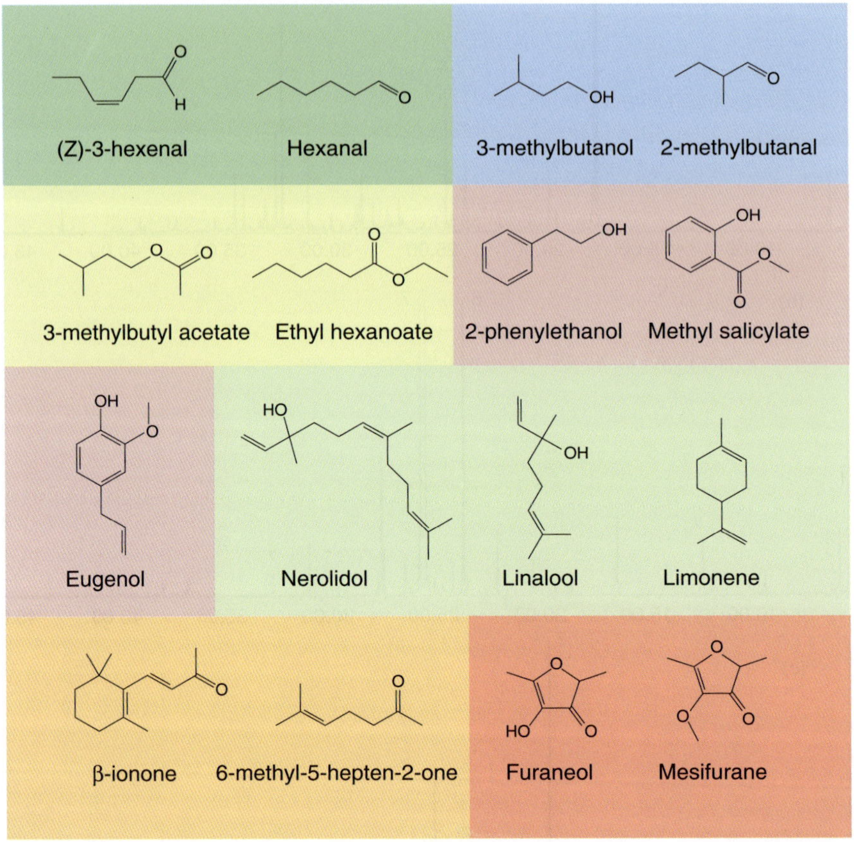

Figure 6.2 Some volatile compounds representing the main metabolic pathways—fatty acid derivatives (Z)-3-hexenal and hexanal; amino-acid-derived 3-methylbutanol and 2-methylbutanal; esters 3-methylbutyl acetate and ethyl hexanoate; benzenoids 2-phenylethanol, methyl salicylate, and eugenol; terpenoids nerolidol, linalool, and limonene; apocarotenoids β-ionone and 6-methyl-5-hepten-2-one; furanones furaneol and mesifurane.

encoding the myeloblastosis (MYB) transcription factor SlODO1 (Orzaez et al., 2009), very little is known of the transcriptional factors regulating specific volatile pathways. An important aspect in volatile regulation seems to be compartmentation, which has a strong influence on substrate availability for the biosynthetic enzymes, and also in the release of volatiles from nonvolatile conjugated forms. The present knowledge on the metabolic pathways leading to the most relevant volatile compounds found in different fruits is described below (Fig. 6.3).

Fatty Acid Derivatives Pathways

Volatiles derived from fatty acids constitute a large class of compounds, which include the so-called green leaf volatiles, a family of C_6 aldehydes, alcohols, and derived esters, that are

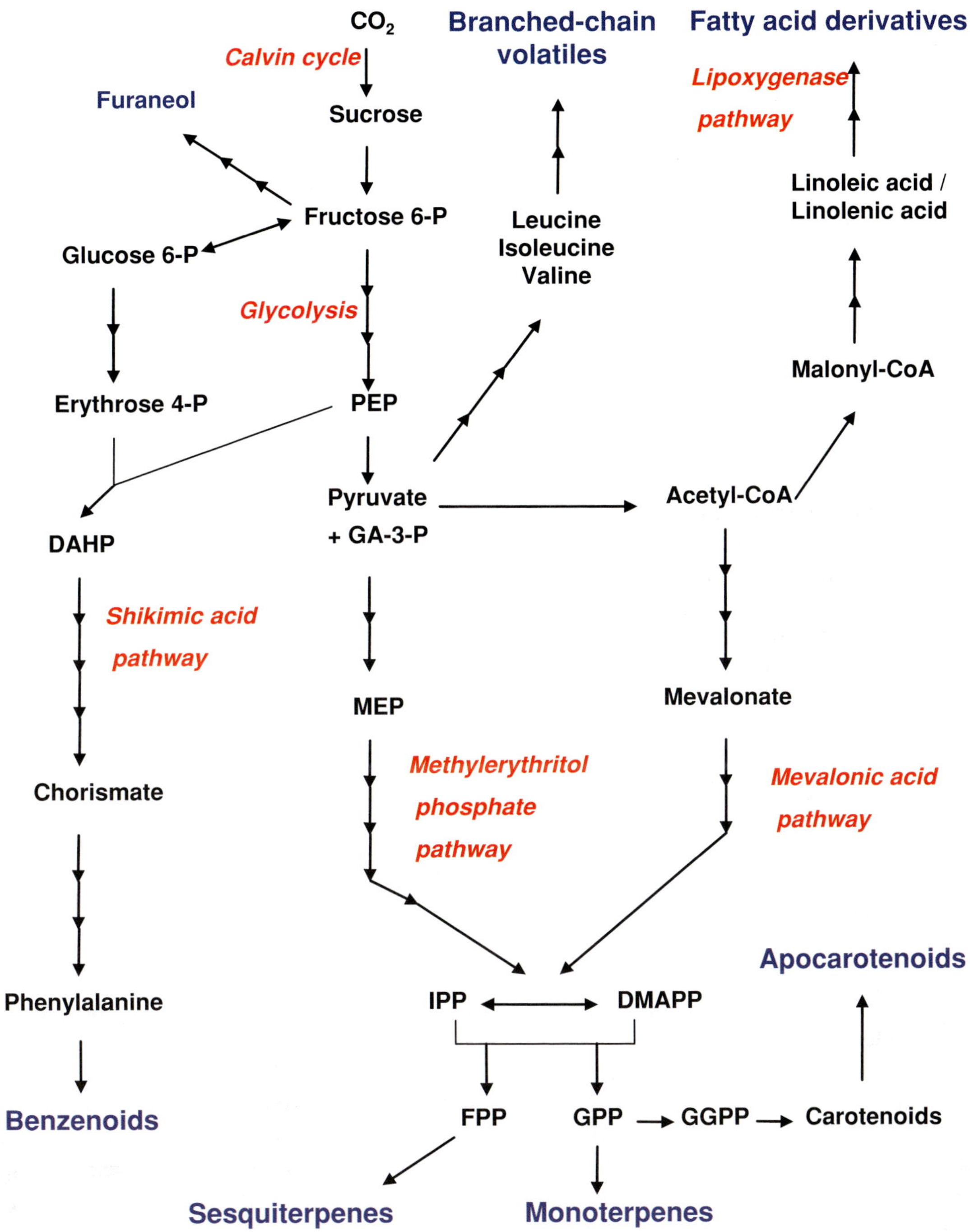

Figure 6.3 Metabolic pathways leading to the biosynthesis of volatile compounds in fruits. The names of the metabolic pathways are in red, and the volatiles are highlighted in blue. DAHP, 3-deoxy-D-arabino-heptulosonate 7-phosphate; DMAPP, dimethylallyl diphosphate; FPP, farnesyl diphosphate; GA-3-P, glyceraldehyde 3-phosphate; GPP, geranyl diphosphate; GGPP, geranyl geranyl diphosphate; IPP, isopentenyl diphosphate; MEP, 2-*C*-methyl-D-erythritol 4-phosphate; PEP, phosphoenolpyruvate.

responsible for the characteristic fresh, "green" aroma typical of leaves when wounded or crushed. Some of them, like (Z)-3-hexenal or hexanal, are abundant compounds in vegetative tissues, particularly when disrupted, and also in some fruits such as tomato where it contributes to their aroma (Buttery et al., 1990). Some of the C_6 aldehydes and alcohols have antibacterial and antifungal properties, and could have a role in inhibiting the invasion of microorganisms into damaged tissues (Ongena et al., 2004). They have also been involved in indirect plant defense against herbivore insects, by attracting predators when infested (Kessler et al., 2004; Allmann and Baldwin, 2010), and it has been hypothesized for a long time that fatty-acid-derived volatiles act as plant-to-plant signals, and therefore have been involved in plant–plant communication (Baldwin et al., 2002).

The catabolism of fatty acids proceeds through different pathways, such as β-oxidation, α-oxidation, or lipoxygenase pathways. In many cases, this catabolism results in the production of a variety of volatile compounds including aldehydes, alcohols, and even lactones and methyl jasmonate. The most thoroughly studied fatty-acid-derived volatile pathway and the most important for fruit aroma is the lipoxygenase–hydroperoxide pathway. The first family of enzymes in this pathway is the lipoxygenase (LOX). These are nonheme iron-containing fatty acid dioxygenases that catalyze the regio- and stereospecific dioxygenation of polyunsaturated fatty acids with a (1Z,4Z)-pentadiene moiety, such as the C_{18} linoleic and linolenic acids, converting them into fatty acid hydroperoxides (Liavonchanka and Feussner, 2006). According to their positional specificity of oxygenation, lipoxygenases can be divided in two groups: 13-LOX, producing 13-hydroperoxides, and 9-LOX, producing 9-hydroperoxides. The hydroperoxides produced are further metabolized by hydroperoxide lyases (HPLs) to produce an aldehyde and an oxoacid. HPL enzymes are members of the cytochrome P450 family, and can also be classified as 13-HPLs or 9-HPLs depending on the type of hydroperoxide they act on, or as 13/9-HPLs when they can act on both 13- and 9-hydroperoxides with a similar efficiency (Matsui, 2006) (Fig. 6.4).

Several isoforms of lipoxygenases exist in most plants, but it seems that only one or a few of them are responsible for the volatile biosynthesis in fruits. This is the case of the tomato TomloxC, the only isoform related to the generation of C_6 volatiles in a family of at least 5-LOX enzymes (Chen et al., 2004). Also in kiwifruit, only AdLox1 and AdLox5 in a family composed of at least 6 isoforms seem to be responsible for the corresponding volatiles released during ripening (Zhang et al., 2006).

13-LOX are the most widespread lipoxygenases. When linolenic acid is their substrate, the products resulting from the sequential action of both LOX and HPL activities are the volatile (Z)-3-hexenal and the nonvolatile 12-oxo-(Z)-9-dodecenoic acid. It is assumed that (Z)-3-hexenal can be further transformed into (E)-2-hexenal, either nonenzymatically or enzymatically by means of a 3Z,2E-enal isomerase (Z3/E2-ISO) (Noordermeer et al., 1999), although this enzyme has not been identified yet. When the substrate is linoleic acid, the volatile product obtained after LOX and HPL activities is hexanal. This volatile can also be produced by means of a 2-alkenal reductase (ALH), which can reduce (E)-2-hexenal to hexanal, as has been described in Arabidopsis (Mano et al., 2002). Apart from the mentioned sources, 13-HPLs have also been identified in fruits such as guava (Tijet et al., 2000), tomato (Howe et al., 2000) and peach (Zhang et al., 2010).

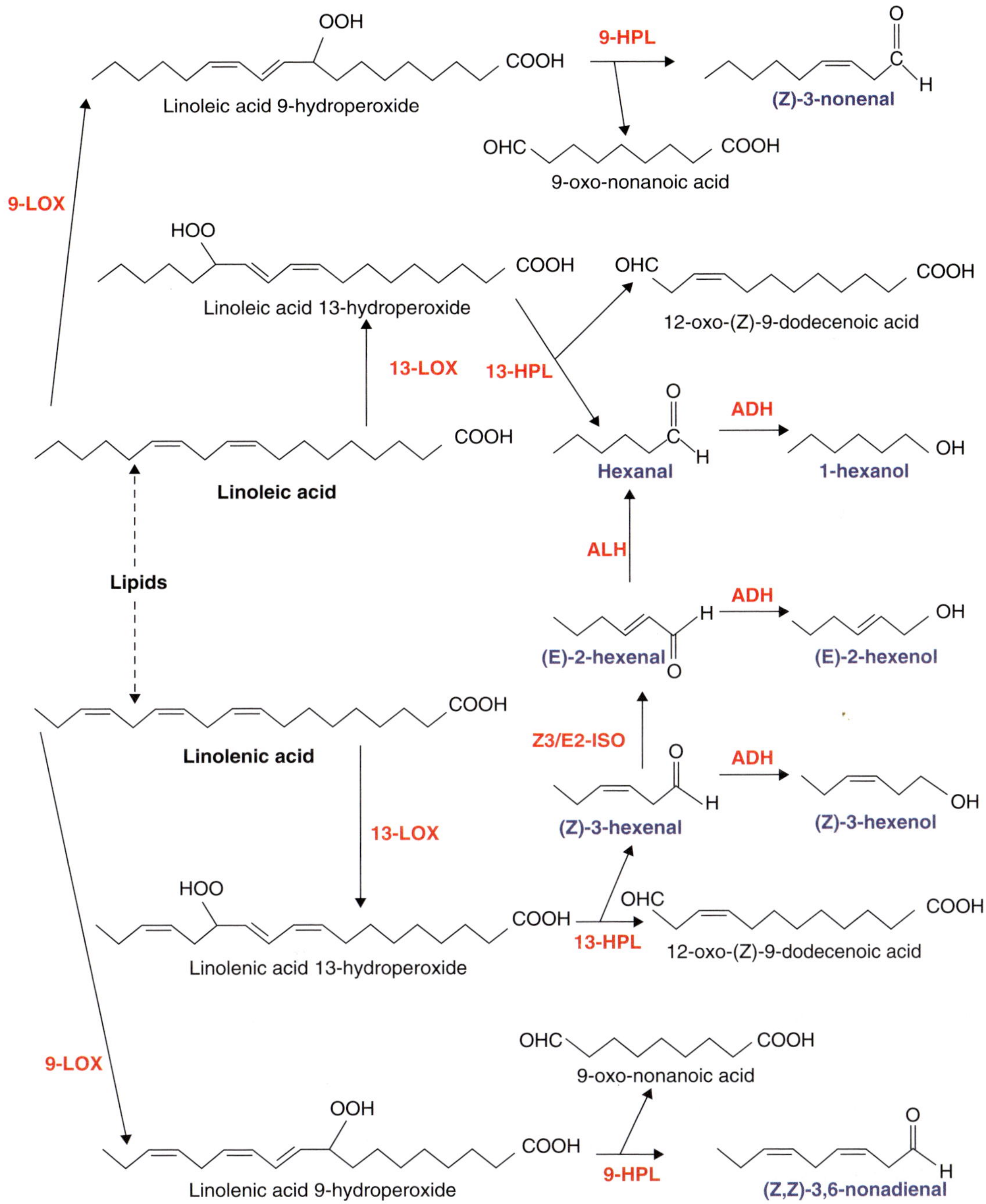

Figure 6.4 Biosynthetic pathway of fatty-acid-derived volatiles in the lipoxygenase–hydroperoxide pathway. Solid arrows indicate reactions in which the enzyme has already been identified; dashed arrows are used when the enzyme has not been identified yet. Volatile compounds are in blue, nonvolatile compounds are in black, and the abbreviated enzymes are in red. ADH, alcohol dehydrogenase; ALH, 2-alkenal reductase; 9-HPL, fatty acid 9-hydroperoxide lyase; 13-HPL, fatty acid 13-hydroperoxide lyase; 9-LOX, 9-lipoxygenase; 13-LOX, 13-lipoxygenase; Z3/E2-ISO, 3Z,2E-enal isomerase.

The action of 9-LOX activities on the same substrates produces 9-hydroperoxides. 9-HPL and 9/13-HPL catalyze the fatty acid oxidative cleavage to produce a volatile C_9 aldehyde and a C_9 oxoacid. The first enzyme with 9-HPL activity was identified in cucumber (Matsui et al., 2000a), where the volatiles produced are responsible for the characteristic cucumber aroma, but ulterior studies identified the same enzymes in many other species (Kuroda et al., 2005; Mita et al., 2005).

The C_6 and C_9 aldehydes produced, as other aldehydes obtained from the different metabolic pathways, like the catabolism of amino acids, can be reduced to alcohols by means of alcohol dehydrogenases (ADHs). These enzymes are involved in the interconversion of aldehydes and alcohols, as described for ADH2 in ripe tomato fruit (Speirs et al., 1998), Cm-ADH1 and Cm-ADH2 in melon (Manríquez et al., 2006), and MiADH1 and MiADH2 in mango (Singh et al., 2010).

Regulation of the biosynthesis of green leaf volatiles is still not completely understood. In plants, fatty acids are not stored as free molecules, but as acylglycerides. The substrates for lipoxygenases are the free fatty acids, but their concentration in intact tissues is negligible. Therefore, LOX activity has to be preceded by the action of a lipase (or lypolytic acyl hydrolase, LAH) that liberates the substrates, which become available for the volatile synthesis pathway. This is supported by an experiment with *Arabidopsis* leaves, where a positive correlation was found between the rapidly decreasing levels of galactolipids following homogenization and the production of (Z)-3-hexenal, in a process that could be repressed by means of a lipase inhibitor (Matsui et al., 2000b). Other studies showed that systemic green leaf volatile production is also increased following wounding or jasmonic acid treatment (Blée, 2002). Also, the synthesis of these volatiles is activated in the ripening of fruits such as tomato, probably due to the loss of integrity of membranes (Klee, 2010). Therefore, regulation of the volatiles produced by the lipoxygenase pathway is considered to rely on a still unidentified lipase, whose activity is enhanced by regulatory signals, fruit ripening, and most notably by tissue disruption, as is produced by wounding or the chewing of the fruit.

Phenylpropanoids and other Benzenoids Pathways

In this group, we have included a set of aromatic compounds derived from the amino acid phenylalanine which constitute a large class of diverse volatiles. These benzenoid compounds play an important role in plant defense against pathogens (Koeduka et al., 2006) and in plant communication (Effmert et al., 2005). They are key components of floral scents typical of roses (Knudsen et al., 1993) and important attractants of pollinators (Schwab et al., 2008). Included in this pathway is vanillin, the most used flavor compound in the world. Some of the volatiles in this pathway have an impact on the aroma of fruits such as tomato, either positively as for 2-phenylethanol (Baldwin et al., 2008) or negatively as for methyl salicylate (Klee, 2010) or guaiacol (Zanor et al., 2009).

According to the chemical structure of the intermediates in the pathway, they could be classified into three subfamilies: phenylpropanoids (with a C_6-C_3 carbon skeleton), benzenoids C_6-C_2, and benzenoids C_6-C_1. Significant progress has been made in the last few years in the understanding of the biosynthesis of C_6-C_2 compounds. Studies on tomato, where high 2-phenylethanol levels are considered desirable, led to the identification of *LeAADC1A*, *LeAADC1B*, and *LeAADC2*, a small family of genes encoding aromatic amino acid decarboxylases that catalyze the conversion of phenylalanine to phenethylamine. Phenethylamine would then require a deamination by a still unknown amine oxidase to be converted to the volatile phenylacetaldehyde, or else a series of uncharacterized reactions to be transformed into 1-nitro-2-phenylethane and phenylacetonitrile (Tieman et al., 2006a). An alternate pathway for the synthesis of phenylacetaldehyde has been described in *Petunia hybrida* by means of a bifunctional phenylacetaldehyde synthase enzyme (PAAS), catalyzing both the decarboxylation of phenylalanine and the posterior amine oxidation (Kaminaga et al., 2006), although there is no evidence for this pathway in fruits. A recent paper suggested a third possible pathway in melon fruits, similar to that described in yeasts (Dickinson et al., 2003), consisting of an initial deamination by an aromatic amino acid transaminase (ArAT), like CmArAT1 described in melon, and ulterior decarboxylation by still unknown enzymes (Gonda et al., 2010).

Phenylacetaldehyde produced from any of these pathways is then reduced to 2-phenylethanol by means of phenylacetaldehyde reductases (PARs). Two enzymes have been identified in tomato, LePAR1 and LePAR2, unidirectionally catalyzing the conversion of aldehyde to alcohol, which could also be implied in the synthesis of benzyl alcohol and cinnamyl alcohol from their respective aldehydes (Tieman et al., 2007).

What we know about the biosynthesis of volatile compounds from the other two subfamilies (phenylpropanoids and benzenoids C_6-C_1) has been obtained in systems different from fruits, but it is reasonable that they provide a picture of the situation that may also occur in fruits (Fig. 6.5).

The first step in the synthesis of phenylpropanoids is the conversion of phenylalanine into (E)-cinnamic acid by means of an L-phenylalanine ammonia-lyase (PAL). The next part of the pathway is shared with that of lignin biosynthesis, and includes a series of hydroxylation and methylation reactions (Humphreys and Chapple, 2002). Although most of the carbon in this pathway is leading to the synthesis of lignin, some of these compounds can also be substrates for the conversion to volatile compounds, as has been described in petunia and sweet basil for eugenol and isoeugenol, both formed from coniferyl acetate by means of an eugenol synthase (EGS) and an isoeugenol synthase (IGS), respectively (Koeduka et al., 2006).

Volatile C_6-C_1 compounds also originate from (E)-cinnamic acid, by shortening of the C_2 moiety in their side chain. Two alternate pathways have been proposed for this, a CoA-dependent β-oxidative pathway and a CoA-independent nonoxidative pathway. Results obtained suggest that both may contribute to the formation of benzenoid volatiles at least in petunia flowers (Boatright et al., 2004). All these volatile compounds synthesized can suffer further modifications such as hydroxylation, acetylation, or methylation, leading to a variety of compounds including alcohols and esters.

Other Amino Acid-derived Volatile Pathways

Amino acids other than phenylalanine are also precursors of volatile compounds in fruits. Particularly important are the branched-chain amino acids leucine, isoleucine, and valine,

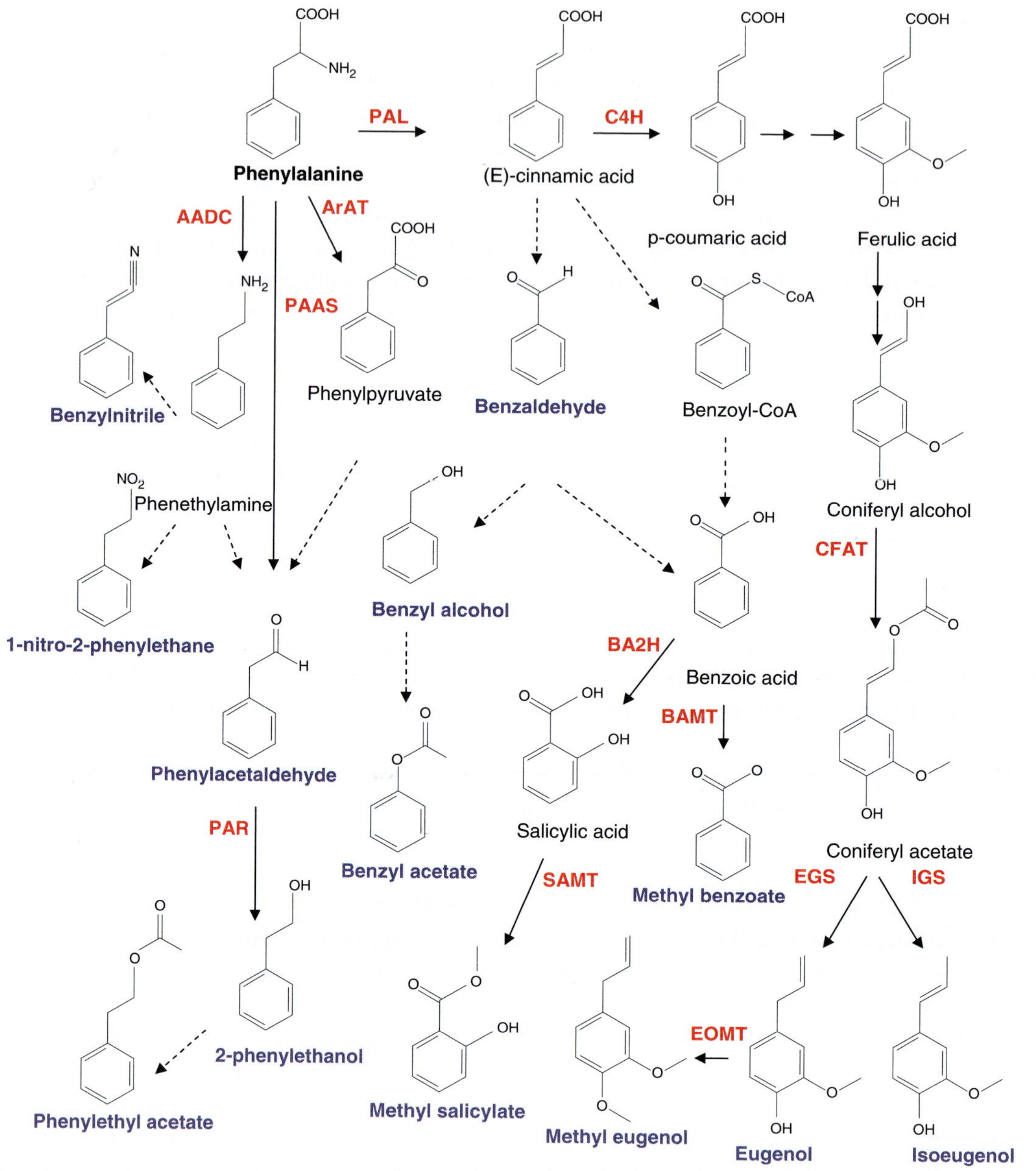

Figure 6.5 Biosynthetic pathway of benzenoids. Solid arrows indicate reactions in which the enzyme has already been identified; dashed arrows indicate probable reactions. Multiple arrows indicate several biochemical steps. Volatile compounds are in blue, nonvolatile compounds are in black, and the abbreviated enzymes are in red. AADC, aromatic amino acid decarboxylase; PAAS, phenylacetaldehyde synthase; ArAT, aromatic amino acid transaminase; PAR, phenylacetaldehyde reductase; PAL, L-phenylalanine ammonia lyase; C4H, cinnamate-4-hydroxylase; CFAT, coniferyl alcohol acyltransferase; BA2H, benzoic acid 2-hydroxylase; EGS, eugenol synthase; IGS, isoeugenol synthase; EOMT, eugenol O-methyltransferase; BAMT, benzoic acid carboxyl methyltransferase; SAMT, salicylic acid methyl transferase.

which are considered precursors of a variety of volatile aldehydes, alcohols, acids, and esters. Many of these compounds, and the esters preeminently, are key constituents of aroma in fruits such as banana (Jordan et al., 2001a), apple (Schumacher et al., 1998), or together with other ester compounds, many varieties of melons (Jordan et al., 2001b).

Their biosynthetic pathway in plants still remains mostly unknown. It has been extensively studied in yeast and bacteria (Marilley and Casey, 2004), and recent studies (Gonda et al., 2010; Maloney et al., 2010) support the hypothesis that their biosynthesis in fruits such as melon or tomato follows a similar pathway. According to this scheme, amino acids would first undergo a transamination reaction to form the corresponding α-keto acids. This is a reversible reaction catalyzed by branched-chain aminotransferases (BCATs), as the CmBCAT1 recently described in melon (Gonda et al., 2010). A group of six BCAT enzymes distributed among chloroplasts, mitochondria, vacuoles, and the cytoplasm have also been characterized in tomato. Apparently, the mitochondria-located SlBCAT1 and SlBCAT2 would be implied in the first step of branched-chain amino acid catabolism, while the enzymes involved in amino acid biosynthesis would be localized to the chloroplast (Maloney et al., 2010). In any case, alterative pathways such as those described for phenylalanine (Kaminaga et al., 2006; Tieman et al., 2006a) could also be possible (Fig. 6.6).

Starting from the α-keto acid substrates and according to what has been described in microorganisms, three alternative biochemical reactions could happen: (1) decarboxylation to form the corresponding aldehydes, (2) reduction to form α-hydroxy acids, or (3) oxidative decarboxylation producing acyl-CoAs which could in turn produce carboxylic acids (Marilley and Casey, 2004). Further reactions from these compounds have already been described, and some of their respective enzymes identified. Aldehydes can be reduced to alcohols by means of ADHs. These alcohols can further react with acyl-CoA molecules by alcohol acyltransferases (AATs) to form esters, which are the compounds with the most odor impact in this group.

The amino acids methionine and cysteine are also substrates for sulfur-containing volatiles, which are key odor compounds in several vegetables (Jones et al., 2004), but have little importance in ripe fruits.

Ester Volatile Pathways

As it has already been mentioned, esters are important compounds with generally low-odor thresholds that provide characteristic aromas which we as humans associate to specific fruits. In general, ester volatiles usually provide "fruity" notes to the aroma, and are among the most abundant compounds in fruits such as apple, strawberry, peach (Seymour et al., 1993), or climacteric varieties of melons (Obando-Ulloa et al., 2008). Furthermore, some may have individually a large impact on the fruit odor, such as 3-methylbutyl acetate and 3-methylbutyl butanoate in banana (Nogueira et al., 2003).

The first committed step in the synthesis of aroma esters is the acyl esterification of alcohols via the corresponding acyl-CoA thioester donor. This reaction is catalyzed by AATs, a family of enzymes within the large protein superfamily named BAHD (D'Auria, 2006). AATs in plants consist of large families of enzymes with different tissue-specific patterns, producing both

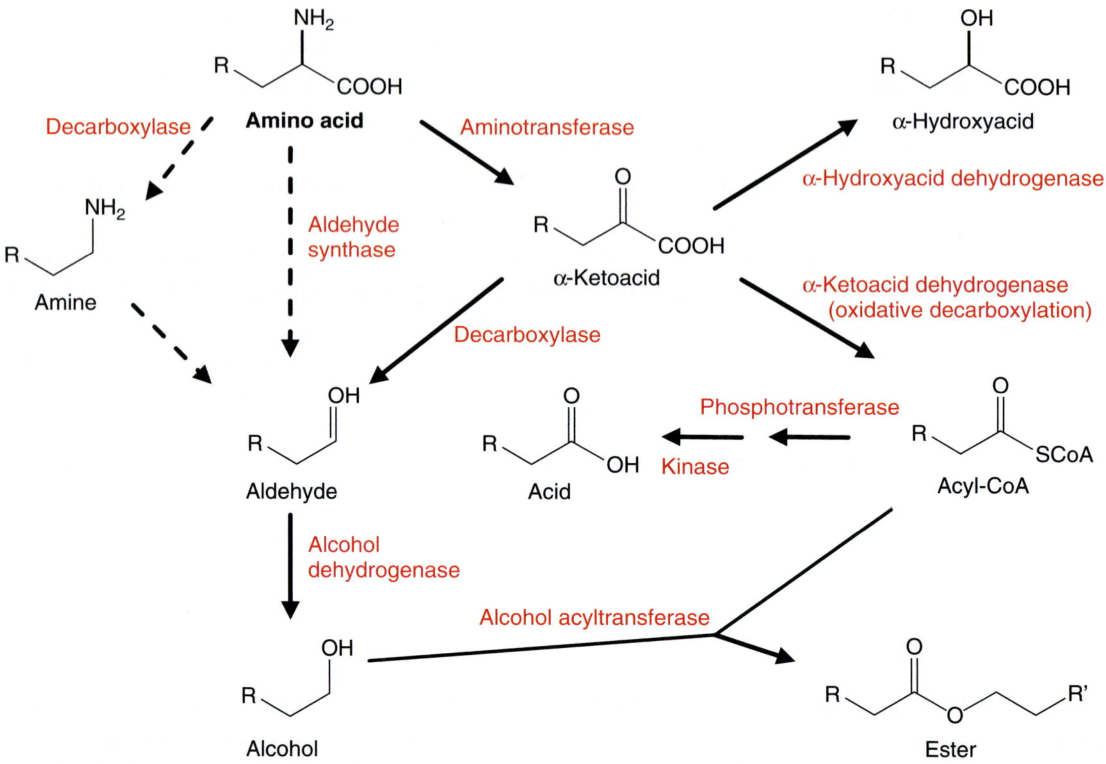

Figure 6.6 Probable biosynthetic pathway of branched amino-acid-derived volatiles. Solid arrows indicate the pathway as described in bacteria (Marilley, and Casey 2004). Dashed arrows indicate other possible reactions in plants. The enzymes implied on each reaction are in red.

volatile and nonvolatile esters, as has been described by genomic and transcriptomic studies in both *Arabidopsis* and *Populus trichocarpa*, where over 60 and 90 putative AAT genes have been identified respectively (Yu et al., 2009). In fruits, a number of AATs involved in the synthesis of volatiles have been reported, such as SAAT in strawberry (Aharoni et al., 2000), CmAAT1, CmAAT3, and CmAAT4 in melon (El-Sharkawy et al., 2005), and VpAAT1 in mountain papaya (Balbontin et al., 2010).

The substrates for the synthesis of esters are either linear or branched-chain or aromatic alcohols and acyl-CoA molecules derived from the catabolism of fatty acids and amino acids. *In vitro* enzyme assays of AAT activity usually show wide substrate specificity, although a given enzyme has differential preferences toward specific acyl-CoAs and alcohols. Nevertheless, *in vitro* substrate preference of the AAT enzymes often is not correlated with the production of the corresponding esters *in vivo*. It has been described that many esters very efficiently produced *in vitro* failed to be detected *in vivo* in the tissues of the species where that particular AAT had been isolated from (Aharoni et al., 2000). Moreover, it has been described that petunia transgenic plants with a strawberry *SAAT* gene were unable to synthesize the esters produced in strawberry.

These results, together with the low correlation observed between substrate preferences and ester production, indicate that often the bottle neck for the biosynthesis of esters *in vivo* is substrate availability (Beekwilder et al., 2004).

A family of enzymes involved in the synthesis of a specific type of esters, the methyl esters, is the SABATH family. This family refers to S-adenosyl-L-methionine-dependent O-methyltransferases, which specifically catalyze the methylation of different small molecular weight substrates such as salicylic acid by SAMT in tomato (Ament et al., 2010; Tieman et al., 2010) producing methyl salicylate, or methyl eugenol and methyl chavicol catalyzed by EOMT and CVOMT from the corresponding alcohols in sweet basil (Gang et al., 2002).

Particularly related to fruit ripening is the FaOMT enzyme from strawberry, which is implicated in the synthesis of vanillin and the methylation of furaneol to mesifurane—key aroma compounds in strawberry (Wein et al., 2002).

Terpenoid Pathway

Volatile terpenoids constitute the largest class of volatile compounds, and have been implicated in diverse scenarios such as plant defense against herbivores and pathogens, plant thermotolerance, and plant reproduction by attracting both pollinators and seed disseminators (Dudareva et al., 2006). Although they only account for a small proportion of the volatile fraction in most of the cultivated fruits, terpenoid volatiles are the compounds responsible for the appreciated floral notes in some aromatic white grape varieties like Muscat or Gewürztraminer (Duchene et al., 2009), are major volatiles in mango fruits (Pino et al., 2005), and are by far the most abundant compounds in most Citrus species, both as part of the essential oils in the peel and in the volatile fraction of the juice (Rouseff et al., 2009).

Volatile terpenoids described in fruits fall into two groups: monoterpenes (C_{10}) and sesquiterpenes (C_{15}), both comprising of a wide range of structurally complex compounds, particularly in the case of sesquiterpenes. All terpenoids are synthesized from the universal five carbon precursors, isopentenyl diphosphate (IPP), and dimethylallyl diphosphate (DMAPP), derived from two alternate biosynthetic pathways localized in different subcellular compartments. In the cytosol, IPP is synthesized from acetyl-CoA by the mevalonic acid (MVA) pathway, and can be further metabolized to DMAPP. In plastids, IPP and DMAPP are derived from pyruvate and glyceraldehyde-3-phosphate via the methylerythritol phosphate (MEP) pathway (Rodríguez-Concepción and Boronat, 2002). Although each of these two pathways has a different subcellular compartmentation, some metabolic cross talk exists between them, particularly in the direction from the plastids to the cytosol (Hemmerlin et al., 2003) (Fig. 6.7).

Once IPP and DMAPP are synthesized, prenyltransferases use them to produce prenyl diphosphates in both plastids and cytosol. The farnesyl diphosphate synthase enzyme (FDS) catalyzes the formation of C_{15} farnesyl diphosphate (FPP), the precursor of all sesquiterpenes, by condensation of two IPP and one DMAPP molecules. Geranyl diphosphate synthase enzyme (GDS) produces C_{10} geranyl diphosphate (GPP), the precursor of monoterpenes, by the condensation of one molecule of IPP and one of DMAPP.

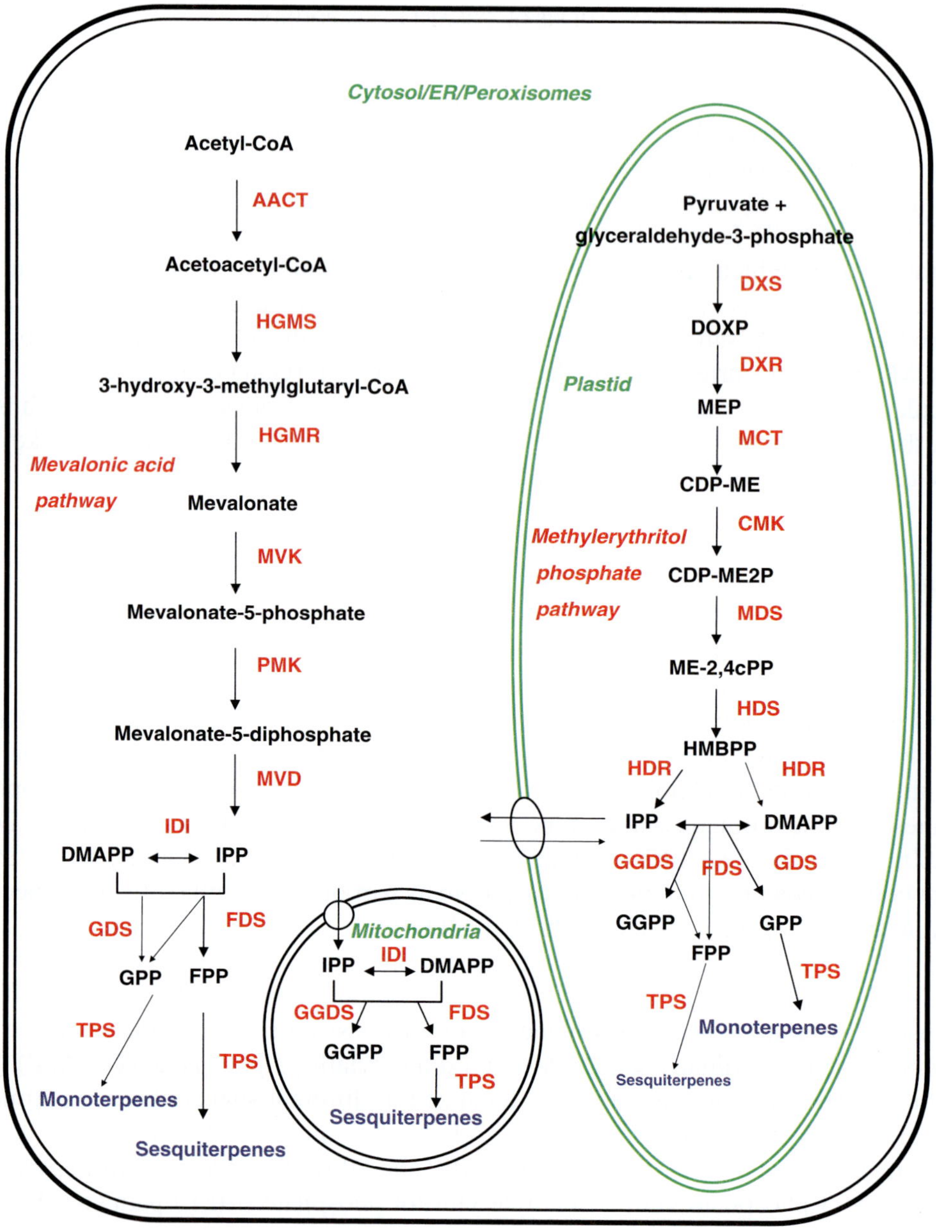

Figure 6.7 Biosynthetic pathways of terpenoid volatiles and their compartmentalization. Volatile compounds are in blue, nonvolatile compounds are in black, and enzymes are in red. AACT, acetoacetyl-CoA thiolase; CDP-ME, 4-(cytidine 5′-diphospho)-2-C-methyl-d-erythritol; CDP-ME2P, 4-(cytidine 5′-diphospho)-2-C-methyl-d-erythritol phosphate; CMK, CDP-ME kinase; DMAPP, dimethylallyl diphosphate; DOXP, 1-deoxy-d-xylulose 5-phosphate; DXR, DOXP reductoisomerase; DXS, DOXP synthase; FDS, farnesyl diphosphate synthase; FPP, farnesyl diphosphate; GDS, geranyl diphosphate synthase; GGDS, geranyl geranyl diphosphate synthase; GGPP, geranyl geranyl diphosphate; GPP, geranyl diphosphate; HDR, (E)-4-hydroxy-3-methylbut-2-enyl diphosphate reductase; HDS, (E)-4-hydroxy-3-methylbut-2-enyl diphosphate synthase; HMBPP, (E)-4-hydroxy-3-methylbut-2-enyl diphosphate; HMGR, 3-hydroxy-3-methylglutaryl-CoA reductase; HMGS, HMG-CoA synthase; IDI, isopentenyl diphosphate isomerase; IPP, isopentenyl diphosphate; MCT, 2-C-methyl-d-erythritol 4-phosphate cytidylyltransferase; MDS, 2-C-methyl-d-erythritol 2,4-cyclodiphosphate synthase; ME-2,4cPP, 2-C-methyl-d-erythritol 2,4-cyclodiphosphate; MEP, 2-C-methyl-d-erythritol 4-phosphate; MVD, mevalonate diphosphate decarboxylase; MVK, mevalonate kinase; PMK, phosphomevalonate kinase; TPS, terpene synthase. Adapted from Nagegowda (2010), with permission from Elsevier.

Using GPP and FPP as substrates, terpene synthases/cyclases (TPSs) produce a great diversity of monoterpenes and sesquiterpenes, respectively. TPSs constitute a large family of enzymes, many of which catalyze the specific formation of a single product. However, many others have the ability to produce multiple compounds from a single prenyl diphosphate substrate (Nagegowda, 2010), or even both mono- and sesquiterpenes from different substrates, as is the case of FaNES1 enzyme in strawberry (Aharoni et al., 2004). The correlation observed between production of terpenoids and the levels of expression of their biosynthetic genes suggests that their biosynthesis is regulated at the level of transcription. This type of regulation has been described in "Valencia" orange in the case of the *Cstps1* gene encoding valencene synthase (Sharon-Asa et al., 2003), in grapes for *VvVal, VvGerD,* and *VvTer* genes, encoding valencene, germacrene D, and monoterpenol synthases, respectively (Lucker et al., 2004), and in strawberry for the *FaNES1* gene which encodes a bifunctional C_{15}/C_{10} nerolidol/linalool synthase (Aharoni et al., 2004).

It was generally accepted that monoterpenes were synthesized in the plastids and sesquiterpenes in the cytosol (Gershenzon and Kreis, 1999). However, this is about to be reexamined, as new evidence indicates that at least some monoterpene synthases can also be located in the cytosol, as described in strawberry for FaNES1 and FvPIN proteins, or in both plastids and mitochondria as described for the bifunctional mono- and sesquiterpene synthase FaNES2 (Aharoni et al., 2004). Also, recently santalene and bergamotene synthase (SBS), a sesquiterpene synthase from the wild tomato species *Solanum habrochaites*, was also found not to follow this general scheme, and was localized in the plastids of glandular trichomes (Sallaud et al., 2009).

Apocarotenoid Pathways

Apocarotenoid volatiles constitute a small family of compounds which can be considered as a group of irregular terpenoids. Unlike the terpenoids previously described, apocarotenoids are derived from the cleavage of carotenoids (C_{40}) by carotenoid cleavage oxygenase enzymes. These carotenoid derivatives appear to participate in a wide range of plant processes. Thus they have recently been implicated in maintaining the functionality of the arbuscural mycorrhiza symbiosis (Walter et al., 2010), they are part of the scent of many flowers like *Freesia hybrida, Rosa hybrida, and Osmanthus fragans* (Dudareva et al., 2006), while some are key compounds in spices like saffron (Moraga et al., 2009), and have been described in many fruits such as tomato (Vogel et al., 2010), grape (Gonzalez-Mas et al., 2009) and starfruit (Fleischman et al., 2003). The human olfactory system is extremely sensitive to these compounds, particularly to cyclic apocarotenoids, which are detectable at very low concentrations (Buttery et al., 1990). Therefore, even if they are minor components of the volatile profile of a fruit, they still can play a major role in its flavor, as has been described in tomato (Vogel et al., 2010).

The key for the biosynthesis of apocarotenoids is the carotenoid cleavage dioxygenases (CCDs). The first CCDs described in fruits were the tomato LeCCD1A and LeCCD1B (Simkin et al., 2004). Apparently, these are promiscuous enzymes recognizing any carotenoid after ζ-carotene in the biosynthesis pathway, and are able to cleave cyclic carotenoids at the 9,10 position and also linear carotenoids at either the 5,6 or 7,8 or 9,10 positions (Vogel et al., 2008; Ilg et al., 2009). Proteins of the CCD1 group are located in the cytosol, and have been identified in fruits of many other species, such as grape (Mathieu et al., 2005), melon (Ibdah et al., 2006), citrus (Kato et al., 2006), and strawberry (Garcia-Limones et al., 2008) among others. Enzymes belonging to subfamilies other than the CCD1 family and located in the plastids have been described, but their implication in the biosynthesis of volatiles in fruit has not been demonstrated at present.

Apocarotenoid biosynthesis has been the subject of intense study during the last few years, and several alternate models considering both one-step and compartmented two-step pathways have been proposed (Walter et al., 2010), but the mechanism of CCD action in fruits is still unknown. A two-step compartmented biosynthesis pathway has been proposed, according to which C_{40} carotenoids synthesized and accumulated in the plastids would first be cleaved inside these organelles, producing a volatile C_{13} and a nonvolatile C_{27} apocarotenoid. The enzymes responsible for this cleavage are not known, but enzymes from the CCD4 and CCD7 subfamilies, located in the plastids and known to have the required 9-10 cleavage activity (Rubio et al., 2008), have been proposed as candidates. Apocarotenoid C_{27} precursors would then be exported to the cytosol, where the cytosolic CCD1 enzymes would cleave the molecules further to produce finally the volatile apocarotenoid compounds. Two alternative one-step biosynthesis models have also been proposed. The first would be the direct 9-10 and 9′-10′ cleavage of the intact C_{40} carotenoids by CCD1 enzymes. This would require access of the enzymes to their substrates, which are protected by the plastidial membrane, and this would imply a previous degradation of the plastid or an enhanced permeability of its membrane. The second model proposes a cleavage of carotenoid fragments previously fragmented as a consequence of physical or chemical damage (Walter et al., 2010).

Regulation of the apocarotenoid synthesis is still unknown, but the existing knowledge suggests that substrate availability is the limiting step. This would be consistent with the fact that, while CCD enzymes are present during all the stages of fruit development, apocarotenoid volatiles are only produced at the latest stages of fruit ripening, in coincidence with the conversion of chloroplasts to chromoplasts (Klee, 2010; Vogel et al., 2010).

Furanone Pathway

Particularly remarkable in this group is 4-hydroxy-2,5-dimethyl-3(2*H*)-furanone (furaneol, HDMF), a compound which deters fungal growth and is a key volatile in the aroma of strawberry (Slaughter, 1999). Furanones are cyclic ester compounds with antioxidant and anticarcinogenic activities and apart from some fresh fruit, are found in many highly cooked foodstuffs and in fruits such as tomato, raspberry, pineapple, and strawberry. Among them, 4-hydroxy-3(2*H*)-furanones, which provide fruity notes, are very important flavor compounds in fruits due to their extraordinarily low-odor thresholds.

Part of the biosynthetic pathway leading to this compound has been described in the last few years. The initial precursor of furaneol seems to be D-fructose-1,6-diphosphate, according to the results of incorporation experiments using labeled precursors (Roscher et al., 1998). It is converted to 4-hydroxy-5-methyl-2-methylene-3(2*H*)-furanone by an unknown enzyme, and then transformed into furaneol by an enone oxidoreductase, like the fruit-ripening-related FaEO isolated from strawberry (Raab et al., 2006) or the SlEO isolated from tomato (Klein et al., 2007). Furaneol can be further metabolized to mesifurane (MDMF) by means of enzymes like FaOMT, a nonspecific O-methyltransferase (Wein et al., 2002), or to conjugate with sugars by means of enzymes such as FaGT2 glucosyltransferase (Landmann et al., 2007).

Conjugation and Emission of Volatiles

As anticipated before, many of the volatile compounds produced in fruit accumulate as non-volatile conjugates. These metabolites are often significantly more abundant than the free volatile forms (Maicas and Mateo, 2005), and do not participate in the aroma. Such compounds have been described in many fruit species (Marlatt et al., 1992), and most often consist of glycosidic derivatives of the volatiles (aglycones), mainly *O*-β-D-glucosides or *O*-diglycosides, although triglycosides have also been described. A wide variety of aglycones has been reported, including monoterpene alcohols, apocarotenoids, benzenoids, and linear alcohols (Ortiz-Serrano and Gil, 2007). In a particular fruit, the pattern of glycosidically bound volatiles has been reported to be similar to that of the free volatiles it produces (Du et al., 2010).

It is considered that these conjugates are not endpoint irreversible products, but can be used as precursors of volatile compounds, and therefore can be an important potential reservoir for aroma production. Therefore, there is not only an increased interest in better understanding the mechanisms of volatile glycosylation in fruits (Birtic et al., 2009), but also on the hydrolysis of those conjugates in order to liberate the volatile aglycons and increase the aroma (Maicas and Mateo, 2005).

A recent study stressed the relevance of conjugates by reporting that the main compounds responsible for the emission of some benzenoid volatiles in tomato fruit upon tissue homogenization were their glycoconjugates. On tissue disruption, glycoconjugates are hydrolyzed by glycosidases (located in different compartments) and the free volatiles, liberated. This process could vary in different genotypes, and it was observed that the genotypes with a low emission of these benzenoid volatiles performed a further glycosylation step during ripening, producing conjugate species of higher complexity in the ripe fruit that the endogenous glycosidases could not hydrolyze. Therefore, contrasting volatile emission was a consequence of a differential conjugation pattern (Tikunov et al., 2010).

Conjugation has the ability to regulate the levels of volatiles emitted by a plant, and sometimes works in the sense of sequestering them dramatically. This mechanism was observed in transgenic Arabidopsis and petunia plants harboring a linalool synthase gene, which either produced preferentially linalool glycosides instead of the free volatile (Aharoni et al., 2003), or produced only S-linalyl-β-D-glucopyranoside but failed to emit the volatile at all (Lucker et al., 2001).

It could be concluded that conjugation and corresponding hydrolysis are mechanisms that can modulate the levels of volatile compounds accumulated in a fruit and those released under certain conditions, but it still remains a poorly studied subject. There is a lack of specific information about the enzymes responsible *in vivo* for both the glycosylation and the corresponding hydrolysis, and how these pathways are controlled. Further knowledge about these processes would be required for their manipulation in order to modify the aroma of fruits and their derived products.

Identification of Quantitative Trait Loci for Volatiles

A useful strategy for improving the volatile content contributing to fruit aroma/flavor is to identify the genes or genome regions that have an influence on each of the volatiles detected as desirable. When such quantitative trait loci (QTL) are mapped and linked to molecular markers, they become useful tools for marker-assisted selection in breeding programs. Based on naturally occurring biological diversity, which is high for volatile compounds in fruits of most cultivated species, many QTL for sensory attributes and volatile production have been identified and mapped in the last few years. Different categories of genes can be hypothesized as responsible for the QTL detected: genes encoding enzymes directly involved in the biosynthetic and metabolic pathways, and genes encoding factors that have an effect at some level in the regulation of the pathway. The latter could include genome regions affecting gene expression to more complex factors like morphology, permeability, etc. So far information on the identity of genes underlying QTL for fruit volatiles is limited. Moreover, known genes involved in volatile biosynthesis do not colocalize with any major QTL, which means that most of the potential for plant breeding relies on still unidentified genes.

Most of the effort in this direction has been performed on tomato, where over 100 QTL have been identified for volatile compounds or their precursors. This QTL information has been obtained by analysis of fruit volatile compound variability in introgression lines, representing genetic variability within the species *Solanum lycopersicum* var. *cerasiforme* (Causse et al., 2001; Causse et al., 2002; Zanor et al., 2009) or between species, as provided by the wild tomato species *Solanum pennellii* (Tieman et al., 2006b), *S. habrochaites* (Mathieu et al., 2009), or *Solanum pimpinellifolium* (Granell et al., unpublished), and this represents just a small sample of the large variability available in the germplasm repositories.

Increasing interest in the mapping of fruit volatile QTL has arisen in the last few years, and QTL mapping has been performed on other species such as grape (Doligez et al., 2006; Battilana et al., 2009; Duchene et al., 2009), apple (Dunemann et al., 2009), raspberry (Woodhead et al., 2010), and melon (Obando-Ulloa et al., 2008). QTL maps for fruit volatiles are in progress in other species such as strawberry (Zorrilla-Fontanesi et al., 2010), peach (own results, unpublished). A limitation of this approach is that, while useful for defining regions for marker-assisted selection, it is difficult to obtain sufficient resolution for gene identification; an interesting possibility is to combine genetics and genomics approaches for the identification of candidate genes. Knowledge of QTL for many different metabolites is also important in defining strategies for crop improvement. For example, although mutants in carotenoid biosynthesis failed to produce the corresponding apocarotenoid volatiles, QTL for these volatile constituents do not colocalize with QTL for the carotenoid precursors, indicating that these normally are not the limiting factors.

Metabolic Engineering of the Fruit Volatile Pathways

In the last 15 years, a number of reports showed success in cases of plants genetically transformed with genes that resulted in an altered volatile profile of their fruits. They were reported either just as a way to demonstrate the capability of a newly developed technique (Endo et al., 2009; Orzaez et al., 2009), as a strategy to prove the function of an identified gene (Chen et al., 2004; Simkin et al., 2004; Simkin et al., 2007), or as a targeted approach to alter the volatile profile of fruit, either directly or indirectly.

Except a recent work describing the silencing of a limonene synthase in the citrus species *Poncirus trifoliata* (Endo et al., 2009), all the other reports on genetic engineering of volatiles in fruit have been performed on tomato. The first successful attempt to genetically engineer volatile production was by means of a Δ-9 desaturase from yeast (Wang et al., 1996). The introduced gene was not directly involved in the synthesis of volatiles, but caused an alteration in the levels of many fatty acids, some of them precursors of volatile compounds. Therefore, the levels of the derived volatiles were correspondingly modified. A similar approach was used recently by the introduction of two ω-3 desaturase genes from *Brassica napus* (Dominguez et al., 2010), which produced an increase in the levels of (Z)-3-hexenal and a dramatic depletion of hexanal.

In most of the cases, the genetic engineering of volatile production has involved the modification of a single gene in the last steps of the biosynthetic pathway. The overexpression of a tomato *ADH* (Speirs et al., 1998) produced an alteration in the aldehyde to alcohol ratio in many fatty acid-derived volatiles correlating to ADH activity, and preliminary taste trials revealed an increase in ripe flavor perception associated to higher alcohol levels. The silencing of the tomato lipoxygenase C gene (Chen et al., 2004) produced a dramatic decrease in the fatty acid derivatives—hexanal, hexenol and hexenal isomers.

Carotenoid-derived volatiles have been modified by means of the silencing of a tomato carotenoid cleavage deoxygenase (Simkin et al., 2004), producing reduced volatile levels, and also indirectly by the introduction of a pepper fibrillin gene (Simkin et al., 2007), which altered the chromoplast structure and increased the production of both carotenoids and their derived volatiles.

Benzenoids have been engineered by the overproduction of tomato aromatic amino acid decarboxylases (Tieman et al., 2006a), which catalyze the first step in their biosynthetic pathway, resulting in three- to tenfold increases of 2-phenylethanol and other related volatiles. This compositional change was detectable by consumers, but resulted in no significant preference between transgenic and wild type fruits (Klee, 2010).

Several attempts to introduce novel compounds in the flavor/aroma of the tomato fruit have been conducted by modifying the terpenoid pathway. The introduction of a linalool synthase from *Clarkia breweri* (Lewinsohn et al., 2001) in tomato caused the accumulation of linalool and 8-hydroxylinalool in fruits at levels detectable by the human nose. A zingiberene synthase from basil has also been introduced in tomato (Davidovich-Rikanati et al., 2008), producing a wide range of sesquiterpenes and monoterpenes, most notably α-zingiberene. Also from basil, a geraniol synthase, the first enzyme in the pathway of monoterpene synthesis, was also introduced in tomato (Davidovich-Rikanati et al., 2007), producing high levels of a variety of monoterpenes and derived compounds, some of them unexpected, such as limonene or myrcene. When tried by a taste panel, most of the panelists could distinguish the transgenic fruit, and 60% of them preferred it to the wild type. Unfortunately, the production of monoterpenes was at the expense of competition with the synthesis of carotenoids, which resulted in reduced levels of lycopene and their derived volatiles.

The experience accumulated to date from the metabolic engineering of volatiles in plants shows that it is not an easy and straightforward matter, and metabolic manipulations performed in the past often yielded unpredictable results. The main reason for this is our limited knowledge about the whole metabolic processes leading to volatile synthesis and the interrelation between the different metabolic pathways. Some of the problems reported have been the absence of a suitable substrate for the enzyme that was introduced (Beekwilder et al., 2004) or the successful synthesis of the target compound but then its modification into nonvolatile conjugates (Lucker et al., 2001). Even when the expected compounds were synthesized and emitted, often other compounds were unexpectedly synthesized or their levels modified (Davidovich-Rikanati et al., 2007; Orzaez et al., 2009). This reveals the degree of unpredictability associated with metabolic engineering at the present time.

Competition between metabolic pathways is another problem observed. To circumvent this, it has been proposed to regulate the expression of the transgenes by wound-inducible promoters, which may work well in raw preparations of the fresh fruit (Sangwan and Sangwan, 2007). Another alternative could be to direct the engineered enzyme to a subcellular location where it has no contact with its substrate. Therefore, enzyme accumulation would not interfere with any metabolic pathway, and high amounts of the enzyme would be ready for the synthesis of volatiles immediately after tissue disruption, similarly as naturally occurs in the synthesis of the fatty acid-derived volatiles.

A new level of difficulty arises when the final objective is to modify volatile contents in order to increase consumer acceptance of fruit. This increased difficulty is a consequence of flavor being the result of a very complex interaction between many compounds, both volatile and nonvolatile, and optimum levels for consumer preference are extremely difficult to set. Therefore, for this purpose, the production of the engineered volatile cannot be considered a success unless it is produced at levels detectable by the human nose, its presence is not masked by other compounds, and the resulting modified aroma induces an increased preference on the consumer toward the modified fruit.

Future Perspectives

Much progress has been achieved over the last few years in the identification of enzymes and genes involved in the biosynthesis of volatile compounds and in defining their biosynthetic pathways. This has allowed us to initiate the metabolic engineering of volatiles, but still much work has to be done to better understand the metabolic networks and their regulation during ripening before we are in a position to change the volatile profile of a fruit in a reasonably predictable and favorable way for the consumer. The development of new tools will presumably increase our ability to understand the whole process in the coming years.

Metabolomics has become a valuable source of information, but dealing with the extraordinary amount of data generated is a major challenge, which has driven attention recently. The approach for non-targeted data analysis for metabolomics proposed by Tikunov and collaborators (2005) is a valuable contribution for dealing with the complex data obtained by gas chromatography–mass spectrometry from the analysis of volatiles. Advanced statistical processing of the data is necessary including creating correlation networks (Ursem et al., 2008) or the use of constraint-based Bayesian networks (Gavai et al., 2009), just to mention two examples. Such techniques are useful to provide a more comprehensive view of the data, and can be used to propose likely metabolic pathways based on the relation between the metabolites analysed.

References

Aharoni, A., Giri, A.P., Deuerlein, S., et al. (2003) Terpenoid metabolism in wild-type and transgenic Arabidopsis plants. *The Plant Cell*, 15, 2866–2884.

Aharoni, A., Giri, A.P., Verstappen, F.W.A., et al. (2004) Gain and loss of fruit flavor compounds produced by wild and cultivated strawberry species. *The Plant Cell*, 16, 3110–3131.

Aharoni, A., Keizer, L.C.P., Bouwmeester, H.J., et al. (2000) Identification of the SAAT gene involved in strawberry flavor biogenesis by use of DNA microarrays. *The Plant Cell*, 12, 647–661.

Allmann, S., and Baldwin, I.T. (2010) Insects betray themselves in nature to predators by rapid isomerization of green leaf volatiles. *Science*, 329, 1075–1078.

Ament, K., Krasikov, V., Allmann, S., et al. (2010) Methyl salicylate production in tomato affects biotic interactions. *The Plant Journal: for cell and molecular biology*, 62, 124–134.

Balbontin, C., Gaete-Eastman, C., Fuentes, L., et al. (2010) *VpAAT1*, a gene encoding an alcohol acyltransferase, is involved in ester biosynthesis during ripening of mountain papaya fruit. *Journal of Agricultural and Food Chemistry*, 58, 5114–5121.

Baldwin, E.A., Goodner, K., and Plotto, A. (2008) Interaction of volatiles, sugars, and acids on perception of tomato aroma and flavor descriptors. *Journal of Food Science*, 73, S294–S307.

Baldwin, I.T., Kessler, A., and Halitschke, R. (2002) Volatile signaling in plant-plant-herbivore interactions: what is real? *Current Opinion in Plant Biology*, 5, 351–354.

Battilana, J., Constantini, L., Emanuelli, F., et al. (2009) The 1-deoxy-D-xylulose 5-phosphate synthase gene co-localizes with a major QTL affecting monoterpene content in grapevine. *Theoretical and Applied Genetics*, 118, 653–669.

Beekwilder, J., Alvarez-Huerta, M., Neef, E., et al. (2004) Functional characterization of enzymes forming volatile esters from strawberry and banana. *Plant Physiology*, 135, 1865–1878.

Birtic, S., Ginies, C., Causse, M., et al. (2009) Changes in volatiles and glycosides during fruit maturation of two contrasted tomato (*Solanum lycopersicum*) lines. *Journal of Agricultural and Food Chemistry*, 57, 591–598.

Blée, E. (2002) Impact of phyto-oxylipins in plant defense. *Trends in Plant Sciences*, 7, 315–321.

Boatright, J., Negre, F., and Chen, X.L., et al. (2004) Understanding in vivo benzenoid metabolism in petunia petal tissue. *Plant Physiology*, 135, 1993–2011.

Buttery, R.G., Teranishi, R., Ling, L.C., et al. (1990) Quantitative and sensory studies on tomato paste volatiles. *Journal of Agricultural and Food Chemistry*, 38, 336–340.

Causse, M., Saliba-Colombani, V., Lecomte, L., et al. (2002) QTL analysis of fruit quality in fresh market tomato: a few chromosome regions control the variation of sensory and instrumental traits. *Journal of Experimental Botany*, 53, 2089–2098.

Causse, M., Saliba-Colombani, V., Lesschaeve, I., et al. (2001) Genetic analysis of organoleptic quality in fresh market tomato. 2. Mapping QTLs for sensory attributes. *Theoretical and Applied Genetics*, 102, 273–283.

Chen, G.P., Hackett, R., Walker, D., et al. (2004) Identification of a specific isoform of tomato lipoxygenase (TomloxC) involved in the generation of fatty acid-derived flavor compounds. *Plant Physiology*, 136, 2641–2651.

D'Auria, J.C. (2006) Acyltransferases in plants: a good time to be BAHD. *Current Opinion in Plant Biology*, 9, 331–340.

Davidovich-Rikanati, R., Lewinsohn, E., Bar, E., et al. (2008) Overexpression of the lemon basil α–zingiberene synthase gene increases both mono- and sesquiterpene contents in tomato fruit. *The Plant Journal: for cell and molecular biology*, 56, 228–238.

Davidovich-Rikanati, R., Sitrit, Y., Tadmor, Y., et al. (2007) Enrichment of tomato flavor by diversion of the early plastidial terpenoid pathway. *Nature Biotechnology*, 25, 899–901.

Dickinson, J.R., Eshantha, L., Salgado, J., et al. (2003) The catabolism of amino acids to long chain and complex alcohols in Saccharomyces cerevisiae. *Journal of Biological Chemistry*, 278, 8028–8034.

Doligez, A., Audiot, E., Baumes, R., et al. (2006) QTLs for muscat flavor and monoterpenic odorant content in grapevine (*Vitis vinifera* L.). *Molecular Breeding*, 18, 109–125.

Dominguez, T., Hernandez, M.L., Pennycooke, J.C., et al. (2010) Increasing ω-3 desaturase expression in tomato results in altered aroma profile and enhanced resistance to cold stress. *Plant Physiology*, 153, 655–665.

Du, X., Finn, C.E., and Qian, M.C. (2010) Bound volatile precursors in genotypes in the pedigree of 'marion' blackberry (*Rubus* Sp.). *Journal of Agricultural and Food Chemistry*, 58, 3694–3699.

Duchene, E., Butterlin, G., Claudel, P., et al. (2009) A grapevine (*Vitis vinifera* L.) deoxy-D: -xylulose synthase gene colocates with a major quantitative trait loci for terpenol content. *Theoretical and Applied Genetics*, 118, 541–552.

Dudareva, N., Negre, F., Nagegowda, D.A., et al. (2006) Plant volatiles: recent advances and future perspectives. *Critical Reviews in Plant Sciences*, 25, 417–440.

Dudareva, N., Pichersky, E., and Gershenzon, J. (2004) Biochemistry of plant volatiles. *Plant Physiology*, 135, 1893–1902.

Dunemann, F., Ulrich, D., Boudichevskaia, A., et al. (2009) QTL mapping of aroma compounds analysed by headspace solid-phase microextraction gas chromatography in the apple progeny 'Discovery' x 'Prima'. *Molecular Breeding*, 23, 501–521.

Effmert, U., Saschenbrecker, S., Ross, J., et al. (2005) Floral benzenoid methyltransferases: from in vitro to in planta function. *Phytochemistry*, 66, 1211–1230.

El-Sharkawy, I., Manríquez, D., Flores, F.B., et al. (2005) Functional characterization of a melon alcohol acyltransferase gene family involved in the biosynthesis of ester volatiles. Identification of the crucial role of a threonine residue for enzyme activity. *Plant Molecular Biology*, 59, 345–362.

Endo, T., Shimada, T., Fujii, H., et al. (2009) Development of a CiFT co-expression system for functional analysis of genes in citrus flowers and fruit. *Journal of the Japanese Society for Horticultural Science*, 78, 74–83.

Fleischmann, P., Watanabe, N., and Winterhalter, P. (2003) Enzymatic carotenoid cleavage in star fruit (*Averrhoa carambola*). *Phytochemistry*, 63, 131–137.

Gang, D.R., Lavid, N., Zubieta, C., et al. (2002) Characterization of phenylpropene O-methyltransferases from sweet basil: facile change of substrate specificity and convergent evolution within a plant O-methyltransferase family. *The Plant Cell*, 14, 505–519.

Garcia-Limones, C., Schnabele, K., Blanco-Portales, R., et al. (2008) Functional characterization of FaCCD1: a carotenoid cleavage dioxygenase from strawberry involved in Lutein degradation during fruit ripening. *Journal of Agricultural and Food Chemistry*, 56, 9277–9285.

Gavai, A.K., Tikunov, Y., Ursem, R., et al. (2009) Constraint-based probabilistic learning of metabolic pathways from tomato volatiles. *Metabolomics*, 5, 419–428.

Gershenzon, J., and Kreis, W. (1999) Biochemistry of terpenoids: monoterpenes, sesquiterpenes, diterpenes, sterols, cardiac glycosides and steroid saponins. In: *Biochemistry of Plant Secondary Metabolism* (ed. M. Wink), pp. 222–299. CRC Press, Boca Raton, FL.

Goff, S.A., and Klee, H.J. (2006) Plant volatile compounds: sensory cues for health and nutritional value? *Science*, 311, 815–819.

Gonda, I., Bar, E., Portnoy, V., et al. (2010) Branched-chain and aromatic amino acid catabolism into aroma volatiles in Cucumis melo L. fruit. *Journal of Experimental Botany*, 61, 1111–1123.

Gonzalez-Mas, M.C., Garcia-Riano, L.M., Alfaro, C., et al. (2009) Headspace-based techniques to identify the principal volatile compounds in red grape cultivars. *International Journal of Food Science and Technology*, 44, 510–518.

Hemmerlin, A., Hoeffler, J.F., Meyer, O., et al. (2003) Cross-talk between the cytosolic mevalonate and the plastidial methylerythritol phosphate pathways in Tobacco Bright Yellow-2 cells. *Journal of Biological Chemistry*, 278, 26666–26676.

Howe, G.A., Lee, G.I., Itoh, A., et al. (2000) Cytochrome P450-dependent metabolism of oxylipins in tomato. Cloning and expression of allene oxide synthase and fatty acid hydroperoxide lyase. *Plant Physiology*, 123, 711–724.

Humphreys, J.M., and Chapple, C. (2002) Rewriting the lignin roadmap. *Current Opinion in Plant Biology*, 5, 224–229.

Ibdah, M., Azulay, Y., Portnoy, V., et al. (2006) Functional characterization of CmCCD1, a carotenoid cleavage dioxygenase from melon. *Phytochemistry*, 67, 1579–1589.

Ilg, A., Beyer, P., and Al-Babili, S. (2009) Characterization of the rice carotenoid cleavage dioxygenase 1 reveals a novel route for geranial biosynthesis. *FEBS Journal*, 276, 736–747.

Jones, M.G., Hughes, J., Tregova, A., et al. (2004) Biosynthesis of the flavor precursors of onion and garlic. *Journal of Experimental Botany*, 55, 1903–1918.

Jordan, M.J., Tandon, K., Shaw, P.E., et al. (2001a) Aromatic profile of aqueous banana essence and banana fruit by gas-chromatography-mass spectrometry (GC-MS) and gas chromatography-olfactometry (GC-O). *Journal of Agricultural and Food Chemistry*, 49, 4813–1817.

Jordan, M.J., Shaw, P.E., and Goodner, K.L. (2001b) Volatile components in aqueous essence and fresh fruit of *Cucumis melo* cv. Athena (Muskmelon) by GC-MS and GC-O. *Journal of Agricultural and Food Chemistry*, 49, 5929–5933.

Kaminaga, Y., Schnepp, J., Peel, G., et al. (2006) Plant phenylacetaldehyde synthase is a bifunctional homotetrameric enzyme that catalyzes phenylalanine decarboxylation and oxidation. *Journal of Biological Chemistry*, 281, 23357–23366.

Kato, M., Matsumoto, H., Ikoma, Y., et al. (2006) The role of carotenoid cleavage dioxygenases in the regulation of carotenoid profiles during maturation in citrus fruit. *Journal of Experimental Botany*, 57, 2153–2164.

Kessler, A., Halitschke, R., Baldwin, I.T. (2004) Silencing the jasmonate cascade: induced plant defenses and insect populations. *Science*, 305, 665–668.

Klee, H.J. (2010) Improving the flavor of fresh fruits: genomics, biochemistry, and biotechnology. *New Phytologist*, 187, 44–56.

Klein, D., Fink, B., Arold, B., et al. (2007) Functional characterization of enone oxidoreductases from strawberry and tomato fruit. *Journal of Agricultural and Food Chemistry*, 55, 6705–6711.

Knudsen, J.T., Tollsten, L., and Bergstrom, L.G. (1993) Floral scents—a checklist of volatile compounds isolated by headspace techniques. *Phytochemistry*, 33, 253–280.

Koeduka, T., Fridman, E., Gang, D.R., et al. (2006) Eugenol and isoeugenol, characteristic aromatic constituents of spices, are biosynthesized via reduction of a coniferyl alcohol ester. *Proceedings of the National Academy of Sciences of the United States of America*, 103, 10128–10133.

Kuroda, H., Oshima, T., Kaneda, H., et al. (2005) Identification and functional analyses of two cDNAs that encode fatty acid 9-/13-hydroperoxide lyase (CYP74-C) in rice. *Bioscience Biotechnology and Biochemistry*, 69, 1545–1554.

Landmann, C., Fink, B., and Schwab, W. (2007) FaGT2: a multifunctional enzyme from strawberry (*Fragaria* x *ananassa*) fruits involved in the metabolism of natural and xenobiotic compounds. *Planta*, 226, 417–428.

Lewinsohn, E., Schalechet, F., Wilkinson, J., et al. (2001) Enhanced levels of the aroma and flavor compound s-linalool by metabolic engineering of the terpenoid pathway in tomato fruits. *Plant Physiology*, 127, 1256–1265.

Liavonchanka, A., and Feussner, I. (2006) Lipoxygenases: occurrence, functions and catalysis. *Journal of Plant Physiology*, 163, 348–357.

Lucker, J., Bouwmeester, H.J., Schwab, W., et al. (2001) Expression of Clarkia S-linalool synthase in transgenic petunia plants results in the accumulation of S-linalyl-beta-D-glucopyranoside. *The Plant Journal: for cell and molecular biology*, 27, 315–324.

Lucker, J., Bowen, P., and Bohlmann, J. (2004) Vitis vinifera terpenoid cyclases: functional identification of two sesquiterpene synthase cDNAs encoding (+)-valencene synthase and (−)-germacrene D synthase and expression of mono- and sesquiterpene synthases in grapevine flowers and berries. *Phytochemistry*, 65, 2649–2659.

Luft, S., Curio, E., and Tacud, B. (2003) The use of olfaction in the foraging behaviour of the golden-mantled flying fox, Pteropus pumilus, and the greater musky fruit bat, Ptenochirus jagori (Megachiroptera: Pteropodidae). *Naturwissenschaften*, 90, 84–87.

Maicas, S., and Mateo, J.J. (2005) Hydrolysis of terpenyl glycosides in grape juice and other fruit juices: a review. *Applied Microbiology and Biotechnology*, 67, 322–335.

Maloney, G.S., Kochevenko, A., Tieman, D.M., et al. (2010) Characterization of the branched-chain amino acid aminotransferase enzyme family in tomato. *Plant Physiology*, 153, 925–936.

Mano, J., Torii, Y., Hayashi, S., et al. (2002) The NADPH : quinone oxidoreductase P1-zeta-crystallin in Arabidopsis catalyzes the alpha, beta-hydrogenation of 2-alkenals: detoxication of the lipid peroxide-derived reactive aldehydes. *Plant and Cell Physiology*, 43, 1445–1455.

Manríquez, D., El-Sharkawy, I., Flores, F.B., et al. (2006) Two highly divergent alcohol dehydrogenases of melon exhibit fruit ripening-specific expression and distinct biochemical characteristics. *Plant Molecular Biology*, 61, 675–685.

Marilley, L., and Casey, M.G. (2004) Flavours of cheese products: metabolic pathways, analytical tools and identification of producing strains. *International Journal of Food Microbiology*, 90, 139–159.

Marlatt, C., Ho, C.T., and Chien, M. (1992) Studies of aroma constituents bound as glycosides in tomato. *Journal of Agricultural and Food Chemistry*, 40, 249–252.

Mathieu, S., Dal Cin, V., Fei, Z., et al. (2009) Flavour compounds in tomato fruits: identification of loci and potential pathways affecting volatile composition. *Journal of Experimental Botany*, 60, 325–337.

Mathieu, S., Terrier, N., Procureur, J., et al. (2005) A carotenoid cleavage dioxygenase from *Vitis vinifera* L.: functional characterization and expression during grape berry development in relation to C-13-norisoprenoid accumulation. *Journal of Experimental Botany*, 56, 2721–2731.

Matsui, K. (2006) Green leaf volatiles: hydroperoxide lyase pathway of oxylipin metabolism. *Current Opinion in Plant Biology*, 9, 274–280.

Matsui, K., Ujita, C., Fujimoto, S.H., et al. (2000a) Fatty acid 9- and 13-hydroperoxide lyases from cucumber. *FEBS Letters*, 481, 183–188.

Matsui, K., Kurishita, S., Hisamitsu, A., et al. (2000b) A lipid-hydrolysing activity involved in hexenal formation. *Biochemical Society Transactions*, 28, 857–860.

Mita, G., Quarta, A., Fasano, P., et al. (2005) Molecular cloning and characterization of an almond 9-hydroperoxide lyase, a new CYP74 targeted to lipid bodies. *Journal of Experimental Botany*, 56, 2321–2333.

Moraga, A.R., Rambla, J.L., Ahrazem, et al. (2009) Metabolite and target transcript analyses during *Crocus sativus* stigma development. *Phytochemistry*, 70, 1009–1016.

Nagegowda, D.A. (2010) Plant volatile terpenoid metabolism: biosynthetic genes, transcriptional regulation and subcellular compartmentation. *FEBS Letters*, 584, 2965–2973.

Nogueira, J.M.F., Fernandes, P.J.P., and Nascimento, A.M.D. (2003) Composition of volatiles of banana cultivars from Madeira island. *Phytochemical Analysis*, 14, 87–90.

Noordermeer, M.A., Veldink, G.A., and Vliegenthart, J.F.G. (1999) Alfalfa contains substantial 9-hydroperoxide lyase activity and a 3Z: 2E-enal isomerase. *FEBS Letters*, 443, 201–204.

Obando-Ulloa, J.M., Moreno, E., García Mas, J., et al. (2008) Climacteric or non-climacteric behaviour in melon fruit. 1. Aroma volatiles. *Postharvest Biology and Technology*, 49, 27–37.

Ongena, M., Duby, F., Rossignol, F., et al. (2004) Stimulation of the lipoxygenase pathway is associated with systemic resistance induced in bean by a nonpathogenic Pseudomonas strain. *Molecular Plant-Microbe Interactions*, 17, 1009–1018.

Ortiz-Serrano, P., and Gil, J.V. (2007) Quantitation of free and glycosidically bound volatiles and effect of glycosidase addition on three tomato varieties (*Solanum lycopersicum* L.) *Journal of Agricultural and Food Chemistry*, 55, 9170–9176.

Orzaez, D., Medina, A., Torre, S., et al. (2009) A visual reporter system for virus-induced gene silencing in tomato fruit based on anthocyanin accumulation. *Plant Physiology*, 150, 1122–1134.

Pickersky, E., and Gershenzon, J. (2002) The formation and function of plant volatiles: perfumes for pollinator attraction and defense. *Current Opinion in Plant Biology*, 5, 237–243.

Pickersky, E., Noel, J.P., and Dudareva, N. (2006) Biosynthesis of plant volatiles: nature's diversity and ingenuity. *Science*, 311, 808–811.

Pino, J.A., Mesa, J., Muñoz, Y., et al. (2005) Volatile components from mango (*Mangifera indica* L.) cultivars. *Journal of Agricultural and Food Chemistry*, 53, 2213–2223.

Raab, T., Lopez-Raez, J.A., Klein, D., et al. (2006) *FaQR*, required for the biosynthesis of the strawberry flavor compound 4-hydroxy-2,5-dimethyl-3(2H)-furanone, encodes an enone oxidoreductase. *The Plant Cell*, 18, 1023–1037.

Reinhard, J., Srinivasan, M.V., and Zhang, S.W. (2004) Olfaction: scent-triggered navigation in honeybees. *Nature*, 427, 411–411.

Rodríguez-Concepción, M., and Boronat, A. (2002) Elucidation of the methylerythritol phosphate pathway for isoprenoid biosynthesis in bacteria and plastids. A metabolic milestone achieved through genomics. *Plant Physiology*, 130, 1079–1089.

Roscher, R., Bringmann, G., Schreier, P., et al. (1998) Radiotracer studies on the formation of 2,5-dimethyl-4-hydroxy-3(2H)-furanone in detached ripening strawberry fruits. *Journal of Agricultural and Food Chemistry*, 46, 1488–1493.

Rouseff, R.L., Perez-Cacho, P.R., and Jabalpurwala, F. (2009) Historical review of citrus flavor research during the past 100 years. *Journal of Agricultural and Food Chemistry*, 57, 8115–8124.

Rubio, A., Rambla, J.L., Santaella, M., et al. (2008) Cytosolic and plastoglobule-targeted carotenoid dioxygenases from *Crocus sativus* are both involved in beta-ionone release. *Journal of Biological Chemistry*, 283, 24816–24825.

Sallaud, C., Rontein, D., Onillon, S., et al. (2009) A novel pathway for sesquiterpene biosynthesis from Z,Z-farnesyl pyrophosphate in the wild tomato Solanum habrochaites. *The Plant Cell*, 21, 301–317.

Sangwan, N.S., and Sangwan, R.S (2007) Metabolic engineering for flavour enhancement in tomato—path setting for opportunities and strategies. *Current Science*, 93, 899–901.

Schumacher, K., Asche, S., Heil, M., et al. (1998) Methyl-branched flavor compounds in fresh and processed apples. *Journal of Agricultural and Food Chemistry*, 46, 4496–4500.

Schwab, W., Davidovich-Rikanati, R., and Lewinsohn, E. (2008) Biosynthesis of plant-derived flavor compounds. *The Plant Journal: for cell and molecular biology*, 54, 712–732.

Seymour, G., Taylor, J., and Tucker, G. (1993) *Biochemistry of Fruit Ripening*. Chapman & Hall, London.

Sharon-Asa, L., Shalit, M., Frydman, A., et al. (2003) Citrus fruit flavor and aroma biosynthesis: isolation, functional characterization, and developmental regulation of Cstps1, a key gene in the production of the sesquiterpene aroma compound valencene. *The Plant Journal: for cell and molecular biology*, 36, 664–674.

Shepherd, G.M. (2006) Smell images and the flavour system in the human brain. *Nature*, 444, 316–321.

Simkin, A.J., Gaffe, J., Alcaraz, J.P., et al. (2007) Fibrillin influence on plastid ultrastructure and pigment content in tomato fruit. *Phytochemistry*, 68, 1545–1556.

Simkin, A.J., Schwartz, S.H., and Auldridge, M., (2004) The tomato carotenoid cleavage dioxygenase 1 genes contribute to the formation of the flavor volatiles beta-ionone, pseudoionone, and geranylacetone. *The Plant Journal: for cell and molecular biology*, 40, 882–892.

Singh, R.K., Sane, V.A., Misra, A., et al. (2010) Differential expression of the mango alcohol dehydrogenase gene family during ripening. *Phytochemistry*, 71, 1485–1494.

Slaughter, J.C (1999) The naturally occurring furanones: formation and function from pheromone to food. *Biological Reviews of the Cambridge Philosophical Society*, 74, 259–276.

Speirs, J., Lee, E., Holt, K., et al. (1998) Genetic manipulation of alcohol dehydrogenase levels in ripening tomato fruit affects the balance of some flavor aldehydes and alcohols. *Plant Physiology*, 117, 1047–1058.

Tieman, D.M., Loucas, H.M., Kim, J.Y., et al. (2007) Tomato phenylacetaldehyde reductases catalyze the last step in the synthesis of the aroma volatile 2-phenylethanol. *Phytochemistry*, 68, 2660–2669.

Tieman, D., Taylor, M., Schauer, N., et al. (2006a) Tomato aromatic amino acid decarboxylases participate in synthesis of the flavor volatiles 2-phenylethanol and 2-phenylacetaldehyde. *Proceedings of the National Academy of Sciences of the United States of America*, 103, 8287–8292.

Tieman, D.M., Zeigler, M., Schmelz, E.A., et al. (2006b) Identification of loci affecting flavour volatile emissions in tomato fruits. *Journal of Experimental Botany*, 57, 887–896.

Tieman, D., Zeigler, M., Schmelz, E., et al. (2010) Functional analysis of a tomato salicylic acid methyl transferase and its role in synthesis of the flavor volatile methyl salicylate. *The Plant Journal: for cell and molecular biology*, 62, 113–123.

Tijet, N., Waspi, U., Gaskin, D.J.H., et al. (2000) Purification, molecular cloning, and expression of the gene encoding fatty acid 13-hydroperoxide lyase from guava fruit (Psidium guajava). *Lipids*, 35, 709–720.

Tikunov, Y., Lommen, A., Ric de Vos, C.H., et al. (2005) A novel approach for nontargeted data analysis for metabolomics. Large-scale profiling of tomato fruit volatiles. *Plant Physiology*, 139, 1125–1137.

Tikunov, Y.M., de Vos, R.C.H., Gonzalez-Paramas, A.M., et al. (2010) A role for differential glycoconjugation in the emission of phenylpropanoid volatiles from tomato fruit discovered using a metabolic data fusion approach. *Plant Physiology*, 152, 55–70.

Ursem, R., Tikunov, Y., Bovy, A., et al. (2008) A correlation network approach to metabolic data analysis for tomato fruits. *Euphytica*, 161, 181–193.

Vogel, J.T., Tan, B.C., Mccarty, D.R., et al. (2008) The carotenoid cleavage dioxygenase 1 enzyme has broad substrate specificity, cleaving multiple carotenoids at two different bond positions. *Journal of Biological Chemistry*, 283, 11364–11373.

Vogel, J.T., Tieman, D.M., Sims, C.A., et al. (2010) Carotenoid content impacts flavor acceptability in tomato (*Solanum lycopersicum*). *Journal of the Science of Food and Agriculture*, 90, 2233–2240.

Walter, M.H., Floss, D.S., and Strack, D. (2010) Apocarotenoids: hormones, mycorrhizal metabolites and aroma volatiles. *Planta*, 232, 1–17.

Wang, C., Chin, C.K., Ho, C.T., et al. (1996) Changes of fatty acids and fatty acid-derived flavor compounds by expressing the yeast Δ-9 desaturase gene in tomato. *Journal of Agricultural and Food Chemistry*, 44, 3399–3402.

Wein, M., Lavid, N., Lunkenbein, S., et al. (2002) Isolation, cloning and expression of a multifunctional O-methyltransferase capable of forming 2,5-dimethyl-4-methoxy-3(2H)-furanone, one of the key aroma compounds in strawberry fruits. *The Plant Journal: for cell and molecular biology*, 31, 755–765.

Woodhead, M., Weir, A., Smith, K., et al. (2010) Functional markers for red raspberry. *Journal of the American Society for Horticultural Science*, 135, 418–427.

Yu, X.H., Gou, J.Y., and Liu, C.J. (2009) BAHD superfamily of acyl-CoA dependent acyltransferases in *Populus* and *Arabidopsis*: bioinformatics and gene expression. *Plant Molecular Biology*, 70, 421–442.

Zanor, M.I., Rambla, J.L., Chaib, J., et al. (2009) Metabolic characterization of loci affecting sensory attributes in tomato allows an assessment of the influence of the levels of primary metabolites and volatile organic contents. *Journal of Experimental Botany*, 60, 2139–2154.

Zhang, B., Chen, K.S., Bowen, J., et al. (2006) Differential expression within the LOX gene family in ripening kiwifruit. *Journal of Experimental Botany*, 57, 3825–3836.

Zhang, B., Shen, J.Y., Wei, W.W., et al. (2010) Expression of genes associated with aroma formation derived from the fatty acid pathway during peach fruit ripening. *Journal of Agricultural and Food Chemistry*, 58, 6157–6165.

Zorrilla-Fontanesi, Y., Rambla, J.L., Cabeza, A., et al. (2010) Towards the identification of QTLs involved in the aroma of strawberry (*Fragaria x ananassa*). In: *28th International Horticultural Congress (IHC)*, abstract S04.051. Lisbon.

7 Cell Wall Architecture and Metabolism in Ripening Fruit and the Complex Relationship with Softening

Eliel Ruiz-May and Jocelyn K.C. Rose

Introduction

The ripening of fleshy fruits is typically accompanied by changes in a number of organoleptic properties that encourage consumption and subsequent seed dispersal. As described in other chapters of this book, over the last couple of decades much has been learned about the biochemical and molecular bases of such changes. Indeed, the compounds and associated metabolic and regulatory pathways that underlie many color, aroma and taste attributes have been defined to the point that it is now possible to alter such characteristics through targeted genetic engineering, with predictable and quantifiable outcomes. In contrast, the molecular determinants of ripening-associated changes in texture, or what is commonly referred to as "softening", are relatively poorly understood. Thus, while pigment levels can be measured and directly correlated with color, and similarly, sugars and organic acids with taste, no specific molecular changes in ripening fruits can be said to have a direct causal relationship with softening. Notably, even the definition of the term varies and "softening" can be used to refer to changes in the physical properties of an intact fruit, generally determined by resistance to uniaxial compression, or various biomechanical features of the fruit "flesh." Measurements of tensile strength, resistance to compression (firmness) and extensibility can all be taken into account, and all contribute to textural characteristics, including mealiness, melting characteristics, crispness, and juiciness (Harker et al., 1997; Chaïb et al., 2007). However, such parameters likely reflect different underlying molecular structures and interacting forces. Taken together then, the rate of advance in our understanding of, let alone ability to control, fruit softening and texture has been frustratingly slow compared with many other areas of fruit biology.

Nonetheless, there is a general agreement that loss of fruit firmness is fundamentally dependent on cell wall disassembly, which typically involves extensive depolymerization and solubilization of the constituent polysaccharides (Brummell, 2006; Vicente et al., 2007). This results in a loss of biomechanical strength and is sometimes accompanied by substantial wall swelling (Redgwell et al., 1997). In addition to primary cell wall modification, the middle

---

The Molecular Biology and Biochemistry of Fruit Ripening, First Edition.
Edited by Graham B. Seymour, Mervin Poole, James J. Giovannoni and Gregory A. Tucker.

lamella, which cements adjacent cells, also often undergoes substantial degradation during ripening, leading to a loss of intercellular adhesion and cell separation. The relative degree of structural changes in the primary walls or middle lamella appears to vary in different species and this in turn may influence fruit texture and the degree and type of softening (Toivonen and Brummell, 2008).

The sections of this chapter include current perspectives of the composition and architecture of fruit cell walls, the molecular mechanisms involved in disassembly of the wall and the middle lamella, and a review of the growing number of factors that influence wall biology and associated fruit texture.

Building Blocks of Fruit Cell Walls

Most fleshy fruits are composed largely of parenchyma cells with nonlignified primary cell walls that are generally thinner than those of vegetative tissues (Toivonen and Brummell, 2008). As is typical of primary walls, they are comprised mostly of cellulose (15–40% dry weight) and matrix glycan polymer networks of pectins (30–50%) and hemicelluloses (20–30%), together with lesser amounts of structural proteins (Cosgrove and Jarvis, 2012). The relative amounts of the constituent polymers vary between plant taxa; the most commonly cited distinction being between the so-called Type I and Type II walls. Dicotyledons and a subset of monocotyledons are generally described as having Type I walls, where the matrix glycans are predominated by pectins and the hemicellulose xyloglucan (XyG), while grasses and noncommelinid monocotyledons have Type II walls, where glucuronoarabinoxylans (GAXs) provide the major hemicellulose component and the proportion of pectins is less (Carpita and Gibeaut, 1993). It should be noted that a growing number of exceptions to this taxonomic categorization are emerging, and intermediate wall types are being identified. For example, pineapple (*Ananas comosus)* is a monocot with a wall composition that appears to be intermediate between the two designated types (Smith and Harris, 1995). Most commercially important fleshy fruits are derived from dicots and so research on fruit cell walls has focused on "classical" Type I walls, and particularly on tomato (*Solanum lycopersicum*), which is the primary model for studies of fleshy fruit biology (Giovannoni, 2004, 2007; Vicente et al., 2007; Seymour et al., 2012). However, it is important to bear in mind that other fruit species exhibit variations in wall composition and architecture, with relatively higher or lower amounts of specific polysaccharide types (Brownleader et al., 1999). As a general rule though, pectins are particularly abundant in fruit cell walls (Brummell, 2006). Other variants in wall structure include examples of specialized cell types, such as the cells of the vascular tissue that have thicker, secondary, and often lignified walls and the lipid-rich cuticular cell walls of the fruit epidermis (Seymour et al., 2012). Additionally, while the general structures of the major classes of wall polymers have now been elucidated, much remains to be learned about the nature and the frequency of fine structural heterogeneity, down to a cellular level. Nonetheless, the basic structures of the principal classes of wall components have been established, as summarized below.

Cellulose

Approximately a third of the dry mass of primary walls comprises cellulose: long chains of β-1,4-linked glucose (Glc) residues that associate during synthesis to form microfibrils. This chain packing, mediated by inter- and intramolecular hydrogen bonds and van der Waals forces, is sufficiently intimate to allow chain crystallization, resulting in rigid structures with high tensile strength (Somerville, 2006). However, in primary wall microfibrils, the crystallinity is thought to be disrupted by dislocations and interactions between the surface glucan chains and other wall polymers (Rose and Bennett, 1999). Consequently, the microfibrils can show considerable organizational heterogeneity, with a substantial paracrystalline or amorphous surface and a more crystalline core (Viëtor et al., 2002; Šturcová et al., 2004). The factors that influence the degree of crystallinity, or the extent and regulation of interactions with other wall components at the point of wall formation, are poorly resolved, but mechanistic details of cellulose synthesis, chain crystallization, and polymerization are gradually emerging (Endler and Persson, 2011; Harris et al., 2012).

Hemicelluloses

Hemicelluloses comprise a structurally heterogeneous class of polysaccharides, including XyGs, β-(1→3,1→4)-glucans (mixed linkage glucans, or MLGs), xylans, mannans and glucomannans (Scheller and Ulvskov, 2010). They are generally ubiquitous in terrestrial plants, although some taxonomic groups have primary walls that are particularly enriched in certain types; for example, MLG has mostly been detected in the Poales. However, the growing number of exceptions to previously clear-cut ideas of phylogenetic distribution of wall polymers (Sørensen et al., 2008, 2011; Popper et al., 2011) suggests that most plants may have walls with a greater range of hemicelluloses than is typically suggested, but that some may be present at extremely low levels. The increasing use of highly sensitive analytical instruments, together with dramatic increases in gene sequence information should soon resolve this issue. It may also be that the fine structures of these polymers show a greater diversity than is currently supposed, possibly including "hybrid" classes. The basic structures are shown in Figure 7.1.

Pectins

Pectic polysaccharides are generally described as galacturonic acid (GalA) rich polymers that are present in the primary cell wall, particularly of dicotyledons, and also represent the major component of the middle lamella (Cosgrove, 2005). They can be divided into three principal classes (Fig. 7.2), the structurally simplest of which is homogalacturonan (HG), a linear polymer of GalA residues. Many of these residues are methylesterified at the C_6 carboxyl group at the time of secretion to the cell wall, and ester linkages to other moieties have also been suggested (Mohnen, 2008). The second class, rhamnogalacturonan I (RGI), is

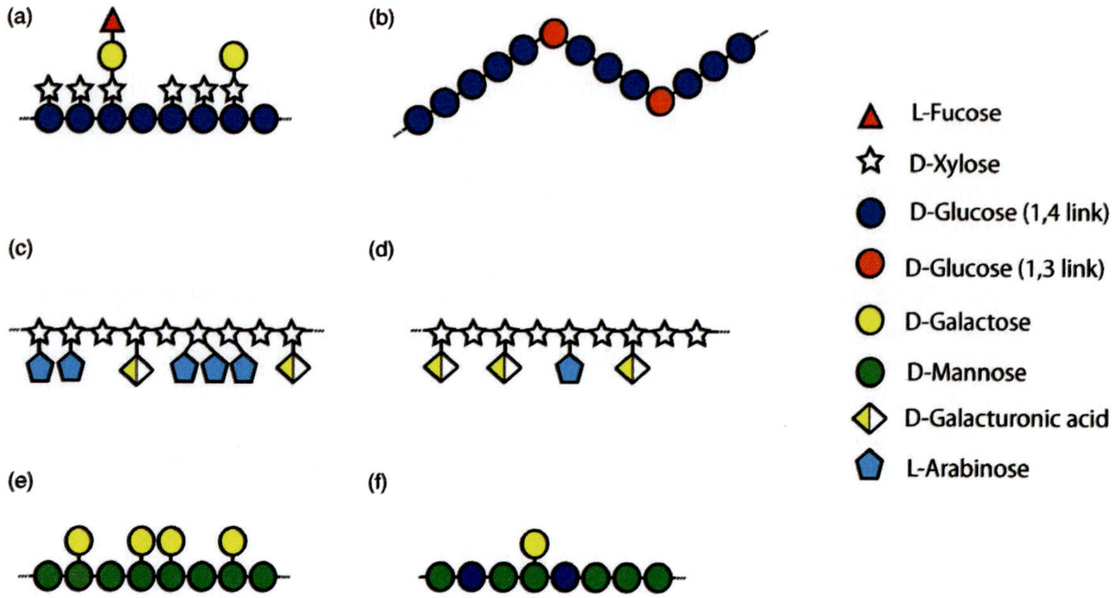

Figure 7.1 Schematic illustration of classes of hemicelluloses found in plant cell walls (taken from Scheller and Ulvskov, 2010). The structure of the hemicelluloses varies greatly in different plant species and tissue types. (a) The backbone xyloglucan [β-D-Glc*p*-(1→4)]$_n$ (XyG) substituted with side chains, as seen in pea and arabidopsis. Variants in side-chain structure are present in XyG from some taxonomic classes, such as the presence of arabinose in the side chains of XyG in the Solanaceae. (b) Mixed linkage β-glucan [β-D-Glc*p*-(1→4)]$_n$-β-D-Glc*p*-(1→3)-[β-D-Glc*p*-(1→4)]$_m$ (MLG), where n or m are 3 or 4—typical of MLG in the Poales. (c) Glucuronoarabinoxylan (GAX), typical of commelinid monocots. (d) A typical dicot GAX structure. (e) Galactomannan, typical of Fabaceae seeds. (f) Galactoglucomannan, characteristic of conifer wood.

structurally highly heterogeneous, with a backbone of alternating GalA and rhamnose (Rha) residues. The RGI polymer can be extensively branched and 20–80% of the Rha moieties can be substituted with linear or branched side chains of arabinose (Ara) or galactose (Gal) residues, forming arabinan, galactan, or hybrid arabinogalactan chains. Other less abundant sugar residues may also be present in the RGI side chains (Mohnen, 2008). The third class is the polymer rhamnogalacturonan II (RGII), which has an HG backbone that is substituted with structurally complex side chains comprising 12 different sugars and more than 20 different glycosyl linkages. RGII is usually present in the cell wall as a dimer that is formed by a borate-diol ester linkage between the apiose residues of two RGII monomers (O'Neill et al., 2004). The structure of RGII is conserved across the plant kingdom (Matsunaga et al., 2004), suggesting an essential role in growth and development. However, as with most cell wall polymers, the functional significance of the structural heterogeneity or conservation is not known at either a biological or a physicochemical level. In addition to these three main classes, two other types of pectic polymers have been identified, both of which have substituted HG backbones and appear to be less ubiquitous. The first is xylogalacturonan (XGA), a polymer with an HG backbone that is heavily decorated with xylose (Xyl) residues and that has been found in several fruits, including watermelon (*Citrullus lanatus*) and apple (*Malus domestica*) (Thibault and Ralet, 2001). The second, apiogalacturonan (AP), comprising HG substituted at O-2 or O-3 with

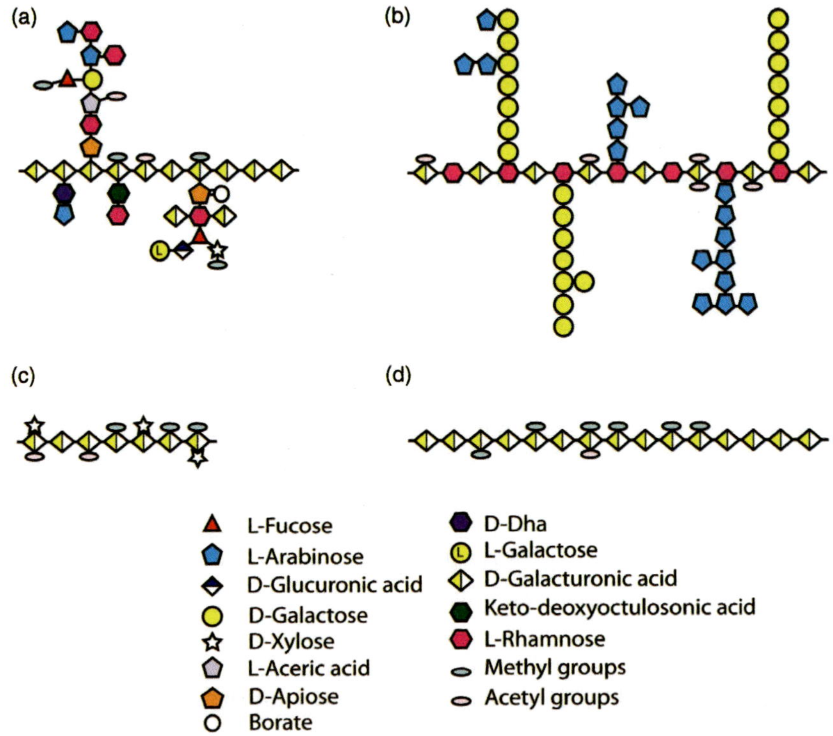

Figure 7.2 Schematic illustration of classes of pectins commonly found in plant cell walls. (a) Rhamnogalacturonan II (RGII), the structure of which is highly conserved across plant taxa. (b) A rhamnogalacturonan I (RGI) polymer, the side-chain structure of which is highly variable. (c) Xylogalacturonan (XGA). (d) Homogalacturonan (HG).

apiose, is apparently even more taxonomically restricted, and has been studied mostly in the aquatic monocot *Lemna* (Golovchenko et al., 2002).

Structural Proteins

In addition to the polysaccharide polymers described above, plant primary cell walls also contain a variety of so-called structural proteins, including hydroxyproline-rich glycoproteins (HRGPs), proline-rich proteins (PRPs), glycine-rich proteins (GRPs), and arabinogalactan proteins (AGPs). Some of these are mostly proteinaceous, while others, such as the AGPs, comprise mostly arabinogalactan polymers with a small protein core. They have been attributed with a wide variety of roles, including cell wall reinforcement, defense against pathogens, and signal transduction (Ringli et al., 2001; Seifert and Roberts, 2007; Deepak et al., 2010). The involvement of these protein/proteoglycan classes in fruit ripening has not been widely reported, although there are several suggestions that AGPs may contribute to some aspect of ripening-related wall metabolism. For example, an analysis of transcript expression in ripening tomato revealed that five fruit-related fasciclin-like arabinogalactan proteins (FLAs) show a substantial decrease in transcript abundance during ripening (The Tomato Genome Consortium, 2012). FLAs have been proposed as cell wall cross-linking molecules or as pectin plasticizers (Ellis et al., 2010), which may have significance in wall rheology, but this has yet to be tested. Similarly, a study of transcripts that are upregulated in grape berry ripening identified several HRGPs, although again their function in the context of ripening is unknown (Davies and Robinson, 2000). Their involvement in fruit texture will not be discussed further in this chapter, but this represents an interesting area for future investigation.

The preceding sections provide a basic compendium of cell wall polymer constituents, but it is important to reiterate that these mostly represent broad generalizations of the structures of the typical classes and do not take into account the tremendous microheterogeneity that is increasingly being reported in the fine structures of most polymer classes (Burton et al., 2010), or the variations seen among different fruit species. A long-term and extremely challenging goal will be to better resolve this heterogeneity, relate it to function, and to determine whether that functionality is conserved between different plant taxa. This will be aided by the development of tools to identify specific polysaccharide structures, and one valuable approach will doubtless be the use of a growing toolbox of antibodies and carbohydrate-binding modules that recognize specific glycan epitopes (Moller et al., 2007; Lee et al., 2011), and that can be used to visualize and even quantify the presence of those epitopes in plant walls. The detailed structural characterization of cell wall polysaccharides is in itself technically challenging; however, an even more difficult task is to develop an understanding of how these building blocks are incorporated into the three-dimensional architecture of the plant wall, and then to determine the dynamics of the resulting structures.

The Architecture of Fruit Cell Walls

Most recent depictions of plant primary cell wall architecture are based on the models of Carpita and Gibeaut (1993) and while some differ slightly from the originals, or highlight specific features (Carpita, 1996; Somerville et al., 2004; Cosgrove, 2005; Doblin et al., 2010; Albersheim et al., 2011), they share the same basic structural elements and emphasize similar structure–function relationships. The basic model, sometimes referred to as the "tethered network", involves helical arrays of cellulose microfibrils embedded and cross-linked in a highly hydrated matrix of hemicelluloses and pectins (Fig. 7.3a). A core element of the model is that the microfibrils are coated and cross-linked by hemicelluloses (XyG in Type I walls and GAX in Type II walls) and that this collectively provides the load-bearing component of the primary wall. For example, current models generally describe XyG as extensively hydrogen bonded to the microfibril surfaces, which prevents them from aggregating, as well as being present in an extended conformation, spanning and thus tethering adjacent microfibrils. It has also been suggested that XyG is present as partially attached strands or embedded within the paracrystalline regions of the microfibrils (Pauly et al., 1999; Rose and Bennett, 1999). The tethered network model therefore attributes a critical structural role to the cross-linking glycans and most models of wall loosening during cell expansion involve modification of this matrix, either through the loosening of hydrogen bonds or cleavage of the intermicrofibrillar hemicellulose

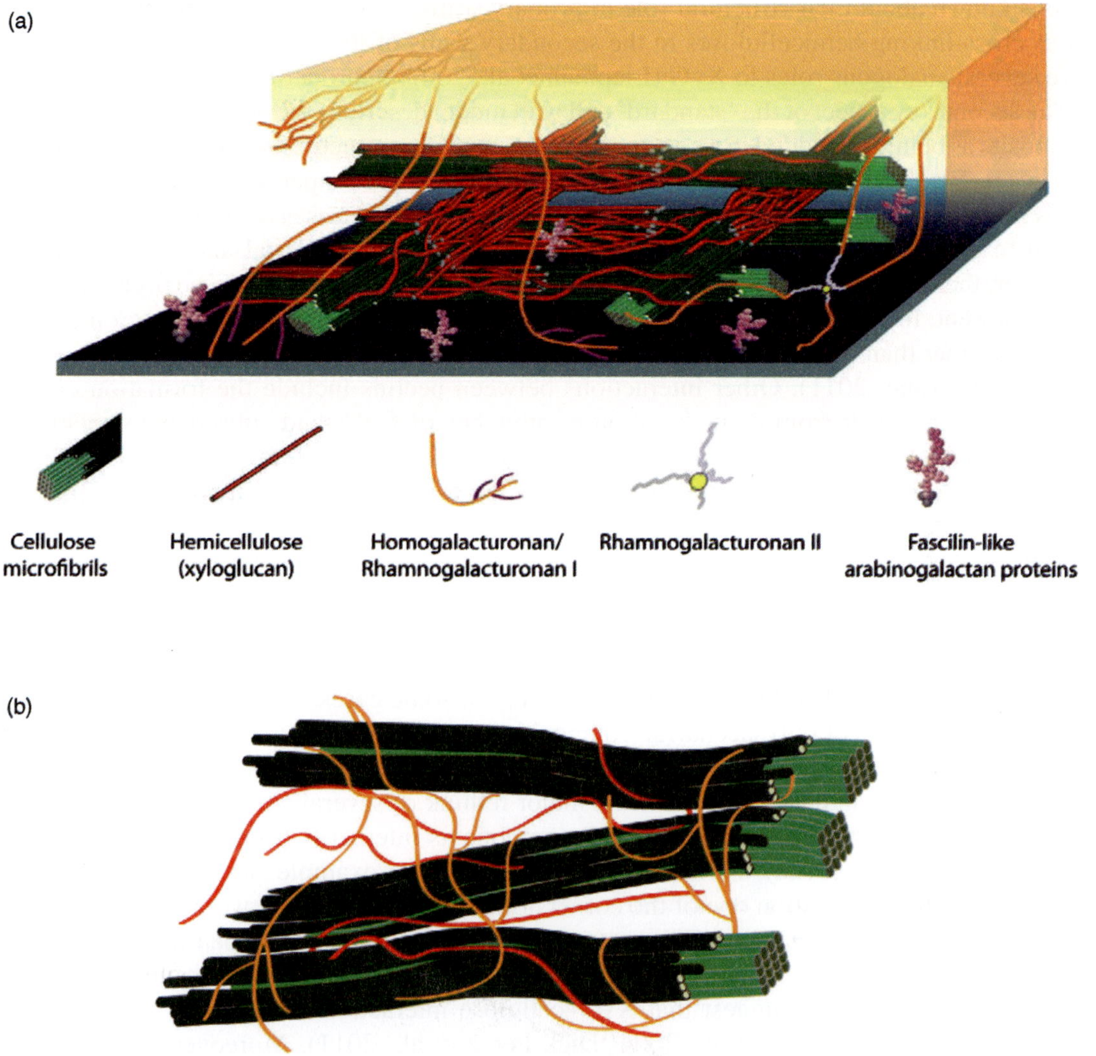

Figure 7.3 (a) Model of the "Type I" primary cell wall, highlighting the tethered network concept of a heavily cross-linked load-bearing network of cellulose microfibrils and xyloglucan (XyG) embedded in a gel of pectins. (b) A revised model (based on Park and Cosgrove, 2012b and Dick-Pérez et al., 2012) of the interactions between primary wall components, emphasizing potential direct cellulose–cellulose interactions, fewer XyG cross-links, and a more extensive and direct association between cellulose and pectins.

chains. A similar phenomenon has been proposed to occur in ripening fruits, where the consequent loss of wall mechanical strength contributes to loss of flesh firmness (Rose and Bennett, 1999). In addition to XyG, mannans are also a common component of fruit cell walls, although they are typically present at far lower concentrations than XyG and show greater structural heterogeneity as a consequence of a range of substitution patterns with Gal and Glc residues, giving rise to galactomannans and galactoglucomannans (Moreira and Filho, 2008). Relatively little is known about their structural role in primary walls, although they bind to cellulose and act as cross-linking hemicelluloses in the secondary walls of gymnosperms, and so they may have several analogous roles to XyG (Liepman et al., 2007; Schröder et al., 2009).

The second key aspect of the 'standard' cell wall model describes the hemicellulose–cellulose composite as being embedded in a coextensive, but somewhat structurally independent, hydrated pectin gel (Fig. 7.3a), which contributes important rheological properties to the wall and influences porosity (Cosgrove, 2005). It is thought that the various classes of pectins are covalently linked to each other to form a large macromolecular structure (Caffal and Mohnen, 2009). However, the frequency and distribution of the various subclasses appear to be highly complex and at present, little is known of the higher-order structure of the pectin supermacromolecular structure, other than that it is highly heterogeneous (Mohnen, 2008; Burton et al., 2010; Round et al., 2010; Yapo, 2011). Other interactions between pectins include the formation of ionic cross-links that result from demethylation of stretches of GalA and subsequent chelation of calcium ions, resulting in the formation of "egg-box" structures (Grant et al., 1973). Pectin chains are also ionically cross-linked through the boron-mediated RGII dimers, which may act as key tethering points between long pectin chains.

This widely adopted model of the primary wall has been of great value in shaping our understanding of wall dynamics in the context of cell expansion and differentiation. However, such models are designed to be updated and discoveries on a number of fronts, in part driven by the development of new analytical technologies, suggest that our understanding of several aspects of primary wall architecture needs revising. In some cases, these new insights relate to the finer scale details of polysaccharide composition or the identification of new intermolecular bonds, such as pectin–XyG and pectin–xylan covalent linkages (Mohnen, 2008). However, some recent studies indicate the need for a major rethink of several core themes of the model, and in particular how some of the key wall components interact (Cosgrove and Jarvis, 2012; Park and Cosgrove, 2012a). Perhaps, the most significant example has come from several reports that call into question one of the cornerstones of the tethered network hypothesis: XyG molecules that bridge adjacent cellulose microfibrils act as crucial load-bearing bonds and so loosening of those tethers is a rate-limiting and integral aspect of cell expansion. Contrary to this idea, NMR analyses suggest that XyG–cellulose interactions are far less extensive than previously thought (Booten et al., 2004; Dick-Pérez et al., 2011). Moreover, biomechanical studies of wild type and XyG-deficient cell walls were interpreted to suggest that only a small fraction of the total XyG is part of a load-bearing network, and that such regions may represent "hotspots" for wall loosening (Fig. 7.3b). These regions of the wall also appear to be somewhat enzyme inaccessible (Park and Cosgrove, 2012b). The authors were commendably cautious with their interpretations and highlight a number of caveats, but nonetheless, this represents an important discovery. Another recently reported solid-state NMR analysis targeted the dynamics and interactions of the pectin network in semi-intact cell walls (Dick-Pérez et al., 2012). The results were interpreted to suggest that pectins are highly mobile in the wall and are more extensively associated with cellulose than is shown in current models; indeed to the same degree as XyG–cellulose interactions.

Such observations underline the need for reevaluating primary wall architecture, even concerning some of the most widely accepted molecular interactions. Equally important, the heterogeneity in wall structure and composition between species, organs, and cell or tissue types requires that caution be used when making generalizations about wall organization and dynamics. Indeed, immunological studies using antisera to specific polysaccharide epitopes suggest considerable variation in wall microarchitecture, even within the face of a single wall (Casero and Knox, 1995; Orfila and Knox, 2000; Ordaz-Ortiz et al., 2009). In the context of this chapter, such wholesale changes in our current appreciation of wall architecture will inevitably have consequences for understanding the mechanisms of wall modification during ripening. Moreover, in some cases new models of wall composition, interpolymeric interactions, or higher-order architecture may help explain the unexpected outcomes or counterintuitive data from previously reported studies. The results of such studies may now be reinterpreted in the light of the new architectural models, examples of which are described below.

Cell Wall Dynamics in Ripening Fruit

A considerable number of comprehensive reviews over the last few years have described the structural and compositional changes that occur in the cell walls of a diverse range of ripening fruits, as well as the numerous classes of proteins that have been associated with ripening-related wall disassembly (Rose et al., 2003; Brummell, 2006; Vicente et al., 2007; Toivonen and Brummell, 2008; Payasi et al., 2009; Li et al., 2010). Collectively, they provide a detailed overview of the field and so the goal of this section is to provide a broad synopsis, with a focus on new insights, unresolved questions and emerging themes.

Ripening-related Disassembly of Cell Wall Polysaccharides

Decades of research have revealed that, generally speaking, most polysaccharide classes undergo structural modification during ripening. This wall disassembly appears to be temporally regulated such that different polymers are sequentially targeted, although the sequence and degree of modification can vary between species, or even between cultivars of the same species (Brummell, 2006). The most evident changes are typically extensive depolymerization of the matrix glycans, involving both cleavage of pectin and hemicellulose backbones, as well as removal or trimming of sugar side chains from those polymers. As a consequence, wall components are solubilized and become more readily extractable from the wall matrix. Other wall modifications include demethylation of HG, and possibly also polymer deacetylation, although this has yet to be confirmed in ripening fruit. Less is known about changes to the structure of the cellulose microfibrils, and while the amount of cellulosic glucan has been reported to be relatively constant during ripening in several species (Rose et al., 2003) an accurate assessment is complicated by the considerable technical challenges of simultaneously evaluating both crystalline and paracrystalline regions. New developments in solid-state NMR may prove to be valuable in this regard, and such a study of kiwifruit found no evidence for a substantial change in cellulose structure during ripening, even in substantially softened fruit

(Newman and Redgwell, 2002). It remains to be seen whether this applies broadly to many fruit species.

As detailed in several of the aforementioned reviews, a number of studies have suggested that one of the first cell wall changes is the loss of the galactan and arabinan side chains from RGI at the onset of, or even slightly prior to, ripening and this coincides with a commonly observed substantial loss of Gal. Given the proposed role of pectins in regulating wall porosity, it may be that this then facilitates access of other wall-modifying proteins to their substrates. Pectin de-esterification and depolymerization of loosely bound polymers then proceed with continuing and often substantial loss of RGI side chains, followed by depolymerization of tightly bound matrix glycans later in softening. As already mentioned, cell walls often swell during ripening and this is likely the result of some of these many changes in wall architecture, although at present this cannot be attributed to any specific structural or compositional changes and no direct causal relationship has been established.

The basic pattern of changes in wall architecture is well established and has been observed, albeit with many minor variations, in many species. Specific quantitative and qualitative changes have also been associated with textural differences, such as fruits that show a soft melting versus firmer crisp texture when fully ripe. However, there are clear exceptions and key distinguishing features have yet to be identified, although the extent of cell–cell separation appears to be important (Brummell, 2006).

It should be noted that all these analyses represent the most substantial and more readily detectable aspects of wall disassembly, typically involving the homogenization of bulk pericarp tissues, containing many cell types. This results in an averaging of wall composition, a complete lack of spatial resolution, and furthermore may prevent the detection of minor and potentially critical wall components of interpolymeric linkages due to dilution effects. This may be particularly significant in the light of some of the newer models of wall architecture, where "hotspots" of susceptibility to wall loosening have been proposed (Park and Cosgrove, 2012b).

Mechanisms of Cell Wall Modification

In parallel with the great progress in defining the many structural modifications that typify softening fruit, the last couple of decades have seen tremendous strides in identifying fruit-cell-wall-modifying enzyme activities, as well as the associated proteins and genes in numerous species. As described in the reviews mentioned above, among others, a large spectrum of proteins has been identified that collectively appear to target almost all cell wall polymers. These include enzymes whose catalytic activity, mode of action substrate, and even three-dimensional structure have been well characterized, as well as those that have no clear catalytic activity and whose *in vitro* substrates have not yet been well defined. Most of these proteins can be assigned to specific glycoside hydrolase families (GH; http://www.cazy.org) and likely act *in vivo* as hydrolases that cleave covalent linkages in the interior of polysaccharide chains (*endo*-glycanases) or remove single sugars from the ends of polymer backbones or side chains (*exo*-glycanases or glycosidases). However, other activities that are involved in wall loosening include transglycosylases, esterases and lyases. Some of these can be expected to directly reduce the tensile strength of the wall by acting on load-bearing bonds, while others are more likely to influence wall viscosity, rheology, porosity and other features that affect different aspects of fruit texture.

It is now apparent that a similar cocktail of wall-modifying proteins and activities is present in essentially all fleshy fruits, but that some can be considerably more pronounced in some species than in others (Brummell, 2006). From this, it may be concluded that the basic mechanisms of wall disassembly are ubiquitous. However, much remains to be learned about how the variability in the extent of depolymerization of a particular polymer and the activity levels of the associated degradative enzymes, translate into differences in the rates of softening or textural changes among different fruits. Indeed, while our understanding of the complement of genes and proteins related to wall modification proteins has significantly improved, there are several major hurdles that still need to be overcome order to address biochemical and biological functions, some of which are listed below.

Confirmation and characterization of *in vivo* enzyme activities

First, while gene sequences can be indicative of a particular wall-modifying activity, based on sequence homology, individual glycan-modifying proteins can catalyze different biochemical reactions depending on factors such as substrate availability and chemical environment. One well-studied example is that of the xyloglucan transglucosylase-hydrolases (XTHs), from GH family 16 (GH16), which specifically catalyze XyG hydrolysis (xyloglucan endohydrolase activity; XEH), resulting in net XyG depolymerization, or XyG transglycosylation (xyloglucan endotransglycosylase activity; XET), which involves grafting of two XyG molecules and need not involve depolymerization (Rose et al., 2002). Different XTH subclasses exhibit either predominant XEH or XET activity *in vitro*, and the structural basis of this has been characterized (Elköf and Brumer, 2010). In principle then, the balance of XEH and XET activities in ripening fruits should be predictable. However, the expression patterns of XTH genes in ripening fruit are complex (Saladié et al., 2006) and the modes of action *in vivo* may also be influenced by factors such as the concentration and availability of donor and acceptor substrates.

A similar situation appears to be the case for mannanases (GH5), which are encoded by large gene families (Rodriguez-Gacio et al., 2012), some of which contain members that appear to contribute to wall disassembly in ripening fruit (Wang et al., 2009). Originally considered to be mannan endo-β-hydrolases, it was subsequently shown that a ripening-related isoform has mannan transglycosylase activity *in vitro* (Schröder et al., 2004) and the protein is therefore more accurately designated as a mannan endotransglycosylase-hydrolase (MTH) enzyme. However, there are discrepancies between *in vitro* activities and an apparent lack of change in the molecular mass of the proposed mannan substrate during ripening (Schröder et al., 2009).

These observations further underline the need for caution when making inferences about the modes of enzyme action *in planta* based on observed activities *in vitro*. For example, it has been noted that factors such as ionic environment, as well as substrate concentration and availability, can have major effects on the preferred activity of dual-action enzymes such as these (Schröder et al., 2006). MTHs and XTHs are just two examples where apparent hydrolases may have other activities in the cell wall and it may be that this is the case for many other GH classes. Another interesting example of novel wall-remodeling activities is the discovery, in a wide variety of plants, of new kinds of hetero-transglycosylating activities, where glycosidic linkages are formed between two different classes of polysaccharides (Hrmova et al., 2007; Franková and Fry, 2011). None of these activities has been examined in fruit, but it is entirely possible that the modification of heteropolymer linkages by transglycosylation or hydrolysis may be an integral part of ripening-related wall disassembly.

In addition to the identification of wall-modifying activities, for which there is not yet a known enzyme or a corresponding gene, there are also the instances of proteins whose substrates or modes of action in the wall have not been well defined. One example is the endo-1,4-β-glucanase (EGase, GH9) family of proteins. EGase activities, as determined using artificial β-1,4-glucan polymeric substrates, as well as the expression of EGase genes, correlate with fruit softening in many species (Brummell et al., 1994; Urbanowicz et al., 2007), but the substrate(s) *in planta* are still not known. Contrary to earlier suggestions, it appears that XyG is not readily hydrolyzed by these enzymes, at least *in vitro* (Urbanowicz et al., 2007), and experiments with transgenic pepper lines also suggest that they act on other, currently unidentified, matrix glycans (Harpster et al., 2002a, 2002b).

Another example of a wall-modifying protein with no well-defined substrate is expansin, which has a potent wall loosening action but no apparent catalytic activity. Expansins are thought to disrupt the hydrogen bonds between cellulose microfibrils and matrix glycans, although their exact molecular targets are unclear (Brummell et al., 1999; Georgelis et al., 2011; Tabuchi et al., 2011), and altering the expansin expression in transgenic tomato fruit has not provided any further clarification (Brummell et al., 1999).

Comparative studies of cell-wall-texture associations in different fruit species

Over the last 20 years, the functions of a number of cell-wall-modifying proteins have been assessed through suppressing their activities in transgenic fruits (see Table 1 of Li et al., 2010), mostly targeting tomato. The first- and best-studied example is that involving the suppression of polygalacturonase (PG), thought to be the primary enzyme involved in the depolymerization of HG in both the primary wall and the middle lamella. Fruits with substantially reduced PG levels still showed significant softening, but also prolonged cell–cell adhesion and a delay in textural deterioration later in ripening (Kramer et al., 1992; Langley et al., 1994), suggesting that modification of the middle lamella and cell separation are significant factors in softening, particularly in the later stages. Nonetheless, the fruits still undergo essentially normal softening from the pre-ripe stage until reaching an edible texture. A similar conclusion was reached through a comparative study of peach cultivars, where PG levels correlated with aspects of texture but not loss of fruit firmness (Ghiani et al., 2011). However, the opposite appears to be the case for apple, where suppression of a PG gene not only prolonged cell–cell adhesion, but also resulted in firmer fruits (Atkinson et al., 2012). Interestingly, a number of pleiotropic effects were noted including smaller and more tightly packed cells, and reduced transpirational water loss, which complicates interpretation of the basis of the underlying firmness phenotype.

Additionally, PG appears to be important for strawberry fruit firmness (Quesada et al., 2009), as does another endo-pectinase, pectate lyase (PL; Jimenez-Bermudez et al., 2002; Marin-Rodriguez et al., 2002; Santiago-Domenech et al., 2008). The take-home message from such studies is that conclusions drawn from studies in one species do not automatically translate to others. While tomato is by far the best-studied model for fleshy fruit ripening (Giovannoni, 2007; Seymour et al., 2012), it is important to evaluate each fruit type on a case-by-case basis and develop an appreciation for factors such as the relative degree of primary cell wall or middle lamella modification, as well as some of the often minor differences in wall structure and textural attributes.

Evaluation of coordinated and synergistic mechanisms of cell wall disassembly

In addition to PG, other cell-wall-modifying proteins whose functional significance has been assessed in transgenic fruit include enzymes that act directly on pectins (pectin methylesterase, PME; pectate lyase, PL; and an *exo*-galactanase) or hemicelluloses (XTH; Miedes et al., 2010), as well as expansin (see Table 1 of Li et al., 2010). As was the case with PG, suppressing the expression of a single-wall-modifying gene has in some cases resulted in marginal effects on softening, but more often subtle differences in texture that are generally more apparent later in ripening. Even suppressing the expression of both PG and expansin simultaneously in tomato is apparently not sufficient to prevent softening, but rather results in a small increase in firmness and differences in the fruit rheological properties (Kalamaki et al., 2003; Powell et al., 2003). In retrospect, it is hardly surprising that reducing the expression of single-wall-modifying proteins does not block softening, although it is again notable that the effects clearly vary in different species. Such studies reinforce the notion that cell wall metabolism in ripening fruit results from the combined action of multiple-wall-modifying proteins acting on the wall matrix. Such activities are likely to be highly coordinated and indeed there is evidence to suggest that they are also synergistic.

Probably the best-studied example of synergistic wall modification in ripening fruit is that between PG and PME. Newly synthesized HG is incorporated into the cell wall in a highly methylesterified form and during ripening PMEs can catalyze substantial demethylation in blockwise and non-blockwise patterns (Willats et al., 2001a, 2001b). For example, in tomato approximately 90% of GalA residues are methylesterified immediately prior to ripening, but this proportion declines to about 35% in ripe fruit (Koch and Nevins, 1989), and this is thought to have a number of consequences. First, blocks of unesterified HG are able to interact ionically with calcium, forming microdomains in the form of a pectin gel that increases wall stiffness (Jarvis, 1984). The importance of this interaction was further suggested in a report describing the overexpression of a proton/calcium exchanger in transgenic tomato lines, which resulted in firmer fruit with elevated calcium levels (Park et al., 2005). However, it should be noted that the calcium levels in the apoplast were not assessed, nor were any cell wall analyses described, but nonetheless, this represents an interesting correlative study. Other evidence, obtained through a study of wall biomechanics in apple fruit, suggests that calcium is an important factor in tissue firmness and affects whether tissue cracking preferentially occurs between cells along the middle lamella, or as a result of cell wall rupture (Cybulska et al., 2012). Second, PME action produces protons, resulting in local acidification of the apoplast and it has been proposed that this change of pH could reduce the activity of some PME isoforms while activating other cell wall enzymes, such as PG and PL (Moustacas et al., 1991; Bordenave, 1996; Bosch and Hepler, 2005; Krichevsky et al., 2007; Pelloux et al., 2007; Draye and Van Cutsem, 2008). Third, demethylesterification of HG strongly influences its susceptibility to hydrolysis by PG (Koch and Nevins, 1989; Wakabayashi et al., 2003), which was shown in ripe tomato fruit to be primarily localized to domains of cell walls with low degrees of methyl esterification (Steele et al., 1997). This has led to the conclusion that during ripening PME activity is an important prerequisite for the action of PG, and possibly also PL (Brummell and Harpster, 2001; Wakabayashi et al., 2003; Prasanna et al., 2007). This example of synergistic polysaccharide modification appears to be spatially controlled in tomato (Steele et al., 1997) and may also involve other enzymes that act on different backbone or side-chain components of the pectin macromolecule, such as rhamnogalacturonases (Vicente et al., 2007). Such coordinated disassembly likely applies to both the primary wall and the middle lamella and is likely common to most fleshy fruit species, although this has yet to be investigated in detail.

In addition to the synergistic modification of specific wall polymers, there is also some evidence to suggest that the same applies to higher-order features of wall architecture. This was illustrated by studies of transgenic tomato lines with elevated or decreased expression of a ripening-related expansin (Brummell et al., 1999). Overexpressing expansin induced precocious and extensive depolymerization of uncharacterized hemicellulose cell wall components, while suppressing expression resulted in a reduction in polyuronide depolymerization later in ripening, but had no apparent effect on normal ripening-related hemicellulose degradation. One interpretation of the data is that expansin may enhance the accessibility of wall polymers to other enzymes (Brummell and Harpster, 2001), thereby accelerating wall hydrolysis, consistent with a sequential and synergistic mechanism of wall disassembly. The effects on pectin degradation were particularly unexpected given the association between expansins and the disruption of hemicellulose–cellulose interactions. However, they are even more intriguing in the context of newly proposed models of wall organization in which cellulose–pectin interactions are thought to be more extensive than previously supposed (Dick-Pérez et al., 2012). As such architectural models are further refined, data from experiments such as those with the transgenic expansin lines may be more readily incorporated in clearer mechanistic models of wall restructuring.

Another indication that sequential and synergistic enzymatic modification may be an intrinsic part of cell wall disassembly comes from studies showing that access of antibodies that recognize hemicellulosic polysaccharides, namely xylan, XyG, and mannan, to their ligands is often blocked by pectins (Marcus et al., 2008, 2010; Hervé et al., 2009). This apparently widespread phenomenon of masking provides a useful indication of the relative proximity and organization of wall polymers *in muro*. However, in addition, the observation that the application of pectinases results in unmasking suggests that sequential depolymerization of wall components, and particularly pectins, may provide an important means of regulating access of degradative enzymes to their substrates during wall remodeling.

Regulation of cell-wall-modifying enzyme action

Characterization of the enzymatic activities, substrate interactions, and expression of cell-wall-modifying proteins provides important insights into their significance in wall disassembly. However, it is important to note that their actions *in vivo* are influenced by many other factors and so detecting the protein of gene expression alone does not necessarily provide evidence of activity. The chemical environment of the apoplast can be particularly important and factors such as pH and abundance of various ions can influence both the enzyme activity and the physicochemical characteristics of the wall substrate. For example, it has been reported that the pH of apoplastic fluid extracted from tomato fruit decreases from 6.7 at a pre-ripe stage to 4.4 in ripe fruits (Almeida and Huber, 1999). However, such characteristics are difficult to measure *in vivo* and it may be that the microenvironment of the cell wall shows considerable spatial, as well as temporal heterogeneity.

Another means of controlling enzyme activity is through the interaction with proteinaceous inhibitors. For example, PME inhibitors (PMEIs), which have been identified in a range of fruit species (Jolie et al., 2010), form a complex with PMEs (Di Matteo et al., 2005), thereby blocking activity. Localization studies have detected PMEIs in fruit primary walls, the middle lamella and intercellular regions (Vandevenne et al., 2011). Additionally, a recent report described the colocalization of a tomato PMEI with regions of more highly esterified pectins in ripe tomato fruit, further suggesting that they provide an effective posttranslational control mechanism (Reca et al., 2012). Tomato PG also appears to be regulated by a modifying protein; the PG β-subunit, or converter (reviewed in Brummell and Harpster, 2001), which shows strong binding to PG. The function of this protein is still not clear, but experiments with transgenic tomato lines with suppressed expression of the β-subunit suggest that it restricts PG activity, either directly or by limiting its distribution within the cell wall.

The importance of posttranslational regulation of cell-wall-modifying proteins was also suggested by an analysis of transgenic tomato and pepper fruits in which the expression of enzymes associated with the trimming of *N*-glycans attached to glycoproteins was suppressed (Meli et al., 2010; Ghosh et al., 2011). In both cases the fruits showed substantially increased shelf life and a firmer pericarp, and the tomato transgenic lines also showed downregulation of transcripts encoding several glycoproteins associated with the cell wall disassembly. However, cell wall composition and structure were not described and the global inhibition of *N*-glycosylation through this strategy did not provide new insights into any specific deglycosylation events that may have been responsible for the phenotypes. Nonetheless, the authors suggested that modification of the sugar backbone of *N*-glycoproteins, or the generation of free *N*-glycans in the cell wall may have been contributory factors. Clearly, posttranslational regulation of cell wall proteins, including glycosylation and proteolytic processing, is a fertile area for future research.

The Cuticular Cell Wall and Fruit Softening

Most efforts to date to understand the role of the cell wall in fruit softening and accompanying textural changes have focused on the polysaccharide components of the extracellular matrix.

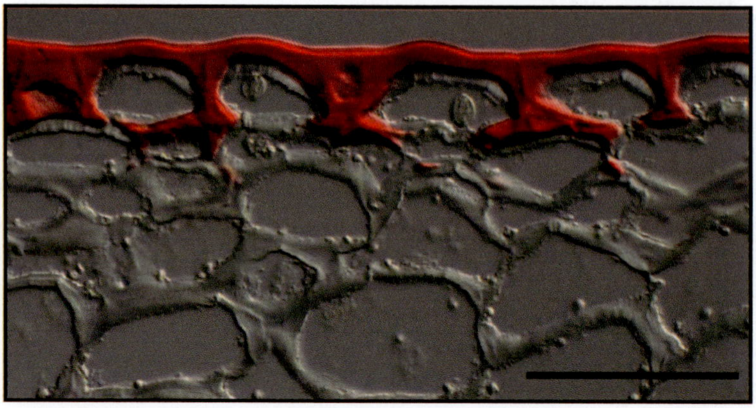

Figure 7.4 Light microscopy image of the cuticle of a mature green stage tomato fruit (cultivar M82) stained with Sudan Red. Scale bar = 50 μm.

However, there is a growing interest in the relative significance of the specialized lipid cell wall of the cuticle (Fig. 7.4), either in the context of its biomechanical properties or its function as the major regulator of transpirational water loss (Petracek and Bukovac, 1995; Riederer and Schreiber, 2001; Bargel et al., 2004; Matas et al., 2004; Bargel and Neinhuis, 2005; Edelmann et al., 2005; Dominguez et al., 2011). Cellular turgor pressure (P) is a critical determinant of tissue firmness and texture, so factors that affect overall fruit water status and the distribution of water in the apoplast and symplast are central to any consideration of fruit softening (Vicente et al., 2007; Seymour et al., 2012). Indeed, while direct measurements of P are relatively scarce, studies in tomato (Shackel et al., 1991; Saladié et al., 2007), cucumber (Sajnin et al., 2003), apple (Tong et al., 1999), and grape (Thomas et al., 2008; Wada et al., 2008, 2009; Matthews et al., 2009) suggest that decreases in P may serve as a major contributing factor to softening in many fleshy fruits.

The cuticle is composed of two main components, the polyester cutin and a variety of hydrophobic compounds that are collectively called waxes (Jeffree, 2006; Nawrath, 2006), and each has been suggested to influence water transpiration and softening in tomato fruit. For example, *Lecer6* is a tomato wax mutant that is deficient in β-ketoacyl-coenzyme A synthase, an important element of the very-long-chain fatty acid elongase complex (Vogg et al., 2004). Chemical characterization of the content of cuticular waxes of *Lecer6* in several stages of tomato ripening showed a major reduction in *n*-alkanes of chain length $>C_{28}$ and an increase of cyclic triterpenoids, which directly correlated with an increase in water transpiration (Leide et al., 2007). Another case study was provided through an analysis of the Delayed Fruit Deterioration (DFD) tomato cultivar, whose fruits undergo normal ripening, but remain firm and show no loss of integrity for many months after reaching the fully ripe stage (Saladié et al., 2007). Physiological and biomechanical analysis showed that DFD fruits lose very little water through transpiration and have a considerably higher cell turgor than normally softening tomato fruit as ripening progresses. This correlates with the increased fruit firmness. These striking characteristics of DFD fruits were accompanied by a higher accumulation of wax and cutin than a normally softening tomato cultivar, although a direct causal relationship has not been established.

Other studies using several cutin-deficient tomato mutants (*cd1, cd2, and cd3*) resulted in the conclusion that the amount of cutin does not always affect water loss, but correlates more with susceptibility to microbial infection (Isaacson et al., 2009). Nevertheless, one of the three *cd* mutants (*cd1*) showed a higher level of water loss as well as a reduction in the amounts of *n*-alkanes of chain length $>C_{30}$ and an increase in triterpenoids content, similar to *Lecer6* (Vogg et al., 2004; Leide et al., 2007). In contrast, the rate of water loss in pepper fruit was found to positively correlate with the amount of cutin monomers, but not with the total amount of surface waxes and was associated with qualitative differences in both cutin and wax composition (Parsons et al., 2012). It was concluded that the chemical composition of the cuticle is a critical factor in determining water loss, rather than the quantitative characteristics. Studies of leaf cuticles indicate that factors other than its role as a diffusion barrier act as selection pressures for cuticle thickness and this may also be the case for fruits (Onoda et al., 2012).

Little is known about the dynamics of the cuticle in ripening fruit and there are many questions regarding cuticle deposition, taxonomic variation, and the structural and functional associations between the cuticle and the underlying polysaccharide cell wall. However, this is likely to be an extremely important field of research in developing a more comprehensive understanding of fruit softening, together with other related factors that influence water status, such as the regulation of solutes in the apoplast (Vicente et al., 2007; Seymour et al., 2012).

Summary

Our perception of the relationship between cell wall biology and fruit ripening is growing increasingly complex and we have progressed a long way from early convictions that "fruit softens primarily as a result of pectin degradation". Many new technologies have helped refine this vision and this has been coupled with an expansion in the breadth and forward genetic approaches that are used to identify and dissect loci that underlie such complex traits as fruit firmness in a variety of species (Saliba-Colombani et al., 2001; Causse et al., 2002; Liebhard et al., 2003; Chaïb et al., 2007; Moreno et al., 2008; Chapman et al., 2012). This is important, as while tomato is still the favored experimental system for fleshy fruit biology it is now obvious that wall architecture, and the mechanisms involved in its restructuring show considerable qualitative and quantitative variation. New emerging models of cell wall structure will likely guide a reevaluation of wall dynamics, but many other factors should also be considered, some of which are shown in Figure 7.5. These range in scale from overall fruit anatomy and tissue morphology, down to the microheterogeneity of cell wall fine structure, the apoplastic environment, and the coordinated temporal and spatial regulation of enzymes that modify wall architecture. To conclude, if "softening" is to be understood then all of these elements, and doubtless others whose connection with cell wall metabolism is still mysterious (Nambeesan et al., 2010; Centeno et al., 2011), must be considered *in toto*— a daunting and exciting prospect.

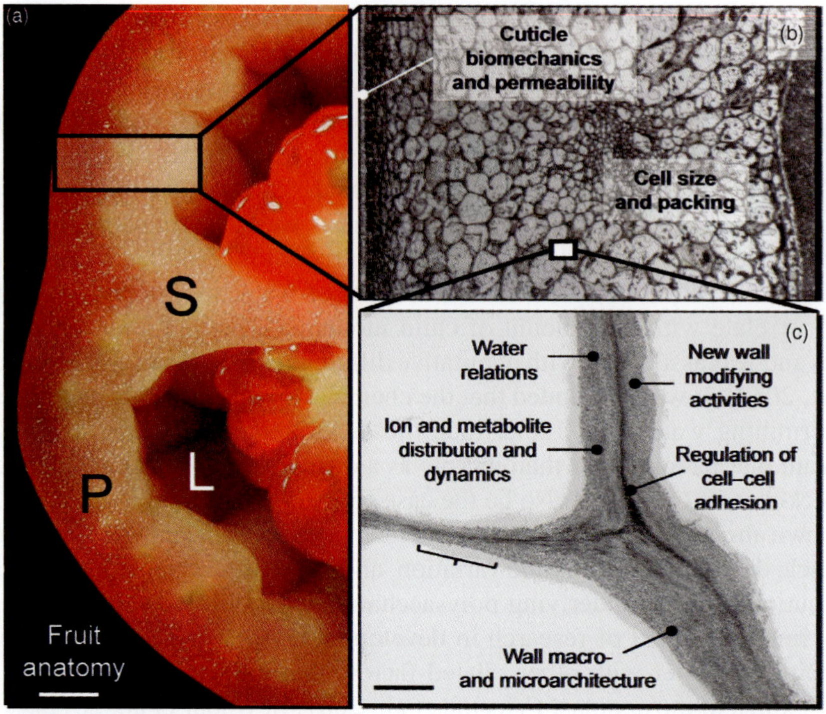

Figure 7.5 An illustration of the many factors that influence fruit softening and the relationship with cell wall disassembly at multiple levels of scale. (a) Gross fruit anatomy of a bisected, ripe tomato fruit showing various anatomical features, including the pericarp (P), locule (L), and septum (S); scale bar = 1 mm. (b) Tissue morphology and properties, showing a light micrograph of a cross section of the tomato pericarp (indicated by a magnification of the box in (a); scale bar = 100 μm). (c) The cell wall and middle lamella, highlighted by a transmission electron micrograph of a junction zone between three tomato fruit pericarp cells (indicated by a magnification of the box in (b); scale bar = 500 nm). All panels are annotated with features that influence aspects of fruit firmness or texture, many of which are still not well understood and that all need to be incorporated into more holistic models.

Acknowledgments

We thank Drs. David Domozych, Gregory Buda, and Antonio Matas for providing photographic images and Drs. Iben Sørensen, Trevor Yeats, and Ana Bertolo for helpful discussion and input. Financial support to JKCR for research in this area is provided by grants from the National Science Foundation (Plant Genome Program; DBI-0606595) and the USDA Cooperative State Research, Education and Extension Service (2011-04197).

References

Albersheim, P., Darvill, A., Roberts, K., et al. (2011) *Plant Cell Walls, From Chemistry to Biology*, pp. 227–272. Garland Science, New York.

Almeida, D.P., and Huber, D.J. (1999) Apoplastic pH and inorganic ion levels in tomato fruit: a potential means for regulation of cell wall metabolism during ripening. *Physiologia Plantarum*, 105: 506–512.

Atkinson, R.G., Sutherland, P.W., Johnston, S.L., et al. (2012) Down-regulation of POLYGALACTURONASE 1 alters firmness, tensile strength and water loss in apple (*Malus* x *domestica*) fruit. *BMC Plant Biology*, 12, 129.

Bargel, H., and Neinhuis, C. (2005) Tomato (*Lycopersicon esculentum* Mill.) fruit growth and ripening as related to the biomechanical properties of fruit skin and isolated cuticle. *Journal of Experimental Botany*, 56, 1049–1060.

Bargel, H., Spatz, H.C., Speck, T., et al. (2004) Two-dimensional tension tests in plant biomechanics-sweet cherry fruit skin as a model system. *Plant Biology*, 6, 432–439.

Booten, T.J., Harris, P.J., Melton, L.D., et al. (2004) ^{13}C-NMR spectroscopy shows that the xyloglucans in the primary cell walls of mung bean (*Vigna radiata* L.) occur in different domains: a new model for xyloglucan-cellulose interactions in the cell wall. *Journal of Experimental Botany*, 55, 571–583.

Bordenave, M. (1996) Analysis of pectin methyl esterases. In: *Plant Cell Wall Analysis* (ed. H.F. Linskens), pp. 165–180. Springer, New York.

Bosch, M., and Hepler, P.K. (2005) Pectin methylesterases and pectin dynamics in pollen tubes. *The Plant Cell*, 17, 3219–3226.

Brownleader, M.D., Jackson, P., Moasheri, A., Pantelides, A.T., Sumar, S., Travan, M., and Dey, P.M. (1999) Molecular aspects of cell wall modifications during fruit ripening. *Critical Reviews in Food Science and Nutrition*, 39, 149–164.

Brummell, D.A. (2006) Cell wall disassembly in ripening fruit. *Functional Plant Biology*, 33, 103–119.

Brummell, D.A., and Harpster, M.H. (2001) Cell wall metabolism in fruit softening and quality and its manipulation in transgenic plants. *Plant Molecular Biology*, 47, 311–340.

Brummell, D.A., Harpster, M.H., Civello, P.M., et al. (1999) Modification of expansin protein abundance in tomato fruit alters softening and cell wall polymer metabolism during ripening. *The Plant Cell*, 11, 2203–2216.

Brummell, D.A., Lashbrook, C.C., and Bennett, A.B. (1994) Plant endo-1,4 β-D glucanase: structure, properties and physiological function. In: *Enzymatic Conversion of Biomass for Fuels Production* (eds. M.E. Himmel, J.O. Baker, and R.P. Overend), pp 100–129. American Chemical Society, Washington, DC.

Burton, R.A., Gidley, M.J., and Fincher, G.B. (2010) Heterogeneity in the chemistry, structure and function of plant cell walls. *Nature Chemical Biology*, 6, 724–732.

Caffal, K.H., and Mohnen, D. (2009) The structure, function and biosynthesis of plant cell wall pectic polysaccharides. *Carbohydrate Research*, 344, 1879–1900.

Carpita, N.C. (1996) Structure and biogenesis of the cell walls of grasses. *Annual Review of Plant Physiology and Plant Molecular Biology*, 47, 445–476.

Carpita, N.C., and Gibeaut, D.M. (1993) Structural models of primary cell walls in flowering plants: consistency of molecular structure with the physical properties of the walls during growth. *The Plant Journal*, 3, 1–30.

Casero, P.J., and Knox, J.P. (1995) The monoclonal antibody JIM5 indicates patterns of pectin deposition in relation to pit fields at the plasma-membrane-face of tomato pericarp cell walls. *Protoplasma*, 188, 133–137.

Causse, M., Saliba-Colombani, V., Lecomte, L., et al. (2002) QTL analysis of fruit quality in fresh market tomato: a few chromosome regions control the variation of sensory and instrumental traits. *Journal of Experimental Botany*, 53, 2089–2098.

Centeno, D.C., Osorio, S., Nunes-Nesi, A., et al. (2011) Malate plays a crucial role in starch metabolism, ripening, and soluble solid content of tomato fruit and affects postharvest softening. *The Plant Cell*, 23, 162–184.

Chaïb, J., Devaux, M.F., Grotte, M.G., et al. (2007) Physiological relationships among physical, sensory, and morphological attributes of texture in tomato fruits. *Journal of Experimental Botany*, 58, 1915–1925.

Chapman, N.H., Bonnet, J., Grivet, L., et al. (2012) High resolution mapping of a fruit firmness-related QTL in tomato reveals epistatic interactions associated with a complex combinatorial locus. *Plant Physiology*, 159, 1644–1657.

Cosgrove, D.J. (2005) Growth of the plant cell wall. *Nature Reviews Molecular Cell Biology*, 6, 850–861.

Cosgrove, D.J., and Jarvis, M.C. (2012) Comparative structure and biomechanics of plant primary and secondary cell walls. *Frontiers in Plant Science*, 3, 204. doi:10.3389/fpls.2012.00204.

Cybulska, J., Pieczywek, P.M., and Zdunek, A. (2012) The effect of Ca^{2+} and cellular structure on apple firmness and acoustic emission. *European Food Research and Technology*, 235, 119–1287.

Davies, C., and Robinson, S.P. (2000) Differential screening indicates a dramatic change in mRNA profiles during grape berry ripening. Cloning and characterization of cDNAs encoding putative cell wall and stress response proteins. *Plant Physiology*, 122, 803–812.

Deepak, S., Shailasree, S., Kini, R.K., et al. (2010) Hydroxyproline-rich glycoproteins and plant defence. *Journal of Phytopathology*, 158, 585–593.

Di Matteo, A., Giovane, A., Raiola, A., et al. (2005) Structural basis for the interaction between pectin methylesterase and a specific inhibitor protein. *The Plant Cell*, 17, 849–858.

Dick-Pérez, M., Wang, T., Salazar, A., et al. (2012) Multidimensional solid-state NMR studies of the structure and dynamics of pectic polysaccharides in uniformly ^{13}C-labeled *Arabidopsis* primary cell walls. *Magnetic Resonance in Chemistry*, 50, 539–550.

Dick-Pérez, M., Zhang, Y., Hayes, J., et al. (2011) Structure and interactions if plant cell-wall polysaccharides by two- and three-dimensional magic-angle-spinning solid state NMR. *Biochemistry*, 50, 989–1000.

Doblin, M.S., Pettolino, F., and Bacic, A. (2010) Plant cell walls: the skeleton of the plant world. *Functional Plant Biology*, 37, 357–381.

Dominguez, E., Heredia-Guerrero, J.A., and Heredia, A. (2011) The biophysical design of plant cuticles: an overview. *New Phytologist*, 189, 938–949.

Draye, M., and Van Cutsem, P. (2008) Pectin methylesterases induce an abrupt increase of acidic pectin during strawberry fruit ripening. *Journal of Plant Physiology*, 165, 1152–1160.

Edelmann, H.G., Neinhuis, C., and Bargel, H. (2005) Influence of hydration and temperature on the rheological properties of plant cuticles and their impact on plant organ integrity. *Journal of Plant Growth Regulation*, 24, 116–126.

Elköf, J.M., and Brumer, H. (2010) The *XTH* gene family: an update on enzyme structure, function, and phylogeny in xyloglucan remodeling. *Plant Physiology*, 153, 456–466.

Ellis, M., Egelund, J., Schultz, C.J., et al. (2010) Arabinogalactan-proteins: key regulators at the cell surface? *Plant Physiology*, 153, 403–419.

Endler, A., and Persson, S. (2011) Cellulose synthases and synthesis in Arabidopsis. *Molecular Plant*, 4, 199–211.

Franková, L., and Fry, S.C. (2011) Phylogenetic variation in glycosidases and glycanases acting on plant cell wall polysaccharides, and the detection of transglycosidase and trans-β-xylanase activities. *The Plant Journal*, 67, 662–681.

Georgelis, N., Tabuchi, A., Nikolaidis, N., et al. (2011) Structure-function analysis of the bacterial expansin EXLX1. *The Journal of Biological Chemistry*, 286, 16814–16823.

Ghiani, A., Onelli, E., Aina, R., et al. (2011) A comparative study of melting and non-melting flesh peach cultivars reveals that during fruit ripening endo-polygalacturonase (endo-PG) is mainly involved in pericarp textural changes, not in firmness reduction. *Journal of Experimental Botany*, 62, 4043–4054.

Ghosh, S., Meli, V.S., Kumar, A., et al. (2011) The *N*-glycan processing enzymes alpha-mannosidase and beta-D-*N*-acetylhexosaminidase are involved in ripening-associated softening in the non-climacteric fruits of capsicum. *Journal of Experimental Botany*, 62, 571–582.

Giovannoni, J.J. (2004) Genetic regulation of fruit development and ripening. *The Plant Cell*, 16(suppl.), S170–S180.

Giovannoni, J.J. (2007) Fruit ripening mutants yield insights into ripening control. *Current Opinion in Plant Biology*, 10, 283–289.

Golovchenko, V.V., Ovodova, R.V., Shashkov, A.S., et al. (2002) Structural studies of the pectic polysaccharide from duckweed *Lemna minor*. *Phytochemistry*, 60, 89–97.

Grant, G.T., Morris, E.R., Rees, D.A., et al. (1973) Biological interactions between polysaccharides and divalent cations: the egg box model. *FEBS Letters*, 32, 195–198.

Harker, F.R., Redgwell, R.J., Hallett, I.C., et al. (1997) Texture of fresh fruit. *Horticultural Reviews*, 20, 121–224.

Harpster, M.H., Brummell, D.A., and Dunsmuir, P. (2002a) Suppression of a ripening-related endo-1,4-beta-glucanase in transgenic pepper fruit does not prevent depolymerization of cell wall polysaccharides during ripening. *Plant Molecular Biology*, 50, 345–355.

Harpster, M.H., Dawson, D.M., Nevins, D.J., et al. (2002b) Constitutive overexpression of a ripening-related pepper endo-1,4-beta-glucanase in transgenic tomato fruit does not increase xyloglucan depolymerization or fruit softening. *Plant Molecular Biology*, 50, 357–369.

Harris, D.M., Corbin, K., Wang, T., et al. (2012) Cellulose microfibril crystallinity is reduced by mutating C-terminal transmembrane region residues CESA1A903V and CESA3T942I of cellulose synthase. *Proceedings of the National Academy of Sciences USA*, 109, 4098–4103.

Hervé, C., Rogowski, A., Gilbert, H.J., et al. (2009) Enzymatic treatments reveal different capacities for xylan recognition and degradation in primary and secondary plant cell walls. *The Plant Journal*, 58, 413–42.

Hrmova, M., Farkas, V., Lahnstein, J., et al. (2007) A barley xyloglucan xyloglucosyl transferase covalently links xyloglucan, cellulosic substrates, and (1,3;1,4)-β-D-glucans. *The Journal of Biological Chemistry*, 282, 12951–12962.

Isaacson, T., Kosma, D.K., Matas, A.J., et al. (2009) Cutin deficiency in the tomato fruit cuticle consistently affects resistance to microbial infection and biomechanical properties, but not transpirational water loss. *The Plant Journal*, 60, 363–377.

Jarvis, M.C. (1984) Structure and properties of pectin gels in plant cell walls. *Plant, Cell and Environment*, 7, 153–164.

Jeffree, C.E. (2006) The fine structure of the plant cuticle. In: *Biology of the Plant Cuticle* (eds. M. Riederer, and C. Muller), pp. 11–125. Blackwell Publishing Ltd., Oxford, UK.

Jimenez-Bermudez, S., Redondo-Nevado, J., Munoz-Blanco, J., et al. (2002) Manipulation of strawberry fruit softening by antisense expression of a pectate lyase gene. *Plant Physiology*, 128, 751–759.

Jolie, R.P., Duvetter, T., Van Loey, A.M., et al. (2010) Pectin methylesterase and its proteinaceous inhibitor: a review. *Carbohydrate Research*, 345, 2583–2595.

Kalamaki, M.S., Harpster, M.H., Palys, J.M., et al. (2003) Simultaneous transgenic suppression of LePG and LeExp1 influences rheological properties and juice and concentrate from a processing tomato variety. *Journal of Agricultural and Food Chemistry*, 51, 7456–7464.

Koch, J.L., and Nevins, D.J. (1989) Tomato fruit cell wall. 1. Use of purified tomato polygalacturonase and pectinmethylesterase to identify developmental changes in pectins. *Plant Physiology*, 91, 816–822.

Kramer, M., Sanders, R., Bolkan, H., et al. (1992) Postharvest evaluation of transgenic tomatoes with reduced levels of polygalacturonase: processing, firmness and disease resistance. *Postharvest Biology and Technology*, 1, 241–255.

Krichevsky, A., Kozlovsky, S.V., Tian, G.W., et al. (2007) How pollen tubes grow. *Developmental Biology*, 303, 405–420.

Langley, K.R., Martin, A., Stenning, R., et al. (1994) Mechanical and optical assessment of the ripening of tomato fruit with reduced polygalacturonase activity. *Journal of the Science of Food and Agriculture*, 66, 547–554.

Lee, K.J.D., Marcus, S.E., and Knox, J.P. (2011) Cell wall biology: perspectives from cell wall imaging. *Molecular Plant*, 4, 212–219.

Leide, J., Hildebrandt, U., Reussing, K., et al. (2007) The developmental pattern of tomato fruit wax accumulation and its impact on cuticular transpiration barrier properties: effects of a deficiency in a beta-ketoacyl-coenzyme A synthase (LeCER6). *Plant Physiology*, 144, 1667–1679.

Li, X., Xu, C., Korban, S.S., and Chen, K. (2010) Regulatory mechanisms of textural changes in ripening fruits. *Critical Reviews in Plant Sciences*, 29, 222–243.

Liebhard, R., Kellerhals, M., Pfammatter, W., et al. (2003) Mapping quantitative physiological traits in apple (*Malus x domestica* Borkh.). *Plant Molecular Biology*, 52, 511–526.

Liepman A.H., Nairn C.J., Willats W.G., et al. (2007) Functional genomic analysis supports conservation of function among cellulose synthase-like a gene family members and suggests diverse roles of mannans in plants. *Plant Physiology*, 143: 1881–1893.

Marcus, S.E., Blake, A.W., Benians, T.A.S., et al. (2010) Restricted access of proteins to mannan polysaccharides in intact plant cell walls. *The Plant Journal*, 64, 191–201.

Marcus, S.E., Verhertbruggen, Y., Hervé, C., et al. (2008) Pectic homoglacturonan masks aundant sets of xyloglucan epitopes in plant cell walls. *BMC Plant Biology*, 8, 60.

Marin-Rodriguez, M.C., Orchard, J., and Seymour, G.B. (2002) Pectate lyases, cell wall degradation and fruit softening. *Journal of Experimental Botany*, 53, 2115–2119.

Matas, A.J., Cobb, E.D., Bartsch, J.A., et al. (2004) Biomechanics and anatomy of *Lycopersicon esculentum* fruit peels and enzyme-treated samples. *American Journal of Botany*, 91, 352–360.

Matsunaga, T., Ishii, T., Matsumoto, S., et al. (2004) Occurrence of the primary cell wall polysaccharide rhamnogalacturonan II in pteridophytes, lycophytes, and bryophytes. Implications for the evolution of vascular plants. *Plant Physiology*, 134, 339–351.

Matthews, M.A., Thomas, T.R., and Shackel, K.A. (2009) Fruit ripening in *Vitis vinifera* L.: possible relation of veraison to turgor and berry softening. *Australian Journal of Grape and Wine Research*, 15, 278–283.

Meli, V.S., Ghosh, S., Prabha, T.N., et al. (2010) Enhancement of fruit shelf life by suppressing *N*-glycan processing enzymes. *Proceedings of the National Academy of Sciences USA*, 107, 2413–2418.

Miedes, E., Herbers, K., Sonnewald, U., et al. (2010) Overexpression of a cell wall enzyme reduces xyloglucan depolymerization and softening of transgenic tomato fruits. *Journal of Agricultural and Food Chemistry*, 58, 5708–5713.

Mohnen, D. (2008) Pectin structure and biosynthesis. *Current Opinion in Plant Biology*, 11, 266–277.

Moller, I., Sørensen, I., Bernal, A.J., et al. (2007) High-throughput mapping of cell-wall polymers within and between plants using novel microarrays. *The Plant Journal*, 50, 1118–1128.

Moreira, L.R.S., and Filho, E.X.F. (2008) An overview of mannan structure and mannan degrading enzyme systems. *Applied Microbiology and Biotechnology*, 79, 165–178.

Moreno, E., Obando, J.M., Dos-Santos, N., et al. (2008) Candidate genes and QTLs for fruit ripening and softening in melon. *Theoretical and Applied Genetics*, 116, 589–602.

Moustacas, A.M., Nari, J., Borel, M., et al. (1991) Pectin methylesterase, metal ions and plant cell-wall extension. The role of metal ions in plant cell-wall extension. *Biochemical Journal*, 279, 351–354.

Nambeesan, S., Datsenka, T. Ferruzzi, M.G., et al. (2010) Overexpression of yeast spermidine synthase impacts ripening, senescence and decay symptoms in tomato. *The Plant Journal*, 63, 863–847.

Nawrath, C. (2006) Unraveling the complex network of cuticular structure and function. *Current Opinion in Plant Biology*, 9, 281–287.

Newman, R.H., and Redgwell, R.J. (2002) Cell wall changes in ripening kiwifruit: ^{13}C solid state NMR characterization of relatively rigid cell wall polymers. *Carbohydrate Polymers*, 49, 121–129.

O'Neill, M.A., Ishii, T., Albersheim, P., et al. (2004) Rhamnogalacturonan II: structure and function of a borate cross-linked cell wall pectic polysaccharide. *Annual Review of Plant Biology*, 55, 109–139.

Onoda, Y., Richards, L., and Westoby, M. (2012) The importance of leaf cuticle for carbon economy and mechanical strength. *New Phytologist*. doi:10.1111/j.1469-8137.2012.04263.x.

Ordaz-Ortiz, J.J., Marcus, S.E., and Knox, J.P. (2009) Cell wall microstructure analysis implicates hemicellulose polysaccharides in cell adhesion in tomato fruit pericarp parenchyma. *Molecular Plant*, 2, 910–921.

Orfila, C., and Knox, J.P. (2000) Spatial regulation of pectic polysaccharides in relation to pit fields in cell walls of tomato fruit pericarp. *Plant Physiology*, 122, 775–782.

Park, S., Cheng, N.H., Pittman, J.K., et al. (2005) Increased calcium levels and prolonged shelf life in tomatoes expressing Arabidopsis H^+/Ca^{2+} transporters. *Plant Physiology*, 139, 1194–206.

Park, B.Y., and Cosgrove, D.J. (2012a) Changes in cell wall biomechanical properties in the xyloglucan-deficient xxt1/xxt2 mutant of Arabidopsis. *Plant Physiology*, 158, 465–475.

Park, B.Y., and Cosgrove, D.J. (2012b) A revised architecture of primary cell walls based on biomechanical changes induced by substrate-specific endoglucanases. *Plant Physiology*, 158, 1933–1943.

Parsons, E.P., Popopvsky, S., Lohrey, G.T., et al. (2012) Fruit cuticle lipid composition and fruit post-harvest water loss in an advanced backcross generation of pepper (*Capsicum* sp.). *Physiologia Plantarum*, 126, 15–25.

Pauly, M., Albersheim, P., Darvill, A., et al. (1999) Molecular domains of the cellulose-xyloglucan network in the cell walls of higher plants. *The Plant Journal*, 20, 629–639.

Payasi A., Mishra N.N., Chaves A.L., et al. (2009) Biochemistry of fruit softening: an overview. *Physiology and Molecular Biology of Plants*, 15:103–113.

Pelloux, J., Rusterucci, C., and Mellerowicz, E.J. (2007) New insights into pectin methylesterase structure and function. *Trends in Plant Science*, 12, 267–277.

Petracek, P.D., and Bukovac, M.J. (1995) Rheological properties of enzymatically isolated tomato fruit cuticle. *Plant Physiology*, 109, 675–679.

Popper, Z.A., Michel, G., Hervé, C., et al. (2011) Evolution and diversity of plant cell walls: from algae to flowering plants. *Annual Review of Plant Biology*, 62, 567–90.

Powell, A.L.T., Kalamaki, M.S., Kurien, P.A., et al. (2003) Simultaneous transgenic suppression of LePG and LeExp1 influences fruit texture and juice viscosity in a fresh market tomato variety. *Journal of Agricultural and Food Chemistry*, 51, 7450–7455.

Prasanna, V., Prabha, T.N., and Tharanathan, R.N. (2007) Fruit ripening phenomena–an overview. *Critical Reviews in Food Science and Nutrition*, 47, 1–19.

Quesada, M.A., Blanco-Portales, R., Posé, S., et al. (2009) Antisense down-regulation of the *FaPG1* gene reveals an unexpected central role for polygalacturonase in strawberry fruit softening. *Plant Physiology*, 150, 1022–1032.

Reca, I.B., Lionetti, V., Camaredella, L., et al. (2012) A functional pectin methylesterase inhibitor protein (SolyPMEI) is expressed during tomato fruit ripening and interacts with PME-1. *Plant Molecular Biology*, 79, 429–442.

Redgwell, R.J., MacRae, E., Hallett, I., et al. (1997) In vivo and in vitro swelling of cell walls during fruit ripening. *Planta*, 203, 162–173.

Riederer, M., and Schreiber, L. (2001) Protecting against water loss: analysis of the barrier properties of plant cuticles. *Journal of Experimental Botany*, 52, 2023–2032.

Ringli, C., Keller, B., and Ryser, U. (2001) Glycine-rich proteins as structural components of plant cell walls. *Cellular and Molecular Life Sciences*, 58, 1430–1441.

Rodriguez-Gacio, M.D., Iglesias-Fernandez, R., Carbonero, P., et al. (2012) Softening-up mannan-rich cell walls. *Journal of Experimental Botany*, 63, 3976–3988.

Rose, J.K.C., and Bennett, A.B. (1999) Cooperative disassembly of the cellulose-xyloglucan network of plant cell walls: parallels between cell expansion and fruit ripening. *Trends in Plant Science*, 4, 176–183.

Rose, J.K.C., Braam, J., Fry, S.C., et al. (2002) The XTH family of enzymes involved in xyloglucan endotransglucosylation and endohydrolysis: current perspectives and a new unifying nomenclature. *Plant and Cell Physiology*, 43, 1421–1435.

Rose, J.K.C., Catalá, C., Gonzalez-Carranza, Z.H., et al. (2003) Plant cell wall disassembly. In: *The Plant Cell Wall*, Annual Plant Reviews Series (ed. J.K.C. Rose), pp. 264–324. Blackwell Publishing Ltd., Oxford, UK.

Round, A.N., Rigby, N.M., MacDougall, A.J., et al. (2010) A new view of pectin structure revealed by acid hydrolysis and atomic force microscopy. *Carbohydrate Research*, 345, 487–497.

Sajnin, C., Gamba, G., Gerschenson, L.N., et al. (2003) Textural, histological and biochemical changes in cucumber (*Cucumis sativus* L) due to immersion and variations in turgor pressure. *Journal of the Science of Food and Agriculture*, 83, 731–740.

Saladié, M., Matas, A.J., Isaacson, T., et al. (2007) A reevaluation of the key factors that influence tomato fruit softening and integrity. *Plant Physiology*, 144, 1012–1028.

Saladié, M., Rose, J.K.C., Cosgrove, D.J., et al. (2006) Characterization of a new xyloglucan endotransglucosylase/hydrolase (XTH) from ripening tomato fruit and implications for the diverse modes of enzymic action. *The Plant Journal*, 47, 282–295.

Saliba-Colombani, V., Causse, M., Langlois, D., et al. (2001) Genetic analysis of organoleptic quality in fresh market tomato. 1. Mapping QTLs for physical and chemical traits. *Theoretical and Applied Genetics*, 102, 259–272.

Santiago-Domenech, N., Jimenez-Bemudez, S., Matas, A.J., et al. (2008) Antisense inhibition of a pectate lyase gene supports a role for pectin depolymerization in strawberry fruit softening. *Journal of Experimental Botany*, 59, 2769–2779.

Scheller, H.V., and Ulvskov, P. (2010) Hemicelluloses. *Annual Review of Plant Biology*, 61, 263–289.

Schröder, R., Atkinson, R.G., and Redgwell, R.J. (2009) Re-interpreting the role of endo-beta-mannanases as mannan endotransglycosylase/hydrolases in the plant cell wall. *Annals of Botany*, 104, 197–204.

Schröder, R., Wegrzyn, T.F., Bolitho, K.M., et al. (2004) Mannan transglycosylase: a novel enzyme activity in cell walls of higher plants. *Planta*, 219, 590–600.

Schröder, R., Wegrzyn, T.F., Sharma, N.N., et al. (2006) LeMAN4 endo-beta-mannanase from ripe tomato fruit can act as a mannan transglycosylase or hydrolase. *Planta*, 224: 1091–1102.

Seifert, G.J., and Roberts, K. (2007) The biology of arabinogalactan proteins. *Annual Review of Plant Biology*, 58, 137–161.

Seymour, G.B., Chapman, N., Chew, B.L., et al. (2012) Regulation of ripening and opportunities for control in tomato and other fruits. *Plant Biotechnology Journal*, doi:10.1111/j.1467-7652.2012.00738.x, Epub ahead of print.

Shackel, K.A., Greve, C., Labavitch, J.M., et al. (1991) Cell turgor changes associated with ripening in tomato pericarp tissue. *Plant Physiology*, 97, 814–816.

Smith, B.G., and Harris, P.J. (1995) Polysaccharide composition of unlignified cell walls of pineapple [*Ananas comosus* (L.) Merr.] fruit. *Plant Physiology*, 107, 1399–1409.

Somerville, C. (2006) Cellulose synthesis in higher plants. *Annual Review of Cell and Developmental Biology*, 22, 53–78.

Somerville, C., Bauer, S., Briminstool, G., et al. (2004) Toward a systems approach to understanding plant cell walls. *Science*, 396, 2206–2211.

Sørensen, I., Pettolino, F.A., Bacic, A., et al. (2011) The charophycean green algae provide insights into the early origins of plant cell walls. *The Plant Journal*, 68, 201–211.

Sørensen, I., Pettolino, F.A., Wilson, S.M., et al. (2008) Mixed-linkage (1 → 3), (1 → 4)-beta-D-glucan is not unique to the poales and is an abundant component of *Equisetum arvense cell* walls. *The Plant Journal*, 54, 510–521.

Steele, N.M., McCann, M.C., and Roberts, K. (1997) Pectin modification in cell walls of ripening tomatoes occurs in distinct domains. *Plant Physiology*, 114, 373–381.

Šturcová, A., His, I., Apperley, D.C., Sugiyama, J., and Jarvis, M.C. (2004) Structural details of crystalline cellulose from higher plants. *Biomacromology*, 5, 1333–1339.

Tabuchi, A., Li, L.-C., and Cosgrove, D.J. (2011) Matrix solubilization and cell wall weakening by beta-expansin (group-1 allergen) from maize pollen. *The Plant Journal*, 68, 546–559.

The Tomato Genome Consortium. (2012) The tomato genome sequence provides insights into fleshy fruit evolution. *Nature*, 485, 635–641.

Thibault, J.-F., and Ralet M.-C. (2001) Pectins, their origin, structure and functions. In: *Advanced Dietary Fibre Technology* (eds. B.V. McCleary, and L. Prosky), pp. 369–378. Blackwell Science, Oxford, UK.

Thomas, T.R., Shackel, K.A., and Matthews, M.A. (2008) Mesocarp cell turgor in *Vitis vinifera* L. berries throughout development and its relation to firmness, growth, and the onset of ripening. *Planta*, 228, 1067–1076.

Toivonen, P.M.A., and Brummell, D.A. (2008) Biochemical bases of appearance and texture changes in fresh-cut fruit and vegetables. *Postharvest Biology and Technology*, 48, 1–14.

Tong, C., Krueger, D., Vickers, Z., et al. (1999) Comparison of softening-related changes during storage of 'Honeycrisp' apple, its parents, and 'Delicious'. *Journal of the American Society for Horticultural Science*, 124, 407–415.

Urbanowicz, B.R., Bennett, A.B., Catalá, C., et al. (2007) Structural organization and a standardized nomenclature for plant endo-1,4-β-glucanases of glycosyl hydrolase family 9. *Plant Physiology*, 144, 1693–1696.

Vandevenne, E., Christiaens, S., Van Buggenhout, S., et al. (2011) Advances in understanding pectin methylesterase inhibitor in kiwi fruit: an immunological approach. *Planta*, 233, 287–298.

Vicente, A.R., Saladié, M., Rose, J.K.C., et al. (2007) The linkage between cell wall metabolism and fruit softening: looking to the future. *Journal of the Science of Food and Agriculture*, 87, 1435–1448.

Viëtor, R.J., Newman, R.H., Ha, M.-A., et al. (2002) Conformational features of crystal-surface cellulose from higher plants. *The Plant Journal*, 30, 721–731.

Vogg, G., Fischer, S., Leide, J., et al. (2004) Tomato fruit cuticular waxes and their effects on transpiration barrier properties: functional characterization of a mutant deficient in a very-long-chain fatty acid beta-ketoacyl-CoA synthase. *Journal of Experimental Botany*, 55, 1401–1410.

Wada, H., Shackel, K.A., and Matthews, M.A. (2008) Fruit ripening in Vitis vinifera: apoplastic solute accumulation accounts for pre-veraison turgor loss in berries. *Planta*, 227: 1351–1361.

Wada, H., Matthews, M.A., and Shackel, K.A. (2009) Seasonal pattern of apoplastic solute accumulation and loss of cell turgor during ripening of *Vitis vinifera* fruit under field conditions. *Journal of Experimental Botany*, 60, 1773–1781.

Wakabayashi, K., Hoson, T., and Huber, D.J. (2003) Methyl de-esterification as a major factor regulating the extent of pectin depolymerization during fruit ripening: a comparison of the action of avocado (*Persea americana*) and tomato (*Lycopersicon esculentum*) polygalacturonases. *Journal of Plant Physiology*, 160, 667–673.

Wang, A., Li, J., Zhang, B., et al. (2009) Expression and location of endo-beta-mannanase during the ripening of tomato fruit, and the relationship between its activity and softening. *Journal of Plant Physiology*, 166, 1672–1684.

Willats, W.G., McCartney, L., Mackie, W., et al. (2001a) Pectin: cell biology and prospects for functional analysis. *Plant Molecular Biology*, 47, 9–27.

Willats, W.G., Orfila, C., Limberg, G., et al. (2001b) Modulation of the degree and pattern of methyl-esterification of pectic homogalacturonan in plant cell walls. Implications for pectin methyl esterase action, matrix properties, and cell adhesion. *The Journal of Biological Chemistry*, 276, 19404–19413.

Yapo, B.M. (2011) Pectic substances: from simple pectic polysaccharides to complex pectins—a new hypothetical model. *Carbohydrate Polymers*, 86, 373–385.

8 Regulatory Networks Controlling Ripening

Betsy Ampopho, Natalie Chapman, Graham B. Seymour, and James J. Giovannoni

A comprehensive understanding of the regulatory networks controlling ripening will be vital for a step change in the development of improved methods for controlling this important developmental process.

Hormonal Control

Ripening fruits have been physiologically characterized as climacteric or nonclimacteric (McMurchie et al., 1972). Climacteric fruit ripening in species such as tomato, banana, apple, pear, and most stone fruits is defined by an increase in respiration often coinciding with a simultaneous increase in the synthesis of the phytohormone, ethylene (Giovannoni, 2004). Storage conditions that minimize ethylene, in addition to gene mutations and targeted repression of ethylene synthesis genes, demonstrate that ethylene is generally required for ripening of climacteric fruits (Hamilton et al., 1990; Klee et al., 1991; Oeller et al., 1991; Watkins et al., 2000). The role of ethylene in ripening has been studied extensively (Chapter 3). However, the roles of other hormones such as auxins, abscisic acid (ABA), gibberellic acid, cytokinins, brassinosteroids, and jasmonic acid are less well understood.

Auxin plays an important role in early fruit development. Fruit set and growth are linked to auxin (Iwahory, 1967; Bottcher et al., 2010) and in tomato, levels of this hormone have been shown to accumulate postanthesis and just prior to the onset of ripening (Mapelli et al., 1978; Bottcher et al., 2010; Kumar et al., 2012). Although exogenous application of auxin to immature fruit delays fruit ripening and results in parthenocarpic fruit set (Vendrell, 1985; Cohen, 1996; Rotino et al., 1997; Jones et al., 2002), auxin is thought to have a pivotal role in initiating fruit ripening. In both tomato and grape (*Vitis vinifera*), levels of free indole-3-acetic acid (IAA) decrease toward the onset of ripening; in contrast, the levels of IAA-aspartic acid (IAA-Asp) increase (Bottcher et al., 2010, 2011; Kumar et al., 2012). The enzyme IAA-amido synthetase, which is encoded by the GH3 gene family, catalyzes the formation of the

The Molecular Biology and Biochemistry of Fruit Ripening, First Edition.
Edited by Graham B. Seymour, Mervin Poole, James J. Giovannoni and Gregory A. Tucker.

IAA-Asp conjugate from IAA, and this process has been reported for the fruits of tomato, grape, and pepper and is associated with ripening initiation (Bottcher et al., 2010, 2011; Kumar et al., 2012). For example, in tomato, two GH3 genes *SlGH3-1* and *SlGH3-2* are upregulated at the onset of ripening and their expression can be correlated with an increase in ethylene production in fruit (Kumar et al., 2012). Additionally, the expression of a pepper (*Capsicum chinense*) GH3 gene (*CcGH3*) is linked to decreased levels of auxin and its expression was shown to be ethylene inducible. Further, when exogenous ethylene was applied to tomato fruit overexpressing *CcGH3*, fruit ripening was accelerated (Liu et al., 2005). These data strongly support the notion that collectively the GH3 genes have a role in initiating ripening through the degradation and modulation of IAA in both climacteric and nonclimacteric fruits (Bottcher et al., 2011).

Members of the auxin response factor (ARF) and Aux/indole-3-acetic acid (Aux/IAA) transcription factor families have important regulatory roles during ripening. The expression of genes encoding ARFs and Aux/IAAs has been shown to be inducible by ethylene in tomato, supporting their role as regulators of the onset of ripening (Jones et al., 2002; Audran-Delalande et al., 2012). Aux/IAAs have differential responses to both auxin and ethylene in tomato and peach fruit, and it is likely that Aux/IAAs may act by modulating crosstalk between the two hormones (Jones et al., 2002; Trainotti et al., 2007; Audran-Delalande et al., 2012). Ethylene response factors (ERFs) have also shown differential response to both hormones, in peach (*Prunus persica*), an ERF gene, which is expressed at the later stages of ripening was upregulated by auxin to a greater extent than ethylene, suggesting that auxin has a role in regulating late ripening in peach (Trainotti et al., 2007).

ABA is a multifunctional plant hormone involved in a diverse range of biological processes including plant growth and development as well as being involved in regulating ripening in both climacteric and nonclimacteric fruits (Jia et al., 2011; Sun et al., 2012). ABA seems to act as a promoter of ripening. In tomato, ABA levels increase in advance of ethylene production at the ripening stage breaker (Sun et al., 2012). Furthermore, when exogenous ABA was applied, fruit firmness decreased and sugar accumulated more rapidly than in untreated fruit (Sun et al., 2012). In both climacteric and nonclimacteric fruits, suppression of 9-cis-epoxycarotenoid dioxygenase (NCED) gene expression which affects ABA biosynthesis, resulted in altered ripening phenotypes. Silencing of *SlNCED1* in tomato led to the downregulation of major genes encoding cell wall enzymes such as polygalacturonase (PG), pectin methylesterase (PME), xyloglucan endotransglycoslyase (XET), endo-1,4-β-cellulose (EGase), and expansins (EXPs). Due to changes in transcription of these genes, *SlNCED1*-RNAi fruits were firmer in texture and had enhanced shelf life (Sun et al., 2012). In *FaNCED1*-RNAi strawberry, receptacle color development was inhibited, but normal pigment accumulation could be rescued by exogenous ABA (Jia et al., 2011). Downregulation of an ABA receptor gene (*FaCHLH/ABAR*) in strawberry also resulted in loss of pigmentation and sugar content was also impaired. Normal ripening could not be rescued by exogenous ABA in this case, suggesting that *FaCHLH/ABAR* is necessary for this process (Jia et al., 2011). In grape, another nonclimacteric fruit, an increase in free levels of ABA is also associated with ripening (Wheeler et al., 2009; Bottcher et al., 2010).

Genetic Networks

Transcription Control Revealed by Analysis of Tomato Ripening Mutations

Tomato has proven to be a useful model for analyses of climacteric ripening, and breeders and geneticists have collected tomato-ripening mutants that reflect defects in the ripening regulatory systems. These include the single-gene mutations: *ripening inhibitor* (*rin*), *nonripening* (*nor*), and *Colorless nonripening* (*Cnr*), which are nonallelic, yet have common physiological characteristics and ripening inhibition, suggesting possible interaction in a common regulatory circuit. The *rin*, *nor*, and *Cnr* mutants do not ripen, but reach the mature green stage, with full-size fruit and mature seeds. All three mutants fail to undergo the climacteric rise in respiration or produce ripening-associated ethylene; do not ripen in response to exogenous ethylene; and respond to ethylene in other tissues and fruits, indicating that the molecular components of ethylene signal transductions remain intact (Eriksson et al., 2004; Manning et al., 2006; Giovannoni, 2007). The physiological properties of these mutations indicate that all play roles in the central ripening phenomena, including the induction of ripening ethylene in addition to activities beyond the regulatory realm of this hormone (Klee and Giovannoni, 2011).

The genes, underlying all three mutations, have been isolated via positional cloning, and found to encode transcription factors. The *rin* locus encodes a MADS-box transcription factor (*LeMADS-RIN*) that is induced at the onset of ripening (Vrebalov et al., 2002). It is located on chromosome 5, next to the tomato *APETALA1* ortholog, *MACROCALYX* (*MC*). *LeMADS-RIN* is a member of the SEPALLATA clade of MADS-box genes, linked to E function in *Arabidopsis* and mediates floral organ growth and development activities (Zahn et al., 2005). The MADS-box genes are developmental regulators conserved in eukaryotes and associated with floral development in plants, most thoroughly described in the context of the ABC(DE) model of floral development (Coen and Meyerowitz, 1991). Tomato plants homozygous for the recessive *rin* mutation produce fruits that develop normally and to full size with viable seed yet do not ripen. *rin/rin* fruits are known to display a large sepal, or macrocalyx, phenotype resulting from the deletion of the C-terminus of the *LeMADS-RIN* gene and approximately 1 kb of the sequence separating *RIN* from *MC*. The mutation manifests in expression of a chimeric *RIN/MC* mRNA with no *RIN* and minimal *MC* function (Vrebalov et al., 2002). The original *rin* allele is widely used in tomato hybrid seed production for heterozygous varieties (*rin/Rin*) and is characterized by a long shelf life and increased fruit firmness. Tomato fruit harboring the recessive *nonripening* (*nor*) mutation, located on tomato chromosome 10, is also inhibited in the normal fruit-ripening process, and displays similar phenotypes and physiology to the *rin* mutation (thus fail to ripen) (Giovannoni et al., 1995). *nor* encodes a transcription factor of the NAC domain family suggesting that both an NAC and a MADS-box gene contribute to the ripening regulatory switch system (Giovannoni, 2004).

COLORLESS NONRIPENING (*CNR*) encodes a tomato *SQUAMOSA*-promoter-binding protein (SBP) residing at the dominant *Cnr* locus on chromosome 2. *Cnr* fruits are inhibited in normal ripening and are colorless, possess a mealy pericarp, and exhibit a substantial loss of cell-to-cell adhesion (Thompson et al., 1999). The *Cnr* phenotype is the result of an unusual spontaneous epigenetic change in its promoter, resulting in a heritable lack of cytosine demethylation, giving rise to one of the few well-characterized epigenetic alleles (epiallele) in any species. Specifically at the *Cnr* locus, high levels of cytosine methylation were detected in a 286 bp contiguous region—2.4 kb upstream from the first ATG of *LeSPL-CNR* (*SPL* = *SQUAMOSA-promoter-binding like*). The DNA in both fruit and leaf tissues were shown to be more highly methylated in this region, in plants exhibiting the *Cnr* phenotype, though said methylation was substantially reduced in normally ripening fruit (Manning et al., 2006). Characterization of the mutation indicates that *CNR*, like *RIN*, plays a major role in the regulatory network that controls fruit ripening (Manning et al., 2006). Comparing *LeSPL-CNR* with SBP-box genes in the *Arabidopsis thaliana* genome, *LeSPL-CNR* is most homologous to *SPL3* (At2g33810). *SPL3* may interact with promoters of the *AP1* or *CAL* MADS-box genes of the SQUAMOSA family *in vivo* as a positive transcriptional regulator controlling floral development (Cardon et al., 1997). There is also an *SPL* gene involved in the early stages of micro- and megasporogenesis (Unte et al., 2003) and with the development of normal plant architecture (Stone et al., 2005). This family of genes has been associated with maize kernel development (Wang et al., 2005), showing a link to the development of dry fruiting structures. The finding that an *SPL* gene resides at the *Cnr* locus and the presence of a MADS-box transcription factor at the *rin* locus (Vrebalov et al., 2002) suggests these as components of a ripening model in which floral development regulatory genes have been recruited to new functions in regulating ripening in both dry and fleshy fruits during the course of angiosperm evolution (Manning et al., 2006). The novel finding of an epiallele at the *Cnr* locus also suggests the tantalizing possibility that the process of fruit development and ripening may include regulatory control elements exerting their influence at the levels of both the DNA and the epigenome.

Most recently, the *uniform ripening* (*u*) mutation has been shown to also encode a putative transcription factor that exerts significant influence on ripening processes, yet is not as global in its impact on fruit maturation as are the genes at the *Cnr, rin,* and *nor* loci. *U* encodes a Golden 2-like (GLK2) transcription factor, *SlGLK2*, regulating plastid development and thus chlorophyll accumulation and distribution in developing fruit. Gene expression analysis of the fruit indicates that normal *U* expression is elevated in the stem end of the fruit as compared to the base and results in the dark green top or "shoulder" typical of unripe wild tomato species, heirloom, and older commercial types. *u* encodes a spontaneous loss-of-function mutation at the *U* locus which has been widely bred into modern tomato varieties due to uniform fruit color, resulting from loss of the green-shoulder phenotype. As an interesting example of unintended consequences, it was shown that selection for *u* may have consumer appeal but the associated reduction in fruit chloroplasts has a negative impact on sugar and carotenoid levels of the mature fruit. There are two *GLK*s found in the tomato genome (The Tomato Genome Consortium, 2012). The same is true for many plants and in tomato, *SlGLK1* and *SlGLK2* are expressed in leaves, but only *SlGLK2* is highly induced in fruit (Powell et al., 2012).

Cloning of the *CNR*, *RIN,* and *NOR* genes defined the first ripening-specific transcription factors and provided insight into a ripening control system upstream of ethylene (production and/or signaling) (Klee and Giovannoni, 2011). RIN is a central player in the developmental regulation of fruit ripening (climacteric respiration and ripening-related ethylene biosynthesis) and acts upstream of both ethylene and nonethylene-mediated ripening control (Osorio et al.,

2011). Interestingly, a *RIN* homolog was shown to be necessary for ripening in nonclimacteric strawberry fruit (Seymour et al., 2011), suggesting conservation of molecular regulatory systems of fruits exhibiting both ripening physiologies. *LeMADS-RIN*, as with many genes in the MADS-box family, has been shown to interact with CArG-box elements in promoters of *ACS2* and *ACS4*, in addition to cell wall hydrolases, signifying that it interacts with promoters of genes representing a range of ripening activities including with promoters of other ripening genes such as those encoding *CNR* and *NOR* in addition to *RIN*'s own promoter suggesting direct regulatory interaction among these genes in addition to feedback control (Ito et al., 2008; Fujisawa et al., 2011; Martel et al., 2011). In the *rin* mutant, *CNR* expression is reduced, indicating that it may act downstream of *MADS-RIN* in the regulatory hierarchy, though clearly the relationship among these regulators and ethylene is not linear as indicated by the complex interactions of RIN with its own and promoters of additional regulatory genes (Martel et al., 2011). Targets of SBP-box genes as defined in *Arabidopsis* include MADS-box genes such as *FRUITFULL* (FUL) (Jaakola et al., 2010). The tomato ortholog of the *FUL*, *TDR4*, has substantially reduced mRNA accumulation in *Cnr* suggesting a similar regulatory interaction between the corresponding tomato genes (Manning et al., 2006). While TDR4 does not have a clear effect on tomato fruit maturation, possibly due to redundancy in the genome (The Tomato Genome Consortium, 2012), repression of a homologous gene in bilberry resulted in ripening inhibition and reduced fruit pigment accumulation suggesting an important regulatory function downstream of *CNR* (Jaakola et al., 2010). A recent report of targeted repression of both tomato *FUL* homologs indeed strongly suggests redundancy of function of these genes in contributing to tomato ripening phenotypes (Bemer et al., 2012).

Additional Transcriptional Regulators of Ripening Identified Through Molecular Studies

Molecular studies emphasizing ripening-related gene expression have revealed multiple additional transcription factors that are likely to influence ripening processes, and a number have been functionally characterized and shown to operate either through direct or indirect regulation of ethylene biosynthesis or signal transduction, while others display broader control similar to the genes revealed through mutation above.

Tomato expression profiling studies suggest that additional MADS-box genes beyond *RIN* contribute to fruit development and maturation (Alba et al., 2005; Osorio et al., 2011). As examples, the *AGAMOUS* clade defined in *Arabidopsis* to encode flower and fruit development (Yanofsky et al., 1990) includes *TAGL1* (*Tomato AGAMOUS-like 1*) and TAG1. Unlike the *SEPALLATA* clade, which may have expanded in tomato versus *Arabidopsis* to enable fleshy-fruit-ripening processes, the *AGAMOUS* (*AG*) clade of MADS-box genes is characterized by four genes in both tomato and *Arabidopsis* (Hileman et al., 2006; Vrebalov et al., 2009). *TAGL1* is induced earlier on in carpel development and later at the onset of ripening. The repression of *TAGL1* via RNAi results in ripening inhibition and reduction in carpel thickness due to reduced cell division during early carpel development. The fact that *TAGL1* repression had no notable impact on *RIN*, *NOR*, or *CNR* expression and that mutations in these genes display normal *TAGL1* expression, even though they do not ripen, suggests that TAGL1 influences ripening through a molecular regulatory network distinct from that mediated by these other transcription factors (Itkin et al., 2009; Vrebalov et al., 2009; Pan et al., 2010). It is unique that the bimodal function of *TAGL1* in early and late fruit development represents a novel molecular link between fleshy fruit expansion and subsequent ripening, the latter a process unique to fleshy fruits.

TAG1, which is the tomato ortholog of *AG*, has been shown to confer a similar homeotic and necessary function in carpel development to its *Arabidopsis* counterpart as repression of *TAG1* resulted in flowers with outer whorl floral structure but lacking in carpels (Pnueli et al., 1994). Interestingly, *TAG1* is expressed in ripening fruit and induced when *TAGL1* is repressed. This suggests a possible compensatory function during ripening. Furthermore, overexpression of *TAG1* in tomato under the 35S promoter resulted in fleshy expansion of, and lycopene accumulation in, tomato sepals (Pnueli et al., 1994). Sepal expansion and carotenoid accumulation also resulted from transgenic overexpression of *TAGL1* suggesting they may have similar and possibly redundant functions in maturing fruits. TAG1 function in mature fruits remains to be tested with a fruit-specific reduction in activity.

As noted above, the *LeSPL-CNR* protein product possibly interacts with sequence motifs in the promoters of MADS-box genes of the SQUAMOSA family. *TDR4,* the ripening-related MADS-box transcription factor and a likely ortholog of the *A. thaliana* FRUITFULL *(FUL)* which regulates silique dehiscence in *A. thaliana* (Liljegren et al., 2004), represents an additional contributor to fruit-ripening control via a MADS-box activity (Jaakola et al., 2010).

HB-1 encodes a tomato HD-Zip homeobox gene shown to interact directly with, and is necessary for, *ACO1* expression and the enzyme that catalyzes the final step in ethylene synthesis (Lin et al., 2008). The repression of *LeHB-1* in transgenic plants results in reduced *ACO1* expression, a concomitant decrease in ethylene synthesis, and delayed ripening (Lin et al., 2008).

Also contributing to tomato ethylene regulation is the tomato *APETALA2a* (*AP2a*) gene, a member of the AP2 subfamily of the *APETALA2/ETHYLENE RESPONSE FACTOR* (*AP2/ERF*) superfamily. As such, *AP2a* harbors two rather than one conserved AP2 domain. AP2a is a transcription factor, unique to plants, that regulates fruit ripening via regulation of ethylene biosynthesis and signaling (Chung et al., 2010; Karlova et al., 2011). Specifically, this gene functions as a negative regulator of ripening via inhibition of ethylene production (Chung et al., 2010). Tomato *AP2*-like genes have been shown to function in floral development as well, similar to *Arabidopsis AP2,* while *AP2a* may have gained a novel function more specific for fleshy fruit development (Karlova et al., 2011). *SlAP2a* is induced during ripening and its repression led to accelerated, elevated ethylene production, and altered carotenoid accumulation. All of these phenotypes were specific to the fruit, suggesting that *SlAP2a* is a fruit-specific and ripening-related suppressor of ethylene production and ripening. Interestingly, *AP2a* expression is induced by ethylene, and repressed in ethylene insensitive fruits of the *Gr* mutant, indicating a role in refining ethylene response possibly to ensure rapid ripening yet sufficient stability at the ripe stage to help ensure consumption and seed dispersal by frugivores. *RIN, NOR,* and *CNR* function upstream of *AP2a* and positively regulate its expression (Chung et al., 2010; Karlova et al., 2011). *AP2a* in turn negatively regulates CNR expression, indicating additional complexity to the network via a negative feedback loop. This regulation is likely to be direct, as CNR is able to bind directly to a promoter element of *AP2a (* Karlova et al., 2011).

Tomato *SlERF6*, a member of the ethylene response factor/ethylene responsive element binding protein (ERF/EREBP) family of transcription factors, was also associated with ripening via initial whole genome expression profiling and inverse correlation with carotenoid levels as determined by transcription factor/carotenoid network analysis (Lee et al., 2012). RNAi-mediated transgene repression revealed that *SlERF6* indeed influences carotenoid biosynthesis and additional ripening phenotypes. To initially identify *SlERF6*, Lee et al. performed ripe fruit transcriptional (microarray) and metabolic profiling to a population of mapped, single subchromosomal introgressions of wild tomato species *Solanum pennellii* into a cultivated *Solanum lycopersicum* parent (Eshed and Zamir, 1995). *SlERF6* has an inverse expression correlation in ripe fruit with trans-lycopene and β-carotene, the predominant carotenoids of ripe tomatoes, imparting upon them their characteristic red-orange color. *SlERF6* is highly expressed in ripe fruit. Reduced expression of *SlERF6* by RNAi enhanced both carotenoid and ethylene levels during fruit ripening, thus supporting a negative regulatory role for the SlERF6 protein. In three independent RNAi lines, the mRNA levels of *DXS*, *HSP21*, *ACS2*, *ACO1*, and *ACO3* were increased in one or more ripening stages. The expression of *ACS* and *DXS* are regulated by ethylene, whereas *HSP21* is not, suggesting that some genes that are affected by *SlERF6* are acting downstream of *SlERF6*-mediated ethylene synthesis, while others are influenced by a more direct regulatory effect. *ACS2* and *ACO1* encode the major enzymes responsible for ethylene biosynthesis in tomato fruit ripening (Barry et al., 1996, 2000), while *ACO3* mRNA is predominantly expressed in flowers, and is slightly induced during ripening and leaf senescence (Barry et al., 1996). The expression of all three genes was increased when SlERF6 was suppressed in ripening fruit, suggesting a regulatory function in ethylene synthesis through the genes, encoding the final steps of the pathway. It remains uncertain whether or not *SlERF6* is a direct regulator of these genes as promoter interaction studies were not reported. The bulk of genes that were substantially influenced by SlERF6 repression were upregulated, supporting the hypothesis that the primary function of *SlERF6* is via negative regulation.

It is clear from recent work on ripening control in tomato that multiple transcription factors influence the ripening process, and analysis of the recently described tomato genome indicates that ancient and more recent polyploidization events made available a reservoir of additional genes that contribute to control and manifestation of fleshy fruit development and ripening phenomena (The Tomato Genome Consortium, 2012). Some regulators such as RIN may be more ancient and well conserved as evidenced by the fact that apparent homologs were identified in the fleshy fruited climacteric monocot, banana (Elitzur et al., 2010). The evolved specific activities of *AGAMOUS* clade MADS-box genes, in particular *Arabidopsis AGL1* and tomato *TAGL1*, suggest that other members of this family have diverged in terms of specific molecular function while retaining their overall roles in seed dispersal via distinct activities in the context of very different (dry versus fleshy) fruit types. *RIN*, *NOR*, *CNR*, and *TAGL1* all play necessary roles in ripening though expression and promoter interaction studies suggest that TAGL1 may operate in a distinct ripening system from the rest (Vrebalov et al., 2009). The early fruit development rope of this gene and the fact that ectopic expression of *TAGL1* complements the *rin* and *nor* mutations suggest a more primary position for this gene in setting up the proper tissue context and developmental program in which the remaining transcription factors can achieve their necessary transcriptional profiles and regulatory contexts. The full complement of fruit-ripening regulators remains to be discovered but those identified to date reveal the beginnings of a complex matrix of regulatory and feedback circuitry necessary to control a critical and terminal process whose expression out of context could lead to tissue loss, lack of seed dispersal, and associated severe negative impact on fitness.

Downstream Ripening Signals and Activities

Fruit ripening represents an irreversible cascade leading to eventual tissue disruption and loss and in tomato and many species is associated with the apparently irreversible transition of photosynthetic chloroplasts into carotenoid accumulating chromoplasts. This transition is associated with the accumulation of carotenoids along with the disappearance of chlorophyll, disassembly of thylakoid structures, and a reduction in the levels of proteins and mRNAs associated with photosynthesis (Egea et al., 2010). These changes convey major influences upon the nutrient and flavor composition of the fruit. Plastoglobules accumulate in the chromoplast, serving as sinks for the buildup of large quantities of carotenoids that serve as both visual and nutritional indicators that the fruit is mature and suitable for consumption (Goff and Klee, 2006).

The chromoplast and plastoglobule proteomes have been studied in multiple species. The tomato chromoplast predominantly accumulates enzymes associated with synthesis and metabolism of carotenoids, amino acids, and fatty acids (Barsan et al., 2010). Even though the tomato plastoglobule proteome has not been described, that of the related pepper has been (Ytterberg et al., 2006), and the enzymes responsible for carotenoid synthesis indeed are prevalent within these structures. Phytoene synthase (PSY), the primary enzymatic regulator of flux into the carotenoid pathway, is under strong positive ethylene control during ripening, showing a link between chromoplast development and the primary ripening hormone. Repression of the gene encoding the lycopene β-cyclase (LYC) enzyme that would otherwise convert red lycopene to orange β-carotene (Alba et al., 2005) is also under ethylene control, supporting a strong role for ethylene in coordinating the activity of the carotenoid pathway toward predominant accumulation of lycopene.

There have been descriptions of few regulatory genes or activities specifically involved in the chloroplast to chromoplast transition during ripening beyond those involved in carotenoid synthesis and chlorophyll degradation (chlorophyllase). However, analysis of fruit pigment mutations revealed some insights. In addition to the *UNIFORM/GLK2* transcription factor described above, one gene that is clearly involved in this transition is the *green-flesh* (*gf*) locus that encodes a STAY-GREEN senescence-related regulator (Barry et al., 2008). *gf* loss-of-function mutation results in reduced plastid conversion during ripening, yielding fruit containing both chloroplasts and chromoplasts and characterized by a brown or "chocolate" color. Mutations in the orthologous pepper gene results in the same phenotype (Barry et al., 2008; Borovsky and Paran, 2008), and these fruits see commercial use due to their novel appearance and flavor characteristics. Additional genes mediating light signal transduction and response, specifically the tomato homologs of *Arabidopsis* DET1, DDB1, and CUL4 impact the number of chloroplasts and chromoplasts of the green and ripe fruits, respectively, though without additional notable impacts on other ripening phenomena (Mustilli et al., 1999; Liu et al., 2004; Wang et al., 2008).

Ethylene signal transduction (see Chapter 3) has been explored through the characterization of *Arabidopsis* ethylene-response mutations, in addition to second site repressors and interacting factors (Kendrick and Chang, 2008). A ripening-impaired dominant tomato mutation proved to result from a lesion in an ERS-like ethylene receptor incapable of binding ethylene confirming the well-known importance of ethylene in mediating ripening (Wilkinson et al., 1995; Rodriguez et al., 1999; O'Malley et al., 2005). While tomato contains at least six ethylene receptors, two (*NR* and *LeETR4*) appear to play predominant and compensatory roles in ripening fruit (Tieman et al., 2000). While most tomato ethylene signal transduction genes have been associated with previously studied *Arabidopsis* homologs, the tomato *Green-ripe* (*Gr*) fruit-specific ethylene-sensitive mutation revealed a novel function in ethylene signal transduction (Barry et al., 2005). Isolation of the gene underlying the *Gr* locus revealed a putative membrane protein of unknown function, yet possessing consensus—copper-binding motifs possibly playing a role in copper homeostasis with the copper-requiring ethylene receptors (Barry and Giovannoni, 2010). A homologous gene was described in *Arabidopsis* and shown to localize in the ER (Resnick et al., 2006), the cellular site to which ethylene receptors have also been localized (Resnick et al., 2006).

SlAP2a as noted above is a negative regulator of ethylene synthesis during ripening (Chung et al., 2010; Karlova et al., 2011) and was shown to also impact the expression of genes contributing to the chloroplast to chromoplast transition. Specifically, the expression of a number of genes implicated in chromoplast differentiation was found to be significantly downregulated. These were the *STAY-GREEN PROTEIN1* (*SGR1*) gene, encoding red chlorophyll catabolite reductase (*RCCR*), a tomato homolog of the cauliflower (*Brassica oleracea*) *ORANGE* (*OR*) gene, and a homolog of a target of the latter, a tomato homolog of the pepper *PFTF* gene (Barry et al., 2008). While additional regulators of general ethylene signal transduction are likely to be most readily identified in the highly tractable and easily screened *Arabidopsis* genetic system, additional fruit-specific regulators of ethylene synthesis and signaling will likely come forth from continued investigations in tomato and in other fleshy fruits.

Regulation of Fruit Quality Attributes

During and as a result of fruit ripening, downstream developmental and biochemical processes have final effects on softening, nutritional quality, accumulation of sugar, and production of chemicals associated with aroma, color, and taste of the ripe fruit. As noted above, when chloroplasts are converted to chromoplasts, large amounts of carotenoids (mainly lycopene and β-carotene) are synthesized and accumulate, providing visual cues to seed dispersers that the fruit is ripe and nutritious (Goff and Klee, 2006).

The change of color from green to red, typical of ripening tomato fruits is largely due to chlorophyll degradation and accumulation of carotenoids, including β-carotene and lycopene which are predominant nutrient metabolites of tomato and many fruits (Klee and Giovannoni, 2011); see Chapter 4. Lycopene, which gives the red color to ripe tomatoes, accounts for 70–90% of the total carotenoids present in the ripe fruit depending on cultivar (Alba et al.,

2005). These compounds are essential in plant photosynthesis and photoprotection and contribute in some cases essential nutrients to the human diet. Whereas carotenoid synthesis is well characterized (Hirschberg, 1999), factors regulating flux through the pathway are poorly understood at the molecular level. Carotenoid regulatory genes include in a general sense the global ripening transcription factors described previously as they exert control over most if not all aspects of ripening, including pigmentation, while *SlERF6* is one of the few genes with more specific impact on carotenoid accumulation. The light signal transduction mutations and genes characterized in tomato also have effects on plastid and carotenoid accumulation but in the broader context of light responsiveness with pleiotropic effects throughout the aerial parts of the plant (Mustilli et al., 1999; Liu et al., 2004; Wang et al., 2008). Additional fruit carotenoid regulatory mutants include *Lutescent 2* (*l2*) that displays a whole-plant chlorophyll-deficient phenotype, with reduced rates of chlorophyll synthesis during de-etiolation and enhanced rates of chlorophyll loss in leaves and fruits as they age. The mutation in *l2* stems from a defect in a plastid-targeted metalloprotease that facilitates chromoplast development and indicates additional regulatory feedback between plastid development and ripening (Barry et al., 2012). It is noteworthy that most described fruit carotenoid regulatory genes represent general components of ripening, plastid development, and light signal transduction, possibly suggesting that either few genetic regulators specific to fruit carotenoid accumulation exist or that they have not yet been successfully targeted.

The flavor of fruits can impact human nutrition with good flavor promoting elevated intake (Mathieu et al., 2009). Sugars, acids, and volatiles are the three major contributors to fruit flavor. The regulation of many of these events is still poorly understood as is their connection to the higher regulatory network. In tomato, sugars accumulate substantially with ripening, primarily glucose and fructose, which can be as much as 4% of the fresh weight and derived at the onset of ripening from starch that accumulated through fruit development (Ho et al., 1983). Citric and malic acids are most abundant in ripe tomato fruit which have a pH in the area of 4. Sugar and acid balance is important for good flavor, although preferences for sweetness and acid balance are individually and culturally biased. The malic acid content of immature fruit has been predicted to have an inverse correlation with the reducing sugar content of the ripe fruit since starch synthesis is positively correlated with reducing sugar content in ripe fruits (Centeno et al., 2011).

Volatiles are also important contributors to fruit palatability and preferences, though to date remain poorly studied though this is changing as quality becomes increasingly important to consumers and in turn fruit breeders. For example, it has recently been shown that a group of 20–30 volatile chemicals positively contribute to tomato flavor and are derived from lipid, amino acid, and carotenoid precursors (Mathieu et al., 2009). It is noteworthy that in many cases these volatiles are derived from an essential metabolite required in the human diet as they cannot be synthesized *de novo*. As examples, volatiles derived from essential amino acids, such as phenylalanine, leucine, and isoleucine in addition to linolenic acid, are common in fruit species at maturity. Indeed the precursor of Vitamin A, β-carotene, is also the immediate precursor of β-ionone, one of the most recognized and important volatiles of tomato to consumers (Goff and Klee, 2006), suggesting that volatiles serve a function in attraction that extends well beyond pleasant or enticing aromas, also as indicators of sources of necessary nutrients.

Interestingly, at the beginning of ripening there is an increase and a peak, either at or shortly before maturation of the levels of most of the flavor volatiles in a tomato fruit. An example is the rate-limiting step for synthesis of several phenylalanine metabolism volatiles including 2-phenylethanol, phenylacetaldehyde, and 1-nitro-2-phenethane, catalyzed by a small family of aromatic amino acid decarboxylases (AADCs) (Tieman et al., 2006). There is upregulation during ripening in the expression of several of the genes encoding these enzymes and the associated flux into the pathway. The regulation of additional volatile synthesis pathways, including *cis*-3-hexenal and hexanal, derived from linoleic and linolenic acids, is less well characterized. Steps in the biosynthetic pathway have been defined including 13-lipoxygenase (LOXC), hydroperoxide lyase, and alcohol dehydrogenase 2 (ADH2), and an increase in their expression during ripening, though genetic regulators of this change in gene expression during fruit development remain mysteries (Speirs et al., 1998; Kovacs et al., 2009).

In the cases of apocarotenoid volatiles (Chapter 6) there is a family of carotenoid cleavage dioxygenases (CCDs) whose metabolic activities lead to volatile products (Vogel et al., 2008). CCD1A and CCD1B are likely contributors to the production of the above-mentioned volatiles though interestingly, there is little change in expression of these genes during ripening suggesting that regulation is posttranscriptional. It is also noteworthy that carotenoids localize in the plastoglobules while CCD1 appears to not be plastid targeted. Interestingly, it has been reported that the accumulation of chromoplasts is associated with structural changes in the plastids, including the emergence of plastoglobules and stromules, suggesting that these physical changes may put substrates and enzymes in association (Pyke and Howells, 2002; Egea et al., 2010). Consistent with this hypothesis is the fact that leaf wounding results in leaf synthesis of carotenoid-derived volatiles (Vancanneyt et al., 2001). The underlying regulatory mechanisms controlling changes in gene expression, cell biology, and posttranscriptional regulation of flavor and volatile synthesis remain a mystery though it is plausible that at least some of this control in ripening fruit is mediated by ethylene and the ripening regulators described above, and experiments directed toward confirming such interactions or identifying novel regulators are the next step in understanding the molecular biology of regulation of flavor and volatile compounds produced during ripening (Klee and Giovannoni, 2011).

Texture is an important fruit quality trait and is an essential driver for consumer acceptance. Texture is made up of a number of sensory characteristics that influence consumer preference including firmness, mealiness, and juiciness (Sinesio et al., 2010). Quality attributes such as lack of aroma and "mealy" texture were the most disagreeable traits reported in a consumer preference study of fresh tomatoes (Sinesio et al., 2010). Shelf life is another agronomically important fruit quality trait and improvements of these attributes are of horticultural and commercial priorities. An understanding of key genes and pathways that affect texture and the rate of fruit softening will allow the manipulation and improvement of these characteristics. In Chapter 7, the complexity of the cell-wall-remodeling process and the role of fruit water relations in governing texture are discussed at length. It is as with other quality characteristics—the precise links to the molecular circuits controlling ripening are still to be determined. However, it is apparent that RIN binds to the promoters of cell-wall-modifying agents such as PG and EXP (Fujisawa et al., 2011; Martel et al., 2011). Sequencing the tomato genome revealed 700 putative genes with cell-wall-related functions and expression profiling identified more than 50 cell-wall-structure-related genes that are differentially expressed during ripening (The Tomato Genome Consortium, 2012). This indicates the complexity of these events and more work is required to reveal the regulatory network controlling fruit texture changes.

Epigenetic Regulation

A wider understanding of the different mechanisms that control gene expression has resulted in the identification of a number of agronomic traits that are directly or indirectly influenced by epigenetic modifiers, for example, cytosine methylation or posttranscriptional modifications of chromatin. In plants, three sequence site types—CG, CHG, and CHH (H = C, T, or A)—exist where cytosine methylation can occur. The most common of these is CHG methylation. DNA methylation may be heritable and stable through many generations and may function in the regulation of expression of transposon elements, repetitive sequences, and other nongenic sequences that can be detrimental to the cell if expressed. Epigenetic modifiers are also central to the control of genes which regulate developmental cell differentiation and reprogramming (Feng et al., 2010).

It is becoming apparent that epigenetic modifiers can play a role in the generation of novel phenotypes and as a source of natural variation that could be exploited in breeding programs. The epigenetic variation in the *CNR* gene that governs ripening has been mentioned already. However, the *Cnr* phenotype has been replicated by either virus-induced gene silencing of this SBP-box gene or by transgene-induced methylation of specific sites in the *CNR* promoter (Manning et al., 2006; Kanazawa et al., 2011). Developmental changes in methylation were apparent in the cultivar Liberto; a loss of methylation was observed in specific nucleotides prior to the onset of ripening, suggesting epigenetic regulation of ripening (Manning et al., 2006). In addition, differences in the levels of methylation at the SBP-box promoter site were observed between cultivars, with higher percentage of methylation in the Liberto background compared to Ailsa Craig.

Recent research has shown that variants for both methylation and small regulatory RNA species exist in *S. pennellii* ILs and hybrids (Shivaprasad et al., 2012). Discovery of naturally occurring epigenetic variation has also been identified amongst other species including ecotypes of *Arabidopsis*, which show somewhat different patterns of methylation (Vaughn et al., 2007).

Small RNAs (sRNAs), such as micro-RNAs (miRNAs), are important in controlling fruit ripening through posttranscriptional regulation of gene expression (Moxon et al., 2008). The most-studied sRNAs are miRNAs which are procured from single-stranded RNA transcripts and contain an imperfect stem-loop secondary structure processed by DICER-LIKE 1 (DCL1) into shorter stem-loop structures (Reinhart et al., 2002). Cleavage of the imperfect stem-loop secondary structure by DCL1 guides the RNA-induced silencing complex (RISC) to the corresponding mRNA containing the complementary miRNA site, resulting in cleavage of the target mRNA (Kidner and Martienssen, 2005). Interestingly, an miRNA-binding site that is complementary to miR156/157 was identified in the *CNR* 3'UTR (Dalmay, 2010). miRNAs were also shown to regulate genes involved in ethylene signaling (Moxon et al., 2008). Genome-wide studies of the sRNAome during fruit development have revealed that sRNAs mapped to fruit-related promoter regions and differential expression of sRNAs were associated with ripening tomato fruit, yet the function of this change is unknown (The Tomato Genome Consortium, 2012). In a recent study describing genome-scale methylation analysis of tomato fruit development it was shown that while the whole genome methylation status is relatively stable, specific regions generally associated with gene promoters become hypomethylated as the fruit matures and ripens (Zhong et al., 2013). A large proportion of genes associated with ripening and altered expression in the rin mutation both displayed such developmentally-associated methylation changes and binding of the RIN protein indicating a significant role for changes in the epigenome as a contributor to ripening control.

Future studies will combine advances in our understanding of gene regulatory networks with the discovery and validation of genes underlying the QTL and epiQTL that control fruit phenotypes. We are at a new frontier in research on fruit ripening that should lead to rational methods for the fine control of this important developmental process, thereby improving quality and reducing wastage in these important products essential for a healthy diet.

References

Alba, R., Payton, P., Fei, Z.J., McQuinn, R., Debbie, P., Martin, G.B., Tanksley, S.D., and Giovannoni, J.J. (2005) Transcriptome and selected metabolite analyses reveal multiple points of ethylene control during tomato fruit development. *The Plant Cell*, 17, 2954–2965.

Audran-Delalande, C., Bassa, C., Mila, I., Regad, F., Zouine, M., and Bouzayen, M. (2012) Genome-wide identification, functional analysis and expression profiling of the Aux/IAA gene family in tomato. *Plant and Cell Physiology*, 53, 659–672.

Barry, C.S., Aldridge, G.M., Herzog, G., Ma, Q., McQuinn, R.P., Hirschberg, J., and Giovannoni, J.J. (2012) Altered chloroplast development and delayed fruit ripening caused by mutations in a zinc metalloprotease at the lutescent2 locus of tomato. *Plant Physiology*, 159, 1086–1098.

Barry, C.S., Blume, B., Bouzayen, M., Cooper, W., Hamilton, A.J., and Grierson, D. (1996) Differential expression of the 1-aminocyclopropane-1-carboxylate oxidase gene family of tomato. *The Plant Journal: for cell and molecular biology*, 9, 525–535.

Barry, C.S., Llop-Tous, M.I., and Grierson, D. (2000) The regulation of 1-aminocyclopropane-1-carboxylic acid synthase gene expression during the transition from system-1 to system-2 ethylene synthesis in tomato. *Plant Physiology*, 123, 979–986.

Barry, C.S., McQuinn, R.P., Thompson, A.J., Seymour, G.B., Grierson, D., and Giovannoni, J.J. (2005) Ethylene insensitivity conferred by the Green-ripe and Never-ripe 2 ripening mutants of tomato. *Plant Physiology*, 138, 267–275.

Barry, C.S., McQuinn, R.P., Chung, M.Y., Besuden, A., and Giovannoni, J.J. (2008) Amino acid substitutions in homologs of the STAY-GREEN protein are responsible for the green-flesh and chlorophyll retainer mutations of tomato and pepper. *Plant Physiology*, 147, 179–187.

Barry, C.S., and Giovannoni, J.J. (2006) Ripening in the tomato Green-ripe mutant is inhibited by ectopic expression of a protein that disrupts ethylene signaling. *Proceedings of the National Academy of Sciences of the United States of America*, 103, 7923–7928.

Barsan, C., Sanchez-Bel, P., Rombaldi, C., Egea, I., Rossignol, M., Kuntz, M., Zouine, M., Latche, A., Bouzayen, M., and Pech, J.C. (2010) Characteristics of the tomato chromoplast revealed by proteomic analysis. *Journal of Experimental Botany*, 61, 2413–2431.

Bemer, M., Karlova, R., Ballester, A., Tukunov, Y., Bovy, A., Wolters-Arts, M., de Barros Rosetto, M., Angenent, G., and Maagd, R. (2012) The tomato FRUITFUL homologs TDR4/FUL1 and FUL2 regulate ethylene independent aspects of fruit ripening. *Plant Cell*. 24, 4437–4451

Borovsky, Y., and Paran, I. (2008) Chlorophyll breakdown during pepper fruit ripening in the chlorophyll retainer mutation is impaired at the homolog of the senescence-inducible stay-green gene. *Theoretical and Applied Genetics*, 117, 235–240.

Bottcher, C., Boss, P.K., and Davies, C. (2011) Acyl substrate preferences of an IAA-amido synthetase account for variations in grape (Vitis vinifera L.) berry ripening caused by different auxinic compounds indicating the importance of auxin conjugation in plant development. *Journal of Experimental Botany*, 62, 4267–4280.

Bottcher, C., Keyzers, R.A., Boss, P.K., and Davies, C. (2010) Sequestration of auxin by the indole-3-acetic acid-amido synthetase GH3-1 in grape berry (Vitis vinifera L.) and the proposed role of auxin conjugation during ripening. *Journal of Experimental Botany*, 61, 3615–3625.

Cardon, G.H., Hohmann, S., Nettesheim, K., Saedler, H., and Huijser, P. (1997) Functional analysis of the *Arabidopsis thaliana* SBP-box gene SPL3: a novel gene involved in the floral transition. *The Plant Journal: for cell and molecular biology*, 12, 367–377.

Centeno, D.C., Osorio, S., Nunes-Nesi, A., Bertolo, A.L.F., Carneiro, R.T., Araujo, W.L., Steinhauser, M.C., Michalska, J., Rohrmann, J., Geigenberger, P., Oliver, S.N., Stitt, M., Carrari, F., Rose, J.K.C., and Fernie, A.R. (2011) Malate plays a crucial role in starch metabolism, ripening, and soluble solid content of tomato fruit and affects postharvest softening. *The Plant Cell*, 23, 162–184.

Chung, M.Y., Vrebalov, J., Alba, R., Lee, J., McQuinn, R., Chung, J.D., Klein, P., and Giovannoni, J. (2010) A tomato (Solanum lycopersicum) APETALA2/ERF gene, SlAP2a, is a negative regulator of fruit ripening. *The Plant Journal: for cell and molecular biology*, 64, 936–947.

Coen, E.S., and Meyerowitz, E.M. (1991) The war of the whorls: genetic interactions controlling flower development. *Nature*, 353, 31–37.

Cohen, J.D. (1996) In vitro tomato fruit cultures demonstrate a role for indole-3-acetic acid in regulating fruit ripening. *Journal of the American Society for Horticultural Science*, 121, 520–524.

Dalmay, T. (2010) Short RNAs in tomato. *Journal of Integrative Plant Biology*, 52, 388–392.

Egea, I., Barsan, C., Bian, W.P., Purgatto, E., Latche, A., Chervin, C., Bouzayen, M., and Pech, J.C. (2010) Chromoplast differentiation: current status and perspectives. *Plant and Cell Physiology*, 51, 1601–1611.

Elitzur, T., Vrebalov, J., Giovannoni, J.J., Goldschmidt, E.E., and Friedman, H. (2010) The regulation of MADS-box gene expression during ripening of banana and their regulatory interaction with ethylene. *Journal of Experimental Botany*, 61, 1523–1535.

Eriksson, E.M., Bovy, A., Manning, K., Harrison, L., Andrews, J., De Silva, J., Tucker, G.A., and Seymour, G.B. (2004) Effect of the Colorless non-ripening mutation on cell wall biochemistry and gene expression during tomato fruit development and ripening. *Plant Physiology*, 136, 4184–4197.

Eshed, Y., and Zamir, D. (1995) An introgression line population of Lycopersicon pennellii in the cultivated tomato enables the identification and fine mapping of yield-associated QTL. *Genetics*, 141, 1147–1162.

Feng, S., Jacobsen, S.E., and Reik, W. (2010) Epigenetic reprogramming in plant and animal development. *Science*, 330, 622–627.

Fujisawa, M., Nakano, T., and Ito, Y. (2011) Identification of potential target genes for the tomato fruit-ripening regulator RIN by chromatin immunoprecipitation. *BMC Plant Biology*, 11, 26.

Giovannoni, J.J. (2004) Genetic regulation of fruit development and ripening. *The Plant Cell*, 16, S170–S180.

Giovannoni, J.J. (2007) Fruit ripening mutants yield insights into ripening control. *Current Opinion in Plant Biology*, 10, 283–289.

Giovannoni, J.J., Noensie, E.N., Ruezinsky, D.M., Lu, X.H., Tracy, S.L., Ganal, M.W., Martin, G.B., Pillen, K., Alpert, K., and Tanksley, S.D. (1995) Molecular-genetic analysis of the ripening-inhibitor and non-ripening loci of tomato - a first step in genetic map-based cloning of fruit ripening genes. *Molecular & General Genetics*, 248, 195–206.

Goff, S.A., and Klee, H.J. (2006) Plant volatile compounds: sensory cues for health and nutritional value? *Science*, 311, 815–819.

Hamilton, A.J., Lycett, G.W., and Grierson, D. (1990) Antisense gene that inhibits synthesis of the hormone ethylene in transgenic plants. *Nature*, 346, 284–287.

Hileman, L.C., Sundstrom, J.F., Litt, A., Chen, M.Q., Shumba, T., and Irish, V.F. (2006) Molecular and phylogenetic analyses of the MADS-Box gene family in tomato. *Molecular Biology and Evolution*, 23, 2245–2258.

Hirschberg, J. (1999) Production of high-value compounds: carotenoids and vitamin E. *Current Opinion in Biotechnology*, 10, 186–191.

Ho, L.C., Sjut, V., and Hoad, G.V. (1983) The effect of assimilate supply on fruit-growth and hormone levels in tomato plants. *Plant Growth Regulation*, 1, 155–171.

Itkin, M., Seybold, H., Breitel, D., Rogachev, I., Meir, S., and Aharoni, A. (2009) TOMATO AGAMOUS-LIKE 1 is a component of the fruit ripening regulatory network. *The Plant Journal: for cell and molecular biology*, 60, 1081–1095.

Ito, Y., Kitagawa, M., Ihashi, N., Yabe, K., Kimbara, J., Yasuda, J., Ito, H., Inakuma, T., Hiroi, S., and Kasumi, T. (2008) DNA-binding specificity, transcriptional activation potential, and the rin mutation effect for the tomato fruit-ripening regulator RIN. *The Plant Journal: for cell and molecular biology*, 55, 212–223.

Iwahory, S. (1967) Auxin of tomato fruit at different stages of its development with a special reference to high temperature injuries. *Plant and Cell Physiology*, 8, 15–22.

Jaakola, L., Poole, M., Jones, M.O., Kamarainen-Karppinen, T., Koskimaki, J.J., Hohtola, A., Haggman, H., Fraser, P.D., Manning, K., King, G.J., Thomson, H., and Seymour, G.B. (2010) A SQUAMOSA MADS box gene involved in the regulation of anthocyanin accumulation in bilberry fruits. *Plant Physiology*, 153, 1619–1629.

Jia, H.F., Chai, Y.M., Li, C.L., Lu, D., Luo, J.J., Qin, L., and Shen, Y.Y. (2011) Abscisic acid plays an important role in the regulation of strawberry fruit ripening. *Plant Physiology*, 157, 188–199.

Jones, B., Frasse, P., Olmos, E., Zegzouti, H., Li, Z.G., Latche, A., Pech, J.C., and Bouzayen, M. (2002) Down-regulation of DR12, an auxin-response-factor homolog, in the tomato results in a pleiotropic phenotype including dark green and blotchy ripening fruit. *The Plant Journal: for cell and molecular biology*, 32, 603–613.

Kanazawa, A., Inaba, J., Shimura, H., Otagaki, S., Tsukahara, S., Matsuzawa, A., Kim, B.M., Goto, K., and Masuta, C. (2011) Virus-mediated efficient induction of epigenetic modifications of endogenous genes with phenotypic changes in plants. *The Plant Journal: for cell and molecular biology*, 65, 156–168.

Karlova, R., Rosin, F.M., Busscher-Lange, J., Parapunova, V., Do, P.T., Fernie, A.R., Fraser, P.D., Baxter, C., Angenent, G.C., and de Maagd, R.A. (2011) Transcriptome and metabolite profiling show that APETALA2a is a major regulator of tomato fruit ripening. *The Plant Cell*, 23, 923–941.

Kendrick, M.D., and Chang, C. (2008) Ethylene signaling: new levels of complexity and regulation. *Current Opinion in Plant Biology*, 11, 479–485.

Kidner, C.A., and Martienssen, R.A. (2005) The developmental role of microRNA in plants. *Current Opinion in Plant Biology*, 8, 38–44.

Klee, H.J., and Giovannoni, J.J. (2011) Genetics and control of tomato fruit ripening and quality attributes. *Annual Review Genetics*, 45, 41–59.

Klee, H.J., Hayford, M.B., Kretzmer, K.A., Barry, G.F., and Kishore, G.M. (1991) Control of ethylene synthesis by expression of a bacterial enzyme in transgenic tomato plants. *Plant Cell*, 3, 1187–1193.

Kovacs, K., Fray, R.G., Tikunov, Y., Graham, N., Bradley, G., Seymour, G.B., Bovy, A.G., and Grierson, D. (2009) Effect of tomato pleiotropic ripening mutations on flavour volatile biosynthesis. *Phytochemistry*, 70, 1003–1008.

Kumar, R., Agarwal, P., Tyagi, A.K., and Sharma, A.K. (2012) Genome-wide investigation and expression analysis suggest diverse roles of auxin-responsive GH3 genes during development and response to different stimuli in tomato (Solanum lycopersicum). *Molecular Genetics and Genomics*, 287, 221–235.

Lee, J.M., Joung, J.G., McQuinn, R., Chung, M.Y., Fei, Z.J., Tieman, D., Klee, H., and Giovannoni, J. (2012) Combined transcriptome, genetic diversity and metabolite profiling in tomato fruit reveals that the ethylene response factor SlERF6 plays an important role in ripening and carotenoid accumulation. *The Plant Journal: for cell and molecular biology*, 70, 191–204.

Lin, Z.F., Hong, Y.G., Yin, M.G., Li, C.Y., Zhang, K., and Grierson, D. (2008) A tomato HD-Zip homeobox protein, LeHB-1, plays an important role in floral organogenesis and ripening. *The Plant Journal: for cell and molecular biology*, 55, 301–310.

K.D., Kang, B.C., Jiang, H., Moore, S.L., Li, H.X., Watkins, C.B., Setter, T.L., and Jahn, M.M. (2005) A GH3-like gene, CcGH3, isolated from Capsicum chinense L. fruit is regulated by auxin and ethylene. *Plant Molecular Biology*, 58, 447–464.

Liu, Y.S., Roof, S., Ye, Z.B., Barry, C., van Tuinen, A., Vrebalov, J., Bowler, C., and Giovannoni, J. (2004) Manipulation of light signal transduction as a means of modifying fruit nutritional quality in tomato. *Proceedings of the National Academy of Sciences of the United States of America*, 101, 9897–9902.

Liljegren, S.J., Roeder, A.H.K., Kempin, S.A., Gremski, K., Ostergaard, L., Guimil, S., Reyes, D.K., and Yanofsky, M.F. (2004) Control of fruit patterning in *Arabidopsis* by INDEHISCENT. *Cell*, 116, 843–853.

Manning, K., Tor, M., Poole, M., Hong, Y., Thompson, A.J., King, G.J., Giovannoni, J.J., and Seymour, G.B. (2006) A naturally occurring epigenetic mutation in a gene encoding an SBP-box transcription factor inhibits tomato fruit ripening. *Nature Genetics*, 38, 948–952.

Mapelli, S., Frova, C., Torti, G., and Soressi, G.P. (1978) Relationship between set, development and activities of growth-regulators in tomato fruits. *Plant and Cell Physiology*, 19, 1281–1288.

Martel, C., Vrebalov, J., Tafelmeyer, P., and Giovannoni, J.J. (2011) The tomato MADS-Box transcription factor RIPENING INHIBITOR interacts with promoters involved in numerous ripening processes in a COLORLESS NONRIPENING-dependent manner. *Plant Physiology*, 157, 1568–1579.

Mathieu, S., Cin, V.D., Fei, Z., Li, H., Bliss, P., Taylor, M.G., Klee, H.J., and Tieman, D.M. (2009) Flavour compounds in tomato fruits: identification of loci and potential pathways affecting volatile composition. *Journal of Experimental Botany*, 60, 325–337.

McMurchie, E.J., McGlasson, W.B., and Eaks, I.L. (1972) Treatment of fruit with propylene gives information about biogenesis of ethylene. *Nature*, 237, 235.

Moxon, S., Jing, R., Szittya, G., Schwach, F., Rusholme Pilcher, R.L., Moulton, V., and Dalmay, T. (2008) Deep sequencing of tomato short RNAs identifies microRNAs targeting genes involved in fruit ripening. *Genome Research*, 18, 1602–1609.

Mustilli, A.C., Fenzi, F., Ciliento, R., Alfano, F., and Bowler, C. (1999) Phenotype of the tomato high pigment-2 mutant is caused by a mutation in the tomato homolog of DEETIOLATED1. *The Plant Cell*, 11, 145–157.

Oeller, P.W., Wong, L.M., Taylor, L.P., Pike, D.A., and Theologis, A. (1991) Reversible inhibition of tomato fruit senescence by antisense RNA. *Science*, 254, 437–439.

O'Malley, R.C., Rodriguez, F.I., Esch, J.J., Binder, B.M., O'Donnell, P., Klee, H.J., and Bleecker, A.B. (2005) Ethylene-binding activity, gene expression levels, and receptor system output for ethylene receptor family members from *Arabidopsis* and tomato. *The Plant Journal: for cell and molecular biology*, 41, 651–659.

Osorio, S., Alba, R., Damasceno, C.M.B., Lopez-Casado, G., Lohse, M., Zanor, M.I., Tohge, T., Usadel, B., Rose, J.K.C., Fei, Z.J., Giovannoni, J.J., and Fernie, A.R. (2011) Systems biology of tomato fruit development: combined transcript, protein, and metabolite analysis of tomato transcription factor (nor, rin) and ethylene receptor (Nr) mutants reveals novel regulatory interactions. *Plant Physiology*, 157, 405–425.

Pan, I.L., McQuinn, R., Giovannoni, J.J., and Irish, V.F. (2010) Functional diversification of AGAMOUS lineage genes in regulating tomato flower and fruit development. *Journal of Experimental Botany*, 61, 1795–1806.

Pnueli, L., Hareven, A.D., Rounsley, S.D., Yanofsky, M.F., and Lifschitz, E. (1994) Isolation of the tomato agamous gene Tag1 and analysis of its homeotic role in transgenic plants. *The Plant Cell*, 6, 163–173.

Powell, A.L.T., Nguyen, C.V., Hill, T., Cheng, K.L., Figueroa-Balderas, R., Aktas, H., Ashrafi, H., Pons, C., Fernandez-Munoz, R., Vicente, A., Lopez-Baltazar, J., Barry, C.S., Liu, Y.S., Chetelat, R., Granell, A., Van Deynze, A., Giovannoni, J.J., and Bennett, A.B. (2012) Uniform ripening encodes a golden 2-like transcription factor regulating tomato fruit chloroplast development. *Science*, 336, 1711–1715.

Pyke, K.A., and Howells, C.A. (2002) Plastid and stromule morphogenesis in tomato. *Annals of Botany*, 90, 559–566.

Reinhart, B.J., Weinstein, E.G., Rhoades, M.W., Bartel, B., and Bartel, D.P. (2002) MicroRNAs in plants. *Genes & Development*, 16, 1616–1626.

Resnick, J.S., Wen, C.K., Shockey, J.A., and Chang, C. (2006) REVERSION-TO-ETHYLENE SENSITIVITY1, a conserved gene that regulates ethylene receptor function in *Arabidopsis*. *Proceedings of the National Academy of Sciences of the United States of America*, 103, 7917–7922.

Rodriguez, F.I., Esch, J.J., Hall, A.E., Binder, B.M., Schaller, G.E., and Bleecker, A.B. (1999) A copper cofactor for the ethylene receptor ETR1 from *Arabidopsis*. *Science*, 283, 996–998.

Rotino, G.L., Perri, E., Zottini, M., Sommer, H., and Spena, A. (1997) Genetic engineering of parthenocarpic plants. *Nature Biotechnology*, 15, 1398–1401.

Seymour, G.B., Ryder, C.D., Cevik, V., Hammond, J.P., Popovich, A., King, G.J., Vrebalov, J., Giovannoni, J.J., and Manning, K. (2011) A SEPALLATA gene is involved in the development and ripening of strawberry (Fragaria x ananassa Duch.) fruit, a non-climacteric tissue. *Journal of Experimental Botany*, 62, 1179–1188.

Shivaprasad, P.V., Dunn, R.M., Santos, B.A., Bassett, A., and Baulcombe, D.C. (2012) Extraordinary transgressive phenotypes of hybrid tomato are influenced by epigenetics and small silencing RNAs. *The EMBO Journal*, 31, 257–266.

Sinesio, F., Cammareri, M., Moneta, E., Navez, B., Peparaio, M., Causse, M., and Grandillo, S. (2010) Sensory quality of fresh French and Dutch market tomatoes: a preference mapping study with Italian consumers. *Journal of Food Science*, 75, S55–S67.

Speirs, J., Lee, E., Holt, K., Yong-Duk, K., Steele Scott, N., Loveys, B., and Schuch, W. (1998) Genetic manipulation of alcohol dehydrogenase levels in ripening tomato fruit affects the balance of some flavor aldehydes and alcohols. *Plant Physiology*, 117, 1047–1058.

Stone, J.M., Liang, X., Nekl, E.R., and Stiers, J.J. (2005) *Arabidopsis* AtSPL14, a plant-specific SBP-domain transcription factor, participates in plant development and sensitivity to fumonisin B1. *The Plant Journal: for cell and molecular biology*, 41, 744–754.

Sun, L., Sun, Y.F., Zhang, M., Wang, L., Ren, J., Cui, M.M., Wang, Y.P., Ji, K., Li, P., Li, Q., Chen, P., Dai, S.J., Duan, C.R., Wu, Y., and Leng, P. (2012) Suppression of 9-cis-epoxycarotenoid dioxygenase, which encodes a key enzyme in abscisic acid biosynthesis, alters fruit texture in transgenic tomato. *Plant Physiology*, 158, 283–298.

The Tomato Genome Consortium (2012) The tomato genome sequence provides insights into fleshy fruit evolution. *Nature*, 485, 635–641.

Thompson, A.J., Tor, M., Barry, C.S., Vrebalov, J., Orfila, C., Jarvis, M.C., Giovannoni, J.J., Grierson, D., and Seymour, G.B. (1999) Molecular and genetic characterization of a novel pleiotropic tomato-ripening mutant. *Plant Physiology*, 120, 383–389.

Tieman, D.M., Taylor, M.G., Ciardi, J.A., and Klee, H.J. (2000) The tomato ethylene receptors NR and LeETR4 are negative regulators of ethylene response and exhibit functional compensation within a multigene family. *Proceedings of the National Academy of Sciences of the United States of America*, 97, 5663–5668.

Tieman, D., Taylor, M., Schauer, N., Fernie, A.R., Hanson, A.D., and Klee, H.J. (2006) Tomato aromatic amino acid decarboxylases participate in synthesis of the flavor volatiles 2-phenylethanol and 2-phenylacetaldehyde. *Proceedings of the National Academy of Sciences of the United States of America*, 103, 8287–8292.

Trainotti, L., Tadiello, A., and Casadoro, G. (2007) The involvement of auxin in the ripening of climacteric fruits comes of age: the hormone plays a role of its own and has an intense interplay with ethylene in ripening peaches. *Journal of Experimental Botany*, 58, 3299–3308.

Unte, U.S., Sorensen, A.M., Pesaresi, P., Gandikota, M., Leister, D., Saedler, H., and Huijser, P. (2003) SPL8, an SBP-Box gene that affects pollen sac development in *Arabidopsis*. *The Plant Cell*, 15, 1009–1019.

Vancanneyt, G., Sanz, C., Farmaki, T., Paneque, M., Ortego, F., Castanera, P., and Sanchez-Serrano, J.J. (2001) Hydroperoxide lyase depletion in transgenic potato plants leads to an increase in aphid performance. *Proceedings of the National Academy of Sciences of the United States of America*, 98, 8139–8144.

Vaughn, M.W., Tanurdzic, M., Lippman, Z., Jiang, H., Carrasquillo, R., Rabinowicz, P.D., Dedhia, N., McCombie, W.R., Agier, N., Bulski, A., Colot, V., Doerge, R.W., and Martienssen, R.A. (2007) Epigenetic natural variation in *Arabidopsis thaliana*. *PLoS Biology*, 5, e174.

Vendrell, M. (1985) Dual effect of 2,4-D on ethylene production and ripening of tomato fruit tissue. *Physiologia Plantarum*, 64, 559–563.

Vogel, J.T., Tan, B.C., McCarty, D.R., and Klee, H.J. (2008) The carotenoid cleavage dioxygenase 1 enzyme has broad substrate specificity, cleaving multiple carotenoids at two different bond positions. *Journal of Biological Chemistry*, 283, 11364–11373.

Vrebalov, J., Pan, I.L., Arroyo, A.J.M., McQuinn, R., Chung, M., Poole, M., Rose, J.K.C., Seymour, G., Grandillo, S., Giovannoni, J., and Irish, V.F. (2009) Fleshy fruit expansion and ripening are regulated by the tomato SHATTERPROOF gene TAGL1. *The Plant Cell*, 21, 3041–3062.

Vrebalov, J., Ruezinsky, D., Padmanabhan, V., White, R., Medrano, D., Drake, R., Schuch, W., and Giovannoni, J. (2002) A MADS-box gene necessary for fruit ripening at the tomato ripening-inhibitor (Rin) locus. *Science*, 296, 343–346.

Wang, S.H., Liu, J.K., Feng, Y.Y., Niu, X.L., Giovannoni, J., and Liu, Y.S. (2008) Altered plastid levels and potential for improved fruit nutrient content by downregulation of the tomato DDB1-interacting protein CUL4. *The Plant Journal: for cell and molecular biology*, 55, 89–103.

Wang, H., Nussbaum-Wagler, T., Li, B.L., Zhao, Q., Vigouroux, Y., Faller, M., Bomblies, K., Lukens, L., and Doebley, J.F. (2005) The origin of the naked grains of maize. *Nature*, 436, 714–719.

Watkins, C.B., Nock, J.F., and Whitaker, B.D. (2000) Responses of early, mid and late season apple cultivars to postharvest application of 1-methylcyclopropene (1-MCP) under air and controlled atmosphere storage conditions. *Postharvest Biology and Technology*, 19, 17–32.

Wheeler, S., Loveys, B., Ford, C., and Davies, C. (2009) The relationship between the expression of abscisic acid biosynthesis genes, accumulation of abscisic acid and the promotion of Vitis vinifera L. berry ripening by abscisic acid. *Australian Journal of Grape and Wine Research*, 15, 195–204.

Wilkinson, J.Q., Lanahan, M.B., Yen, H.C., Giovannoni, J.J., and Klee, H.J. (1995) An ethylene-inducible component of signal-transduction encoded by never-ripe. *Science*, 270, 1807–1809.

Yanofsky, M.F., Ma, H., Bowman, J.L., Drews, G.N., Feldmann, K.A., and Meyerowitz, E.M. (1990) The protein encoded by the *Arabidopsis* homeotic gene agamous resembles transcription factors. *Nature*, 346, 35–39.

Ytterberg, A.J., Peltier, J.B., and van Wijk, K.J. (2006) Protein profiling of plastoglobules in chloroplasts and chromoplasts. A surprising site for differential accumulation of metabolic enzymes. *Plant Physiology*, 140, 984–997.

Zahn, L.M., King, H.Z., Leebens-Mack, J.H., Kim, S., Soltis, P.S., Landherr, L.L., Soltis, D.E., dePamphilis, C.W., and Ma, H. (2005) The evolution of the SEPALLATA subfamily of MADS-Box genes: a preangiosperm origin with multiple duplications throughout angiosperm history. *Genetics*, 169, 2209–2223.

Zhong, S., Fei, Z., Chen, Y., Zheng, Y., Huang, M., Vrebalov, J., McQuinn, R., Gapper, N., Liu, B., Xiang, J., Shao, Y., and Giovannoni, J. (2013) *Nat. Biotech.* 31, 154–159.

Index

AATs. *See* Alcohol acyl-transferases
Abscisic Acid (ABA), 85, 85f, 94, 94f, 190
ACC. *See* 1-Amino-cyclopropane-1-carboxylic acid
ACC oxidase (ACO), 48, 61, 62t
  inhibition of, 45, 48, 48f, 49f, 50f, 63
  in System-2, 50–52, 51f
ACC synthase (ACS), 48, 61, 62t
  classes of, 52
  inhibition of, 48, 48f, 49f, 50f
  in System-2, 50–52, 51f
Achene, 23t, 24f, 26f
Acidity, development of, 6–7
ACO. *See* ACC oxidase
ACO3 activation, 52
ACO1 inhibition, 45, 48, 48f, 49f, 50f
ACO2 promotion, 53
ACS. *See* ACC synthase
ACS2
  activation of, 52–53
  inhibition of, 48, 48f, 49f, 50f, 63–64
S-Adenosylmethionine (SAM)
  in ethylene production, 8–9, 49f
  in spermidine synthesis, 9
ADHs. *See* Alcohol dehydrogenases
ADP-glucose pyrophosphorylase (AGPase)
  malate and, 6
  starch and, 7
AGAMOUS gene clade, 33, 193, 195
Aggregate fruit, 23t, 25f, 28f
AGPase. *See* ADP-glucose pyrophosphorylase
Alcohol acyl-transferases (AATs), 66, 146–47, 146f
Alcohol dehydrogenases (ADHs), 65–66, 142
2-Alkenal reductase (ALH), 142
Alpha-tomatine, 66
Amborella trichopoda, 25f, 32
Amino acids
  flavor role of, 10
  fruit content of, 6
  volatiles derived from, 138f, 145, 146f
1-Amino-cyclopropane-1-carboxylic acid (ACC), 49f
Ampelopsis brevipedunculata, 28f
ANA grade, 22–23, 29f
Angiosperms, 21
  evolution of, 22–23, 29, 29f
    in fossil record, 30–31
    fruit as driver of diversity, 36–38
    fruit development, 33–36
    fruit variation and angiosperm phylogeny, 32–33
  genetic relationships of, 22–23, 29, 29f
  types of, 21–22, 23t, 24f, 25f, 26f, 27f, 28f
Annona squamosa, 25f
ANR. *See* Anthocyanidin reductase
ANS. *See* Anthocyanidin synthase
Antagonists, fruit evolution and, 36–37
Antheraxanthin, 83, 90
Anthocyanidin reductase (ANR), 126
Anthocyanidin synthase (ANS), 126
Anthocyanins
  biosynthesis of, 118–19, 118f, 123f, 124–25, 124f
    cinnamic acids, 118–19
    coumarins, 120–22, 121f
    monolignols, lignans, and lignin, 120
    regulation of, 126–27
    stilbenoids, 122
    transcriptional regulation of, 127–28
  color with, 61
  engineering elevated levels of, 128–29
  in fruit ripening, 44
  therapeutic properties of, 2
Anticlinal cell division, 34

The Molecular Biology and Biochemistry of Fruit Ripening, First Edition.
Edited by Graham B. Seymour, Mervin Poole, James J. Giovannoni and Gregory A. Tucker.

Antioxidants
flavonols, 123
phenylpropanoids, 117
protective effects of, 2
AP. *See* Apiogalacturonan
AP2, 60, 194
APETALA2/ETHYLENE RESPONSE FACTOR (AP2/ERF), 60, 194
APG III angiosperm phylogeny, 22–23, 29, 29f
Apiogalacturonan (AP), 166–67
Apocarotenoids, 84–86, 85f, 138f
synthesis of, 92–95, 93t, 94f, 139f, 149–50, 199
Apocarpy, 22, 23t
Apoplastic invertase, sink size and, 6
Aquilegia coerulea, 26f
Archaefructus, 30–31
ARF. *See* Auxin response factor
Aril, 27f
Aroma volatiles
ethylene and, 48–49, 66
in fruit ripening, 4, 44, 45f
gene expression in, 62t, 65–66
metabolic pathways for, 4, 5f
Ascorbate, protective effects of, 2
Asterids, 23, 29f
Astringency
proanthocyanins, 125
of wild fleshy fruit, 37
Aux/IAA. *See* Aux/indole-3-acetic acid
Auxin, 47
ERF regulation by, 60
in ripening, 189–90
Aux/indole-3-acetic acid (Aux/IAA), 190
Auxin response factor (ARF), 190

BAHD, 146
BCATs. *See* Branched-chain aminotransferases
Benzenoids, 138f, 139f, 142–45, 144f
genetic engineering of, 153–54
Berry, 23t, 24f, 25f, 26f, 28f
capsules compared with, 36
development of, 33–35
Beta-carotene
cancer and, 2
in chloroplasts, 76
ethylene and development of, 61–62
protective effects of, 2
structure of, 77f
bHLH transcription factors, 127
Bignoniaceae, 32
Bitterness. *See* Astringency
Body weight, fruit nutrients and, 2
Branched-chain aminotransferases (BCATs), 145, 146f
Branched-chain volatiles, 139f
Brix index, starch and, 6
Brugueira gymnorrhiza, 27f
Bulnesia sarmientoi, 27f
Buxus latistyla, 26f
bZIP transcription factors, 47

CAD. *See* Cinnamyl-alcohol dehydrogenase
Caffeic acid, 119
Calendula arvensis, 24f
Campanulids, 23, 29f
Capsaicin, 37
Capsanthin, 91, 91f
Capsorubin, 91, 91f
Capsule, 23t, 26f, 27f, 28f
berries compared with, 36
Carbon metabolism. *See* Central carbon metabolism
Carotenoids
biosynthesis of, 76, 86–92
desaturation, cyclization, and oxidation reactions, 87–92, 89f, 91f
early steps of, 87, 88f
future perspectives on, 100–102
nontranscriptional regulation, 98–100
regulation of, 101f, 198
transcriptional regulation, 95–98
cancer and, 2
color with, 61
distribution of, 75–78, 77f
evolutionary development of, 36
function of, 82–84
genetic engineering of, 153
protective effects of, 2
structures of, 76, 77f, 82–84
Carotenoid cleavage dioxygenases (CCDs), 84, 92–93, 92t, 149–50, 199
Carotenoid metabolites. *See* Apocarotenoids
Carpel, 22, 23t
C-function genes and, 33
wall of, 33
Caryopsis, 25f
Catesbeae-Chiococceae complex, 32
CBF/DREB, 60
CCDs. *See* Carotenoid cleavage dioxygenases
CCR. *See* Cinnamoyl-CoA reductase
Cedrela mexicana, 27f
Cell division, 34–35
Cell expansion, 34–35
Cellulose, 12, 164–65
Cell wall, 163–64
architecture of, 168–71, 169f
building blocks of, 164–68
cellulose, 165
hemicellulose, 165, 166f
pectins, 165–67, 167f
structural proteins, 167–68
cuticular, 177–79, 178f

metabolism of, 11–12
enzymes in, 44–45, 45f
ripening dynamics of, 171–77
coordinated and synergistic mechanisms of disassembly, 175–76
in different fruit species, 174–75
disassembly of polysaccharides, 171–72
mechanisms of, 172–73
regulation of enzyme action in, 176–77
in vivo enzyme activities, 173–74
summary for, 179, 180f
Cell wall hydrolases, 12
Central carbon metabolism, 4–7
acidity, 6–7
starch, 6
sucrose, glucose, and fructose, 4–6
TCA cycle structure, 7
C-function genes, 33
C4H. *See* Cinnamate 4-hydroxylase
Chalcone flavanone isomerase (CHI), 126
Chalcone synthase (CHS), 122, 126
CHI. *See* Chalcone flavanone isomerase
Chlorophylls
biosynthesis and regulation of, 78
degradation of, 76, 78–81, 79f, 80f
future perspectives on, 100–102
distribution of, 75–78, 77f
ethylene and loss of, 61
forms of, 77f, 78
Chlorophyllase, 61, 81
Chlorophyll catabolites. *See* Fluorescent chlorophyll catabolites; Nonfluorescent chlorophyll catabolites
Chlorophyllide, 78
Chlorophyll metabolites, recycling of, 81–82, 82f
Chlorophyll retainer (*cl*) mutant, 80–81
Chloroplasts
carotenoid content of, 75–76
to chromoplast transition, 100, 196
Chromoplasts
in carotenoid synthesis, 97, 100, 196
chloroplast to, transition, 100, 196
fruit development of, 61
CHS. *See* Chalcone synthase
Cinnamate 4-hydroxylase (C4H), 119
Cinnamic acids, 118–19, 118f, 143, 144f
Cinnamoyl-CoA reductase (CCR), 120
Cinnamyl-alcohol dehydrogenase (CAD), 120
Citrate, in ripening, 6–7, 198
cl. See Chlorophyll retainer mutant
4CL. *See* 4-Coumarate:CoA ligase
Climacteric fruit
classification of, 7–8
ethylene in, 46–48
malate in, 6
nonclimacteric compared with, 7–8
ripening gene networks in, 60–61, 62t
ripening of, 45f, 189
CmPG1-3, 64
Cnr. *See Colorless nonripening*
Cochlospermum vitifolium, 27f
Cocos nucifera, 25f
Color. *See also* Pigments
evolution of, 36
gene expression in, 61–62, 62t, 63f
Colorless nonripening (*Cnr*), 8–9, 45, 191–92
Coniferyl alcohol, 143, 144f
CONSTITUTIVE TRIPLE RESPONSES (CTR), 8
ethylene receptor interaction with, 57–58
Copper, in ethylene binding, 55–56
Cornus capitata, 28f
4-Coumarate:CoA ligase (4CL), 119
Coumarins, 118f, 120–22, 121f
4-Coumaroyl-CoA, 125–26
Cryptochromes (CRYs), 96
CTR. *See* CONSTITUTIVE TRIPLE RESPONSES
CTR1, 56–58, 57–58
CTR3, 57–58
Cuticular cell wall, 177–79, 178f
Cyanidin (Cy), 123f, 124
Cytosolic isocitrate dehydrogenase (ICDHc), 9

Dehiscent fruits, 22, 23t
development of, 34
evolution of, 32
Delphinidin (Dp), 123f, 124
1-Deoxy-D-xylulose 5-phosphate reductoisomerase (DXR), 88f, 99
1-Deoxy-D-xylulose 5-phosphate synthase (DXS), 88f, 99
DFR. *See* Dihydroflavonol 4-reductase
Dicotyledons, 23, 29f
cell walls of, 164
Dihydroflavonol 4-reductase (DFR), 126
Dimethylallyl diphosphate (DMAPP), 87, 88f, 147, 148f, 149
Diospyros digyna, 28f
Disease, fruit consumption and, 1–2, 123, 125
DMAPP. *See* Dimethylallyl diphosphate
Dp. *See* Delphinidin
Drimys granadensis, 25f
Drupe, 23t, 25f, 28f
Durian theory, 30
DXR. *See* 1-Deoxy-D-xylulose 5-phosphate reductoisomerase
DXS. *See* 1-Deoxy-D-xylulose 5-phosphate synthase

EBF1, 60
EBF2, 60
Ecballium elaterium, 26f
EGase. *See* Endo-1,4-β-glucanase
EGS. *See* Eugenol synthase

EILs, 59–60
EIN2. *See* ETHYLENE INSENSITIVE2
EIN3. *See* ETHYLENE INSENSITIVE3
EIN4, 55
E8 inhibition, 45
Endo-1,4-β-glucanase (EGase), 174
Endoreduplication, 34
Epigenetic regulation, 200–201
9-*cis*-Epoxycarotenoid dioxygenases (NCEDs), 91, 92t
ERF1. *See* ETHYLENE RESPONSE FACTOR1
ERFs. *See* Ethylene response factors
ERS1, 55–56
ERS2, 55–56
Esters, 138f, 145–47, 146f
Ethylene, 43–67
  in carotenoid synthesis, 97
  in climacteric and nonclimacteric fruit, 46–48
  ERF regulation by, 60
  perception and signaling of, 8, 54–60, 55f, 197
    CTR interaction with receptors, 57–58
    downstream of CTR, 59–60
    enzymes and proteins in, 62t
    other protein interaction with receptors, 58–59
    receptors for, 55–57, 56f, 57f
  in ripening, 7–9, 189
  ripening gene networks and, 53–54, 54f, 60–67, 62t
    color, 61–62, 63f
    flavor and aroma volatiles, 65–66
    pathogen susceptibility, 66–67
    texture, 63–65, 63f, 64f
  ripening upstream of, 8–9
  shelf life and, 48–49
  synthesis of, 48–52
    atmosphere modification and, 3–4
    climacteric compared with nonclimacteric, 7–8
    System-1 and System-2 in, 50–52, 51f
ETHYLENE INSENSITIVE2 (EIN2), 59
ETHYLENE INSENSITIVE3 (EIN3), 59–60
Ethylene receptors, 55–57, 56f, 57f
  CTR interaction with, 57–58
  other protein interaction with, 58–59
  in tomato, 8
ETHYLENE RESPONSE FACTOR1 (ERF1), 59–60
Ethylene response factors (ERFs), 60
  ACO regulation by, 52
  auxin and, 47, 190
Ethyl hexanoate, 138f
ETR1, 55–56
ETR2, 55–56
Eucalyptus sp., 28f
Eudicots, 22–23, 29f
  in fossil record, 30
Eugenol, 138f
Eugenol synthase (EGS), 143, 144f
Euonymus macropterus, 27f
Evolution of angiosperms, 22–23, 29, 29f
  in fossil record, 30–31
  fruit as driver of diversity in, 36–38
  fruit development, 33–36
  fruit variation and angiosperm phylogeny, 32–33
Expansin, 174

Fabids, 23, 29f
FaOMT enzyme, 147
Farnesyl diphosphate (FPP), 149
Farnesyl diphosphate synthase enzyme (FDS), 149
Fasciclin-like arabinogalactan proteins (FLAs), 167–68, 169f
Fatty acid derived volatiles, 138f, 139f, 140–42, 141f
FCCs. *See* Fluorescent chlorophyll catabolites
FDS. *See* Farnesyl diphosphate synthase enzyme
Fertilization, changes triggered by, 33–34, 126–27
Ferulic acid, 119
F3H. *See* Flavanone 3-hydroxylase
Firmness, in fruit ripening, 4, 44, 45f, 163–64
FLAs. *See* Fasciclin-like arabinogalactan proteins
Flavan-3-ols, 123, 123f, 125–26
Flavanone 3-hydroxylase (F3H), 126
Flavonoids
  biosynthesis of, 118f, 125–26
    regulation of, 126–27
    transcriptional regulation of, 127–28
  engineering elevated levels of, 128–29
  types of, 122–25, 123f, 124f
Flavonols, 123–24, 123f
Flavonol synthase (FLS), 126
Flavor, 4
  components of, 10, 198
  development of, 44, 45f
  ethylene and, 48–49
  gene expression in, 62t, 65–66
  perception of, 10–11
  volatile compounds and, 135–36, 198
Fleshy fruit
  development of, 33–36
  ethylene production and, 46–48
  evolution of
    angiosperm phylogeny and, 32–33
    in fossil record, 31
  wild, 37
Flowers
  in angiosperm radiation, 37
  ethylene role in, 53–54, 54f
FLS. *See* Flavonol synthase
Fluorescent chlorophyll catabolites (FCCs), 79, 79f, 80f
Follicle, 23t, 25f, 26f
Fossil record, of angiosperm evolution, 30–31
FPP. *See* Farnesyl diphosphate

Fructose. *See* Sugars
D-Fructose-1,6-diphosphate, 151
Frugivores, fruits and adaptation with, 31, 36
Fruit
  angiosperms and, 21
    as driver of diversity in, 36–38
    phylogeny and variation in, 32–33
  definition of, 21–22
  development of, 33–36
  evolution of, 22–23, 29, 29f
    angiosperm phylogeny and variation in, 32–33
    as driver of angiosperm diversity, 36–38
    fossil record of, 30–31
    fruit development, 33–36
  global production, consumption, and net export of, 2–3, 3f
  quality of, 1, 4, 135
    regulation of, 197–200
  therapeutic properties of, 1–2
  types of, 21–22, 23t, 24f, 25f, 26f, 27f, 28f
Fruit metabolism
  of cell wall, 11–12
  central carbon, 4–7
    acidity, 6–7
    starch, 6
    sucrose, glucose, and fructose, 4–6
    TCA cycle structure, 7
  modification of, 3–4
Fruit ripening
  carotenoid production during, 76
  changes associated with, 4
  control of, 44–45, 45f
  ethylene in, 7–9
  gene expression programs in, 43–45
  regulatory networks controlling, 189–201
    epigenetic, 200–201
    genetic, 191–200
    hormones, 189–90
  seed dispersal and, 43
  silver inhibition of, 45, 46f
  upstream of ethylene, 8–9
Fumarase, 6
Furaneol, 138f, 139f, 150–51
Furanocoumarins, 121
Furanones, 138f, 150–51

Galapagos tomatoes, 36
Galposis speciosa, 24f
Gametoheterotopy theory, 22
GAXs. *See* Glucuronoarabinoxylans
GCC box, 60
Gene expression programs, in fruit ripening, 43–45
Genetic engineering
  for flavonoids, 128–29
  for volatile compound pathways, 153–54
Genetic networks, 191–200
  downstream ripening signals and activities, 196–97
  fruit quality attributes, 197–200
  transcriptional regulators, 193–96
  transcription control, 191–93
Genetic relationships, of angiosperms, 22–23, 29, 29f
Genipa americana, 24f
Geraniol synthase, 154
Geranium maculatum, 28f
Geranyl diphosphate (GPP), 149
Geranylgeranyl pyrophosphate (GGPP), 61–62, 82, 89, 89f
gf. *See Green flesh* mutant
GGPP. *See* Geranylgeranyl pyrophosphate
GLK2. *See* Golden 2-like transcription factor
Global production, consumption, and net export of fruit, 2–3, 3f
Glucose. *See* Sugars
Glucuronoarabinoxylans (GAXs), 164
Glyceraldehyde-3-phosphate, 147, 148f
Golden 2-like (GLK2) transcription factor, 192
GPP. *See* Geranyl diphosphate
GR. *See* GREEN-RIPE
Grain, 23t
Green flesh (*gf*) mutant, 79–80
GREEN-RIPE (GR), 58–59
Gynoecium, 22
Gyrocarpus americanus, 25f

HCAs. *See* Hydroxycinnamic acids
HDMF. *See* 4-Hydroxy-2,5-dimethyl-3(2*H*)-furanone
HDR. *See* (E)-4-Hydroxy-3-methyl but-2-enoyl diphosphate reductase
HD-Zip proteins, 53, 64, 194
Heat shock (HS), ripening and, 45–46
Heisteria povedae, 28f
Helicteres isora, 27f
Hemicelluloses, 164–65, 166f, 168, 169f
Hexanal, 65, 138f, 140
Hexenal, 65
(Z)-3-Hexenal, 138f, 140
Hexenol, 65
HG. *See* Homogalacturonan
High-pigment tomato mutant (*hp*), 62
Homogalacturonan (HG), 165–67, 167f, 169f
  demethylation of, 171
Hormones
  in carotenoid synthesis, 97
  controlling ripening, 189–90
  metabolic pathways for, 4, 5f
hp. *See High-pigment* tomato mutant
HPLs. *See* Hydroperoxide lyases
HRGPs. *See* Hydroxyproline-rich glycoproteins
HS. *See* Heat shock
Hydroperoxide lyases (HPLs), 140–42, 141f

Hydroxycinnamic acids (HCAs), 119
4-Hydroxy-2,5-dimethyl-3(2*H*)-furanone (HDMF), 150–51
(E)-4-Hydroxy-3-methyl but-2-enoyl diphosphate reductase (HDR), 88f, 99
Hydroxyproline-rich glycoproteins (HRGPs), 167–68
Hypanthium, 26f
Hypocotyl, 27f

IAA. *See* Indole-3-acetic acid
IAA-aspartic acid (IAA-Asp), 189–90
Iberis umbellata, 27f
ICDHc. *See* Cytosolic isocitrate dehydrogenase
IGS. *See* Isoeugenol synthase
Ilex, 24f
Indehiscent fruits. *See* Nondehiscent fruits
Indole-3-acetic acid (IAA), 189–90
Insects, in diversification of angiosperms, 37
β-Ionone, 138f
IPP. *See* Isopentenyl diphosphate
Isoeugenol synthase (IGS), 143, 144f
Isoflavonoids, 123
Isopentenyl diphosphate (IPP), 87, 88f, 147, 148f, 149
Isorhamnetin, 123, 123f

Kaempferol, 123, 123f
Ketocarotenoids, 91, 91f

Lammiids, 23, 29f
LAR. *See* Leucoanthocyanidin reductase
LeACO1, 52–53, 52f, 54f
LeACS2, 52–53, 52f
Leaf senescence
  chlorophyll degradation during, 78–81, 79f, 80f
  chlorophyll metabolite recycling, 81–82, 82f
LeCTR2, 58
Lecythis pisonis, 28f
Leefructus, 30
LeETR1, 55, 59
LeETR2, 55
LeETR3. *See* NR
LeETR4, 8, 55–56
LeETR5, 55
LeETR6, 8, 55–57
Legume, 23t, 27f
LeHB-1, 52–54, 52f, 64, 64f, 194
LeMADS-RIN gene, 191–93
LeSPL-CNR, 192, 194
Leucoanthocyanidin reductase (LAR), 126
Light
  carotenoid synthesis and, 96–97
  flavonoid biosynthesis and, 126–27
Lignans, 118f, 120
Lignification, 34
Lignin, 118f, 120
Limonene, 138f
LIN5, 5–6
Linalool, 138f
Linalool synthase, 154
Lipoxygenases (LOX), 65–66, 140–42, 141f, 199
Lithospermum arvense, 24f
LOX. *See* Lipoxygenases
Lutein
  in chloroplasts, 76
  structure of, 77f
Lycopene
  biosynthetic fate of, 89f, 90
  cancer and, 2
  ethylene and development of, 61–62
  structure of, 77f

MADS-box transcription factors, 45, 47, 191
Magnolia virginiana, 25f
Magnoliids, 22–23, 29f
Malate
  metabolism of, 6–7
  in ripening, 6–7, 198
Malonyl-CoA, 125–26
Malvids, 23, 29f
Malvidin (Mv), 123f, 124
Mandragora officinalis, 24f
Mannans, 169
Mannanases, 173
Mannan endotransglycosylase-hydrolase (MTH), 173
1-MCP. *See* 1-Methylcyclopropene
Megafaunal dispersal syndrome, 36
Melon, ACS and ACO inhibition in, 48–49, 50f
MEP pathway. *See* Methylerythritol phosphate pathway
Mericarps, 23t, 24f
Mesangiosperms, 22–23, 29f
Mesifurane, 138f
Metabolism of fruit
  cell wall, 11–12
  central carbon, 4–7
    acidity, 6–7
    starch, 6
    sucrose, glucose, and fructose, 4–6
    TCA cycle structure, 7
  modification of, 3–4
Metabolites
  chronic disease and, 2
  in flavor, 10–11
  microbes and, 36–37
  volatile compounds and, 10–11
Metabolomics, 155
3-Methylbutanal, 65
2-Methylbutanol, 138f
3-Methylbutanol, 65, 138f
3-Methylbutyl acetate, 138f, 146

3-Methylbutyl butanoate, 146
1-Methylcyclopropene (1-MCP), 4, 53, 55
Methylerythritol phosphate (MEP) pathway, 87, 88f, 147
6-Methyl-5-hepten-2-one, 138f
Methyl salicylate, 138f
Methyltransferases (MTs), 126
MFOs. *See* Mixed function oxygenases
Mitochondrial malate dehydrogenase (mMDH)
  malate and, 6
  TCA cycle and, 7
Mixed function oxygenases (MFOs), 89f, 90
mMDH. *See* Mitochondrial malate dehydrogenase
Monocotyledons, 22–23, 29f
  cell walls of, 164
Monolignols, 118f, 120
Monoterpenes, 139f, 147, 148f, 154
Mostly male theory, 22
MTH. *See* Mannan endotransglycosylase-hydrolase
MTs. *See* Methyltransferases
Musa velutina, 26f
Mutualists, fruit evolution and, 36–37
Mv. *See* Malvidin
MYB. *See* Myeloblastosis
MYB genes, 127
Myeloblastosis (MYB) transcription factors, 127–28
Myricetin, 123, 123f
Myrrhis odorata, 24f

NAC. *See* No Apical Meristem
Naringenin chalcone, 125–26
NCCs. *See* Nonfluorescent chlorophyll catabolites
NCEDs. *See* 9-*cis*-Epoxycarotenoid dioxygenases
Nelumbo nucifera, 26f
Neoxanthin, 90–91
  in chloroplasts, 76
Nerolidol, 138f
Neverripe (*Nr*), 8, 44–45, 44f
  ethylene binding in, 56–57, 57f
1-Nitro-2-phenethane, 199
No Apical Meristem (NAC)-domain transcription factor, 45
Nonclimacteric fruit
  classification of, 7–8
  climacteric compared with, 7–8
  ethylene in, 46–48
  malate in, 6
  ripening gene networks in, 60–61, 62t
  ripening of, 45f
Nondehiscent fruits, 22, 23t
  development of, 34
  evolution of, 32
Nonfluorescent chlorophyll catabolites (NCCs), 79, 79f, 80f
Nonripening (*nor*), 8–9, 45, 191
Nontranscriptional regulation, of carotenoid synthesis, 98–100
  intracellular location and storage, 99–100
  post-transcriptional modifications, 99
  substrate availability, pathway flux, and turnover, 98–99
nor. *See Nonripening*
NOR, 53
Nr. *See Neverripe*
NR
  in ethylene reception, 56–57, 57f
  inhibition of, 45
  SlTPR1 interaction with, 59
Nutlets, 23t, 24f
Nutrients, 1–2, 44, 97–98
Nuts, 23t, 25f

Odor. *See* Aroma volatiles
Organic acids, flavor role of, 10, 65, 198
Ovary, 33
Ovules, 33

PAAS. *See* Phenylacetaldehyde synthase enzyme
PAL. *See* Phenylalanine ammonia lyase
PAO. *See* Pheophorbide a oxygenase
Parenchyma cells, 164
Paris mairei, 26f
PAs. *See* Proanthocyanins
Pathogens
  benzenoids and, 142–43
  ethylene and, 66–67
PDS. *See* Phytoene desaturase
PE. *See* Pectin esterase
Pectate lyase (PL), 175–76
Pectins, 12, 164, 165–68, 167f, 169f
Pectin esterase (PE), 44
Pectin methyl esterase (PME), 8, 12, 175–76
Pelargonidin (Pg), 123f, 124
Peonidin (Pn), 123f, 124
PEPC. *See* Phosphoenolpyruvate carboxylase
Pericarp, 33–35
Periclinal cell division, 34
Petunidin (Pt), 123f, 124
Pg. *See* Pelargonidin
PG. *See* Polygalacturonase
Phenethylamine, 143, 144f
Phenylacetaldehyde, 143, 144f, 199
2-Phenylacetaldehyde, 11
Phenylacetaldehyde synthase enzyme (PAAS), 143, 144f
Phenylalanine, 117–18, 118f, 142, 144f
Phenylalanine ammonia lyase (PAL), 118–19, 118f, 143, 144f
Phenyl-2-benzopyrylium, 124
2-Phenylethanol, 11, 138f, 143, 144f, 199

Phenylpropanoids, 117
  engineering elevated levels of, 128–29
  functions of, 117
  metabolism of, 118–19, 118f
    cinnamic acids, 118–19
    coumarins, 120–22, 121f
    flavonoids, 122–28, 123f, 124f
    monolignols, lignans, and lignin, 120
    stilbenoids, 122
  synthesis of, 142–45, 144f
Pheophorbide, 78
Pheophorbide a oxygenase (PAO), 78
Pheophytin, 78
Phloem
  amino acid transport via, 6
  sucrose loading into, 4
  translocation of metabolites in, 10
Phosphoenolpyruvate carboxylase (PEPC), 9
Phytochrome-interacting factor 1 (PIF1), 97
Phytochromes (PHYs), 96
Phytoene, 61–62
  desaturation, cyclization, and oxidation reactions from, 87–92, 89f, 91f
  formation of, 96, 98
Phytoene desaturase (PDS), 89, 89f
Phytoene synthase (PSY), 44, 89, 89f, 96
Phytoene synthase 1 (PSY1), 62, 96
Phytoene synthase 2 (PSY2), 96
Phytol, 81–82, 82f
Phytyl diphosphate, 82
PIF1. *See* Phytochrome-interacting factor 1
Pigments
  anthocyanins, 124
  in fruit ripening, 4, 44, 45f
  metabolic pathways for, 4, 5f
  production of, 44, 45f
  study of, 75
PL. *See* Pectate lyase
Plant-insect interactions, in diversification of angiosperms, 37
Plastoglobules, 84, 100, 196
PME. *See* Pectin methyl esterase
PME inhibitors (PMEIs), 177
Pn. *See* Peonidin
Polyamines, 9–10
Polygalacturonase (PG), 12, 44, 61
  in texture development, 63–65, 63f, 64f, 174–76
Polyphenols, 117
  protective effects of, 2
Polysaccharides
  disassembly of, 171–72
  hydrolysis of, 12
  solubilization of, 12
Prenyl diphosphates, 149
Prenyltransferases, 149
Proanthocyanins (PAs), 118f, 125
  upregulation of, 129
Protea sp., 26f
Provitamin A. *See* Beta-carotene
PSY. *See* Phytoene synthase
PSY1. *See* Phytoene synthase 1
PSY2. *See* Phytoene synthase 2
Pt. *See* Petunidin
Pterogyne nitens, 27f
Putrescine, 9–10
Pyrene, 23t, 24f
Pyruvate, 147, 148f
Pyxidium, 28f

QTL. *See* Quantitative trait loci
Quality of fruit, 1, 4, 135
  regulation of, 197–200
Quantitative trait loci (QTL), for volatile compounds, 152–53
Quercetin, 123, 123f

Raf-1, 58
RAN1 protein, 56
RAV, 60
RCCR. *See* Red chlorophyll catabolite reductase
RCC reductase (RCCR), 78, 79f
Reactive oxygen species (ROS), in carotenoid synthesis, 97–98
Recommended consumption, 1–2
Red chlorophyll catabolite reductase (*RCCR*), 197
Regulatory networks controlling ripening, 189–201
  epigenetic, 200–201
  genetic, 191–200
    downstream ripening signals and activities, 196–97
    fruit quality attributes, 197–200
    transcriptional regulators, 193–96
    transcription control, 191–93
  hormones, 189–90
Respiration rate
  atmosphere modification and, 3–4
  climacteric compared with nonclimacteric, 7–8
  ethylene biosynthesis and, 46–47
Resveratrol, 2, 122
REVERSION-TO-ETHYLENE SENSITIVITY1 (RTE1), 58–59
RGI. *See* Rhamnogalacturonan I
RGII. *See* Rhamnogalacturonan II
Rhamnogalacturonan II (RGII), 166–67, 167f, 169f
Rhamnogalacturonan I (RGI), 165–67, 167f
rin. *See Ripening inhibitor*
RIN, 53, 192–93
Ripening. *See* Fruit ripening

Ripening gene networks, ethylene and, 53–54, 54f, 60–67, 62t
  color, 61–62, 63f
  flavor and aroma volatiles, 65–66
  pathogen susceptibility, 66–67
  texture, 63–65, 63f, 64f
Ripening inhibitor (*rin*), 8–9, 44–45, 44f, 191
ROS. *See* Reactive oxygen species
Rosa omeiensis, 26f
Rosids, 23, 29f
R2R3 MYB family members, 127–28
RTE1. *See* REVERSION-TO-ETHYLENE SENSITIVITY1
RTE1 protein, 56
Rubiaceae, 32

SABATH family, 147
SAM. *See* S-Adenosylmethionine
Samara, 23t, 27f
SAM decarboxylase, 9–10
Schizocarps, 23t
Seed dispersal, ripening and, 43
Seed-dispersing animals, co-evolution with, 31
Senescence. *See* Leaf senescence
Sepals, 53, 54f
Sesquiterpenes, 139f, 147, 148f, 154
SGR. *See Stay-green* gene
Shelf life, 48–49, 199
Shikimate pathway, 118
  cinnamic acids, 118–19
  coumarins, 120–22, 121f
  flavonoids, 122–28, 123f, 124f
  monolignols, lignans, and lignin, 120
  stilbenoids, 122
Sideroxylon grandiflorum, 36
SlERF6, 198
Signal transduction, of ethylene, 8, 54–60, 55f, 197
  CTR interaction with receptors, 57–58
  downstream of CTR, 59–60
  other protein interaction with receptors, 58–59
  receptors for, 55–57, 56f, 57f
Silicle, 23t, 27f
Silique, 23t, 33–35
Silver, ripening inhibition by, 45, 46f
Sinapic acid, 119
SlAP2a, 194, 197
SlERF6, 194–95
SlTPR1, 59
Softening. *See* Texture
Softness. *See* Firmness
SPB protein. *See* SQUAMOSA promoter-binding
Spermidine, 9–10
Spermine, 9–10
SQUAMOSA promoter-binding (SPB) protein, 45, 191–92
Starch
  hydrolysis of, 12
  LIN5 and, 6
  during ripening, 6, 198
Stay-green gene (*SGR*), 81, 196
Stilbene, 122
Stilbene synthase (STS), 122
Stilbenoids, 118f, 122
Strigol, 85f, 86
Strigolactones, 85f, 86, 94–95
Structural proteins, 167–68
STS. *See* Stilbene synthase
Sucrose. *See* Sugars
Sugars
  in central carbon metabolism, 4–6
  flavor role of, 10, 65, 198
  LIN5 and, 5–6
Syncarpy, 22, 23t, 36
System-2
  in ethylene synthesis, 50–52, 51f
  switching on of, 53–54
System-1, in ethylene synthesis, 50–52, 51f
Syzigium jambos, 28f

TAG1. *See TOMATO AGAMOUS1*
TAGL1. *See TOMATO AGAMOUS-LIKE1*
Tambalacoque (*Sideroxylon grandiflorum*), 36
Tannins, 117
TCA cycle, 7
TDR4, 53
Temperature
  flavonoids and, 126
  sepals and, 53, 54f
Terpenoids, 138f, 147–49, 148f
  genetic engineering of, 154
Texture. *See also* Firmness
  cuticular cell wall and, 177–79, 178f
  gene expression in, 62t, 63–65, 63f, 64f, 199–200
  in ripening fruit, 163–64, 199
Tocopherols, 81–82, 82f
TOM5, 61
TOM13, 49f, 61
Tomato
  ACS and ACO inhibition in, 48–49, 48f, 49f
  ethylene receptors in, 8
  transcription control in, 191–93
TOMATO AGAMOUS-LIKE1 (*TAGL1*), 35–36, 52–53, 193–95
TOMATO AGAMOUS1 (*TAG1*), 53, 193–94
TOMLOXA-E, 65–66
TOMLOXC, 45

Transcriptional regulation, of carotenoid synthesis, 95–98
chromoplast-specific genes and plastid differentiation, 97
feedback regulation, 98
hormones in, 97
light in, 96–97
reactive oxygen species and nutrient supply, 97–98
Trapopogon porrifolius, 24f
Tree factor, 47
TRP1, 59

u. See Uniform ripening mutation
UDP-glucose:flavonoid-O-glucosyltransferases (UFGTs), 126
Uniform ripening (*u*) mutation, 192

VaHOX1, 53
Vanillin, 142
Violaxanthin, 76, 83–84, 90–91
Vitamin A. *See* Beta-carotene
Vitamin C. *See* Ascorbate
Vitamins, metabolic pathways for, 4, 5f
Viviparous14 (VP14), 91–92, 92t
Volatile compounds, 10–11, 135–36, 137f, 138f. *See also* Aroma volatiles
conjugation and emission of, 151–52
engineering of pathways for, 153–54
fingerprint of, 136, 137f
in flavor, 135–36, 198
future perspectives on, 155
metabolic pathways of, 136–52, 139f
amino-acid-derived, 145, 146f
apocarotenoids, 149–50
esters, 145–47, 146f
fatty acid derived, 140–42, 141f
furanones, 150–51
phenylpropanoids and other benzenoids, 142–45, 144f
terpenoids, 147–49, 148f
production of, 44–45, 45f
quantitative trait loci for, 152–53
type of, 136, 138f
VP14. *See* Viviparous14

XEH. *See* Xyloglucan endohydrolase activity
XET. *See* Xyloglucan endotransglycosylase activity
XGA. *See* Xylogalacturonan
XTHs. *See* Xyloglucan transglucosylase-hydrolases
XyG. *See* Xyloglucan
Xylogalacturonan (XGA), 166, 167f
Xyloglucan endohydrolase activity (XEH), 173
Xyloglucan endotransglycosylase activity (XET), 173
Xyloglucan transglucosylase-hydrolases (XTHs), 173, 175
Xyloglucan (XyG), 164, 168–70, 169f, 173

Yang cycle, 48

Zea mays, 25f
Zeaxanthin, 77f, 83, 90
Zeaxanthin epoxidase (ZEP), 89f, 90
Zinc finger transcription factors, 47
Zingiberene synthase, 154